Modern Sensors and Sensor Systems: Designs, Instrumentation and Applications

Modern Sensors and Sensor Systems: Designs, Instrumentation and Applications

Edited by Marvin Heather

CLANRYE
INTERNATIONAL
www.clanryeinternational.com

Clanrye International,
750 Third Avenue, 9th Floor,
New York, NY 10017, USA

ISBN: 978-1-63240-587-6

Cataloging-in-publication Data

Modern sensors and sensor systems : designs, instrumentation and applications / edited by Marvin Heather.
p. cm.
Includes bibliographical references and index.
ISBN 978-1-63240-587-6
1. Detectors. 2. Sensor networks. 3. Multisensor data fusion. I. Heather, Marvin.
TK7871.674 .M64 2017
681.2--dc23

For information on all Clanrye International publications
visit our website at www.clanryeinternational.com

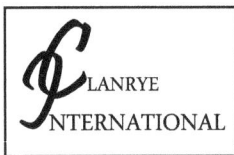

LANRYE
NTERNATIONAL

Printed in the United States of America.

Contents

Permissions

List of Contributors

Index

Preface

This book unravels the recent studies in the field of modern sensors and sensor systems. It presents researches and studies performed by experts across the globe. Sensors and sensor systems are the objects used to detect changes in any given environment. They use electrical and optical signals to find the events and then give outputs based on them. The most basic types of sensors are humidity sensor, touch sensor, pressure sensor, gas sensor, gyro sensor, acceleration sensor, biosensor, chemical sensor, etc. This text explores all the important aspects of sensor systems in the present day scenario. Most of the topics introduced in it cover new techniques and the applications of the subject. While understanding the long-term perspectives of the topics, the book makes an effort in highlighting their impact as a modern tool for the growth of the discipline. Scientists and students actively engaged in this field will find the book full of crucial and unexplored concepts.

Over the recent decade, advancements and applications have progressed exponentially. This has led to the increased interest in this field and projects are being conducted to enhance knowledge. The main objective of this book is to present some of the critical challenges and provide insights into possible solutions. This book will answer the varied questions that arise in the field and also provide an increased scope for furthering studies.

I hope that this book, with its visionary approach, will be a valuable addition and will promote interest among readers. Each of the authors has provided their extraordinary competence in their specific fields by providing different perspectives as they come from diverse nations and regions. I thank them for their contributions.

Editor

Defects and gas sensing properties of carbon nanotube-based devices

S. Baldo[1,2], V. Scuderi[1], L. Tripodi[1,2], A. La Magna[1], S.G. Leonardi[3], N. Donato[3], G. Neri[3], S. Filice[4], and S. Scalese[1]

[1]Istituto per la Microelettronica e Microsistemi, CNR, VIII Strada 5, 95121, Catania, Italy
[2]Dipartimento di Fisica e Astronomia, Università degli Studi di Catania, Via S. Sofia, 95125, Catania, Italy
[3]Dipartimento di Ingegneria Elettronica, Chimica e Ingegneria Industriale, Università degli Studi di Messina, Contrada di Dio, Salita Sperone 31, Messina, Italy
[4]Dipartimento di Chimica e Tecnologie Chimiche, Università della Calabria, Via P. Bucci, cubo 14/D, 87036, Arcavacata di Rende (CS), Italy

Correspondence to: S. Baldo (salvatore.baldo@imm.cnr.it) and S. Scalese (silvia.scalese@imm.cnr.it)

Abstract. In this work we report on the development of back-gated carbon nanotube-field effect transistors (CNT-FETs), with CNT layers playing the role of the channel, and on their electrical characterisation for sensing applications. The CNTs have been deposited by electrophoresis on an interdigitated electrode region created on a SiO_2/Si substrate. Different kinds of CNTs have been used (MWCNTs by arc discharge in liquid nitrogen and MWCNTs by chemical vapour deposition, CVD) and the electrical characterisation of the devices was performed in a NH_3- and NO_2-controlled environment. Preliminary data have shown an increase in the channel resistance under NH_3 exposure, whereas a decrease is observed after exposure to NO_2, and the sensitivity to each gas depends on the kind of CNTs used for the device.

Furthermore, the defect formation by Si ion implantation on CNTs was investigated by high-resolution transmission electron microscopy (TEM) and Raman analysis. The behaviour observed for the different devices can be explained in terms of the interaction between structural or chemical defects in CNTs and the gas molecules.

1 Introduction

Since their discovery, carbon nanotubes (CNTs) have aroused great interest due to their exceptional properties related to their one-dimensional character (Deretzis et al., 2006) like a high current-carrying capacity, high thermal conductivity and reduced charge carrier scattering. The use of both individual nanotubes and CNT networks has been explored, depending on the specific applications (Zhou et al., 2002). Currently they are used in various technological applications, such as microelectronics or nanoelectronics, the automotive field, telecommunications, the aerospace industry, and the biomedical field.

Due to a very high surface-to-volume ratio, high electron mobility, great surface reactivity and high capability of gas adsorption, CNTs can be used as sensitive layers in gas sensors for environmental monitoring, industrial process control and non-invasive biomedical analysis, by exploiting the changes in their electrical characteristics induced by surface chemical modifications (Zhang et al., 2008).

Defects in CNTs play a crucial role in the electronic, optical and mechanical properties (Scuderi et al., 2014), and they can be wanted or unwanted, depending on the kind of application: for example, for sensor applications, defected nanotubes are considered more desirable than ideal CNTs (Robinson et al., 2006; Neophytou et al., 2007), due to a larger interaction of the adsorbing species with defective sites. Of course, in order to take advantage of the defects presence, a suitable control on their generation, kind and amount has to be achieved. Furthermore, depending on the kind of defects, the binding energy should be evaluated and recovery mechanisms after gas molecules have been adsorbed on the CNT wall defects have to be investigated.

In this work, back-gated carbon nanotube-field effect transistors (CNT-FETs) have been produced where a CNT network forms the channel of the FET, and the role of defects in the sensing properties, already present or induced by ion implantation on the CNT walls, has been investigated.

2 Experimental

Two types of CNTs were used to investigate the sensing properties of the devices produced: commercial MWCNTs synthesised by the Sigma Aldrich CoMoCAT® catalytic CVD process, hereafter called MWCNTs-CVD, and MWCNTs synthesised by arc discharge in liquid nitrogen (LN_2) and oxidised in H_2O_2 (diluted at 30 %) for 2 h in an ultrasonic bath (Bagiante et al., 2010; Scalese et al., 2010), hereafter called MWCNTs-LN_2. The MWCNTs were deposited by the electrophoresis technique on an interdigitated electrode region created on a SiO_2/Si n^{++} type substrate (Baldo et al., 2014; Scuderi et al., 2012a), as reported in the scanning electron microscopy (SEM) image in Fig. 1a and b. In Fig. 1c and d, transmission electron microscopy (TEM) images are reported for both kinds of nanotubes, showing relevant differences in the structural order. The conductive Si substrate is used as a back-gate contact. The FET structure is useful in order to investigate any possible changes in the CNT electrical properties induced, for example, by ion implantation processes.

A complete electrical characterisation has been carried out, both in air and in controlled gaseous environments (NO_2 and NH_3), using two source meters (SMU) in a common source configuration, the Keithley 6487 for gate biasing and the Keithley 2400 for source and drain biasing (Fig. 2a). Both instruments, connected to a computer by the general purpose interface bus (GPIB), are totally controlled by MATLAB software. In this way a complete parameter analyser is obtained. The devices have been connected to the instruments with an appropriate test fixture (Fig. 2b) that gives the possibility of performing measurements in air and gaseous environments through an appropriate chamber connected to the gas system (Fig. 2c).

Understanding of the role of defects in the gas sensing properties of CNTs is of great relevance for the improvement of the sensitivity of CNT-based gas sensors. Defect formation by Si ion implantation on MWCNTs-LN_2 was also investigated. The samples were implanted with Si^+ ions at 180 keV. The used energy was fixed so that the silicon implanted profile was fully contained in the SiO_2/Si substrate. The ion doses were 1×10^{13}, 5×10^{13} and 1×10^{14} cm^{-2}.

The number of carboxyl groups has been determined by thermo-gravimetric analysis (TGA) using a Perkin Elmer Pyris 6 TGA thermo-gravimetric analyser. After reaching a thermal equilibrium at 30 °C for 5 min, the samples are heated up to 900 °C with a heating rate of 10 ° min^{-1}. The structural characterisation of the samples was performed by

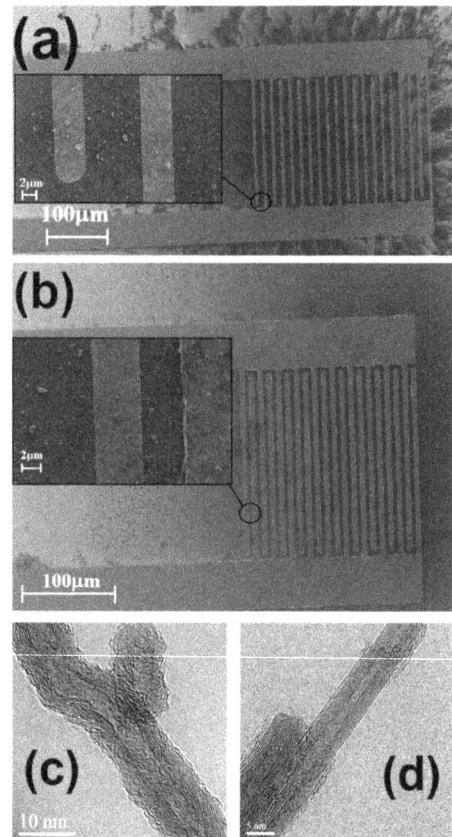

Figure 1. SEM images of devices with MWCNTs-CVD (**a**) and MWCNTs-LN_2 (**b**). In (**c**) and (**d**), TEM images of CNTs produced, respectively, by CVD and arc discharge are reported; the latter were treated by H_2O_2.

SEM, using a ZEISS SUPRA 35 FE-SEM system with a field emission electron gun, and by TEM, using a JEM 2010F JEOL microscope operating with an acceleration voltage of 200 kV.

Raman scattering has been excited by a 514.5 nm radiation coming from an Ar ion laser and the scattered light has been analysed by a single 460 mm monochromator (Jobin–Yvon HR460). Laser power was always kept below 10 mW at the sample to avoid its degradation, and the accumulation time was in the range of a few minutes.

3 Results and discussion

In Fig. 3 we report on the sensitivity of two kinds of devices fabricated, using respectively MWCNTs-LN_2 (black line) or commercial MWCNTs-CVD (red line) as the current channel, when they are exposed to NH_3 (Fig. 3a) and NO_2 (Fig. 3b). The sensitivity is expressed in terms of $\Delta R/R_0$, where R_0 is the channel resistance in dry air (reference resistance) and $\Delta R = R - R_0$, where R is the channel resistance in NH_3 or NO_2, keeping the same bias conditions ($V_{gs} = V_{ds} = 5$ V). Electrical characterisation shows that the

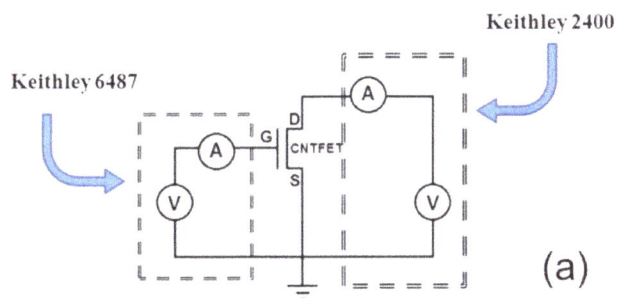

Figure 2. (a) Measurement set-up scheme; (b) test fixture; (c) test fixture and chamber for gas exposure.

Figure 3. Sensitivity values ($\Delta R/R_0$) for MWCNT-CVD and MWCNT-LN$_2$ devices to NH$_3$ (a) and NO$_2$ (b).

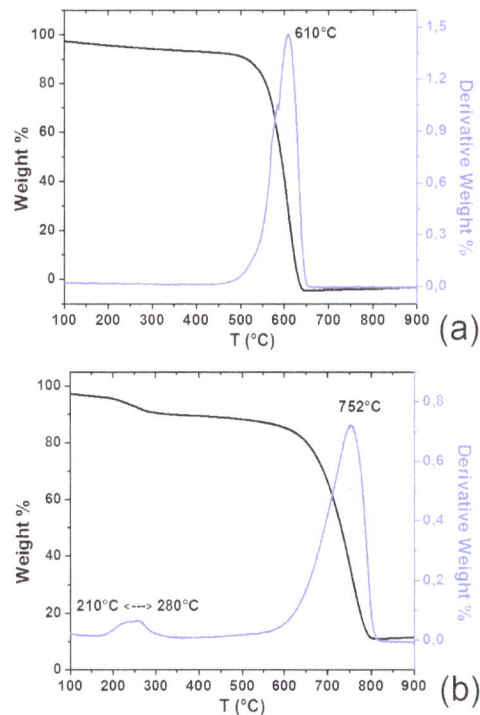

exposure to NH$_3$ induces an increase in the channel resistance as the concentration of ammonia is increased, and vice versa, during the exposure to NO$_2$, a decrease in the channel resistance is noted as the NO$_2$ concentration is increased. The response is attributed to the electrical charge transfer induced on the CNTs by the two kinds of molecules, which are electron donors (NH$_3$) or acceptors (NO$_2$) (Donato et al., 2011). The device with MWCNTs-LN$_2$ shows a better sensitivity to NH$_3$ compared to the device with MWCNTs-CVD (Fig. 3a), while the device with MWCNTs-CVD shows a better sensitivity to NO$_2$ compared to the other one (Fig. 3b).

These results depend on the different interaction of the gaseous species with CNTs. In fact, the presence of the -COOH groups on the CNT walls, in the case of MWCNTs-LN$_2$, favours an acid–base chemical interaction between the oxidised CNT and NH$_3$ molecules. NO$_2$ molecules instead are generally adsorbed by structural defects like vacancies, which are more abundant in MWCNTs-CVD. TGA results are shown in Fig. 4, expressed as weight % or derivative

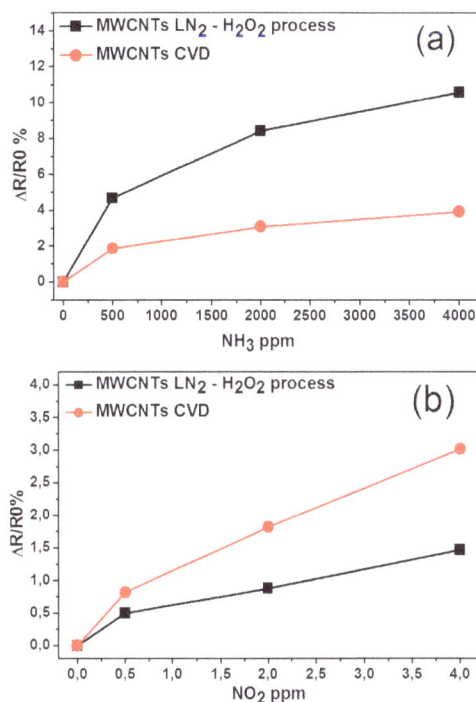

Figure 4. Thermo-gravimetric analysis of MWCNTs-CVD (a) and MWCNTs-LN$_2$ (b). The values are expressed as weight % on the left axis or as derivative weight on the right axis.

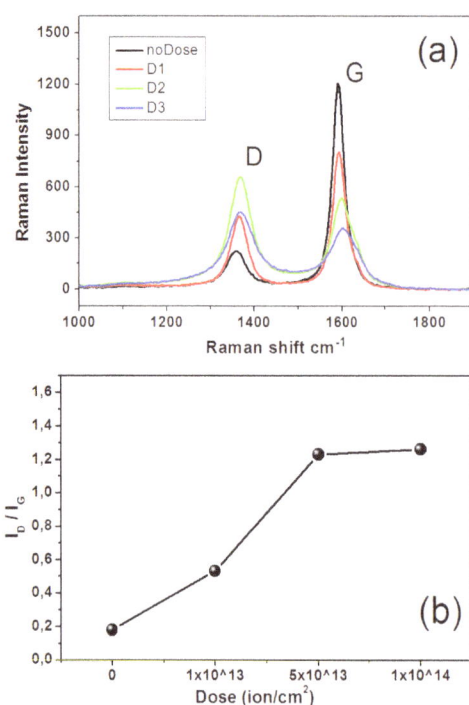

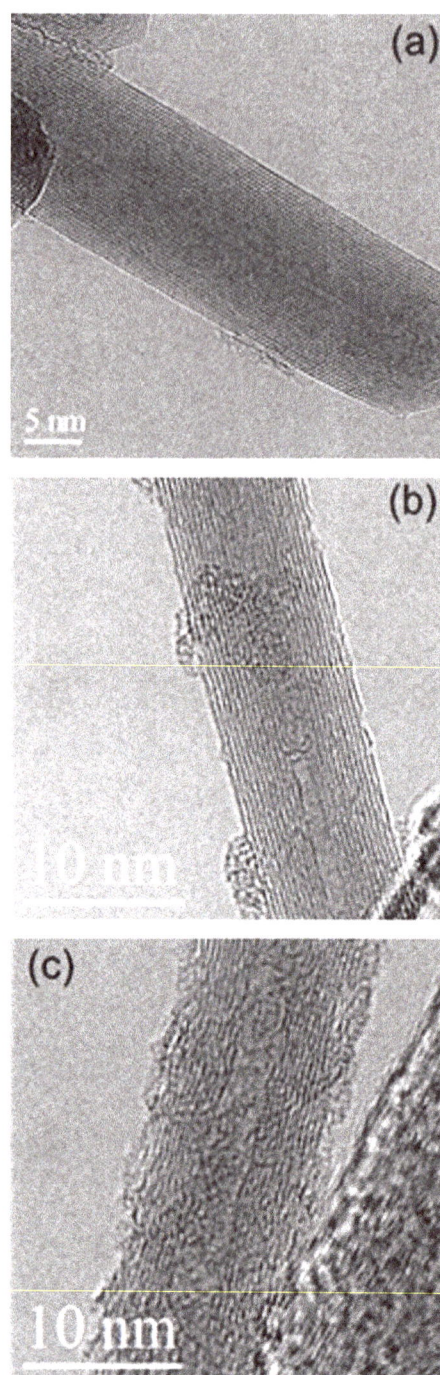

Figure 5. Raman spectra of the sample (**a**) not implanted (black line) and implanted with Si ion doses of 1×10^{13} cm^{-2} (red line), 5×10^{13} cm^{-2} (green line) and 1×10^{14} cm^{-2} (blue line); (**b**) I_D/I_G ratio vs. implant dose.

weight (left and right axes) as a function of temperature. A weight loss of 4 % between 210 and 280 °C (Fig. 4b) is observed in the case of MWCNTs-LN$_2$, confirming the presence of carboxyl groups (Datsyuk et al., 2008). Conversely, in MWCNTs-CVD, no weight loss is observed in that range of temperatures (Fig. 4a). In both materials, no hydroxyl functionalities, usually degrading in the temperature range 350–500 °C, are observed. Furthermore, the TGA shows that MWCNTs-LN$_2$ are inherently less defective and therefore more stable than the other ones, since the total weight loss of carbon is obtained for higher temperatures with respect to the other kinds of nanotubes (752 vs. 610 °C).

In order to investigate the role of defects in the CNT sensing properties, we have introduced structural defects by ion implantation on the CNTs that show a better structural quality (arc discharge synthesis). In particular, Si ion implantation has been performed at 180 keV at room temperature with three doses: 1×10^{13} cm^{-2} (called D1), 5×10^{13} cm^{-2} (called D2) and 1×10^{14} cm^{-2} (called D3). The reference sample, not exposed to ion implantation, is called D0.

After the ion implantation, the samples were analysed by Raman spectroscopy, in order to evaluate the structural order of the nanotubes. The spectra were acquired at different points of each sample in order to verify the uniformity of the material deposited on the area between the electrodes. The obtained spectra showed a very similar I_D/I_G ratio for each

Figure 6. Typical CNT structures and defects observed by TEM for the samples (**a**) not implanted, (**b**) implanted with a Si ion dose of 1×10^{13} cm^{-2}, and (**c**) implanted with a Si ion dose of 1×10^{14} cm^{-2}.

sample, indicating good uniformity in the entire electrode area. In Fig. 5a we report just one of the acquired spectra for each sample. The spectra show the typical D band at around 1361 cm^{-1}, due to defects present in the nanotubes, and the G band at around 1591 cm^{-1}, related to the graphitic order of

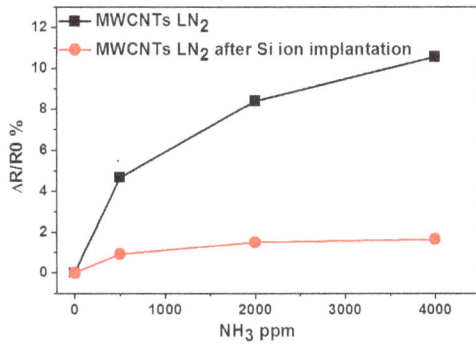

Figure 7. Comparison between $\Delta R/R_0$ values obtained for MWCNTs-LN2 and MWCNTs-CVD in NH_3 before (black line) and after (red line) Si implantation ($V_{gs} = V_{ds} = 5V$).

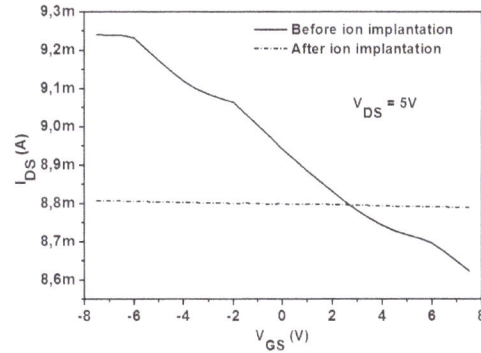

Figure 8. Dependence of I_{ds} on V_{gs} for the CNT-FET device before and after implantation (Si ion dose: 1×10^{13} cm^{-2}), for a fixed V_{ds} value of 5V.

the nanotubes. Starting from the reference spectrum (black curve) related to sample D0, as the ion dose is increased, it is possible to observe an increase in the D peak intensity, a decrease in the G peak intensity and an enlargement of both. The ratio between the D and G band intensities (I_D/I_G) gives information on the graphitic order of C structures: in particular, a lower I_D/I_G ratio means a better structural order. The I_D/I_G ratio calculated from Fig. 5a is reported as a function of the implant dose in Fig. 5b. As expected, the CNT structural quality worsens for higher implant doses: the I_D/I_G value increases from 0.6 for the sample not implanted up to 1.5 for the sample implanted with a dose of 1×10^{14} cm^{-2}.

The samples D0, D1 and D3 were also investigated by transmission electron microscopy in order to compare the damage degrees. The TEM images, reported in Fig. 6, show the effects of ion implantation, like vacancy formation and coalescence, shell burning, and diffusion of carbon atoms inside the structure. In particular, the damage is more evident for the sample implanted with the highest dose (1×10^{14} cm^{-2}), shown in Fig. 6c, in agreement with the results obtained by Raman spectroscopy.

The silicon ions that arrive on carbon nanotubes with an energy of 180 keV can break the outer shell and induce migrations of carbon atoms on the structures. For the highest implant dose, the amorphous carbon arranges itself on the outer shell like a carbon overcoat (Fig. 6c).

The removal of carbon atoms from CNT walls leads to a rearrangement of the network from a hexagonal structure to a coherent structure also containing non-six-membered rings. In particular, the presence of pentagons and/or heptagons in the structure changes the curvature of graphene cylinders (Iijima et al., 1992; Scuderi et al., 2012b) and, furthermore, it is possible to see that external diameter shrink due to the continuous loss of atoms (Banhart et al, 2005).

In order to investigate the effect of defects induced by ion implantation (dose of 1×10^{13} cm^{-2}) on the sensing properties, an electrical characterisation of the CNT network before and after ion implantation was done in an NH_3 gaseous envi-

ronment. In Fig. 7, we compare the sensitivity of the device, before (black line) and after (red line) ion implantation, keeping the same biasing conditions ($V_{gs} = V_{ds} = 5V$) in terms of $\Delta R/R_0$. A worse response of the device is evidenced (red line) after ion implantation. Therefore, it looks like Si implantation has led to a deterioration in the CNT sensitivity to NH_3 molecules.

We guess that the removal of carbon atoms from CNT walls and the consequent rearrangement of the hexagonal lattice, due to ion implantation, in some way also involves the -COOH groups responsible for the NH_3 adsorption. In particular, the removal of or the change in the -COOH groups causes a reduction in the molecules' adsorption on the CNT walls and, therefore, a reduced sensitivity to NH_3.

In Fig. 8 we report a comparison between the $I_{ds} - V_{gs}$ trans-characteristics obtained for the same device before and after ion implantation (Si ion dose: 1×10^{13} cm^{-2}): the initial dependence of I_{ds} on V_{gs} disappears after implantation, thus indicating a transition of the electrical behaviour of CNTs from semiconducting to quasi-metallic. This behaviour is compatible with the removal of -COOH functionalities. Further experiments are still needed to understand these findings deeply.

4 Conclusions

In this work, back-gated CNT-based sensors have been developed and characterised using a CNT network as the FET channel. In particular, we have investigated the role of defects, already present or induced by ion implantation on the CNT walls, in the sensing properties. Electrical characterisation has shown that the exposure to NH_3 induces an increase in the channel resistance as the concentration of ammonia is increased. The device with MWCNTs-LN$_2$ shows a better sensitivity to NH_3 compared to the device with MWCNTs-CVD. TGA shows the presence of carboxyl groups in the MWCNTs-LN$_2$, confirming that the interaction between CNTs and NH_3 occurs by -COOH groups present after treatment in H_2O_2. Vice versa, during the exposure to NO_2, a decrease in the channel resistance was noted as the NO_2 concentration is increased. In this case, the device with MWCNTs-CVD shows a better sensitivity to NO_2 compared to the device with MWCNTs-LN$_2$, showing that structural defects due to the CVD growth improve the NO_2 sensing properties.

Furthermore, defect formation by Si ion implantation on MWCNTs-LN$_2$ has been investigated. Raman analysis shows an increase of the I_D/I_G ratio related to an increase in structural disorder, as expected. TEM analyses confirm these observations, showing the presence of large holes (vacancy agglomerations) on the CNT outer walls and the presence of amorphous carbon layers on the outer surface of the CNTs.

Electrical characterisation indicates that the Si ion implantation has reduced the CNT sensitivity to NH_3. We guess that ion implantation, in some way, removes or alters the -COOH groups responsible for the NH_3 adsorption.

Further studies have to be carried out in order to achieve the complete understanding of the observed behaviours and to tailor the structural properties of CNTs for the improvement of sensing properties of such CNT-based devices.

Acknowledgements. The authors acknowledge expert technical support by A. Marino for ion implantation processes, C. Bongiorno for TEM analysis, S. Di Franco for lithographic processes, N. Godbert for TGA analysis and the group of Prof. G. Compagnini for Raman spectroscopy.

This work has been funded by MIUR by means of the PON R&C 2007-2013 national programme, project "Hyppocrates – Sviluppo di Micro e Nano-Tecnologie e Sistemi Avanzati per la Salute dell'uomo" (PON02 00355).

References

Bagiante, S., Scalese, S., Scuderi, V., D'Urso, L., Messina, E., Compagnini, G., and Privitera, V.: Role of the growth parameters on the structural order of MWCNTs produced by arc discharge in liquid nitrogen, Phys. Stat. Sol. B, 247, 884–887, 2010.

Baldo S., Scalese, S., Scuderi, V., Tripodi, L., La Magna, A., Romano, L., Leonardi, S. G., and Donato, N.: Correlation between structural and sensing properties of carbon nanotube-based devices, Sensors – Proceedings of the Second National Conference on Sensors, Rome, Italy, 19–21 February, 2014.

Banhart, F., Li, J. X., and Krasheninnikov, A. V., Carbon nanotubes under electron irradiation: stability of the tubes and their action as pipes for atom transport, Phys. Rev. B, 71, 241408-241411, 2005.

Datsyuk, V., Kalyva, M., Papagelis, K., Parthenios, J., Tasis, D., Siokou, A., Kallitsis, I., and Galiotis, C.: Chemical oxidation of multiwalled carbon nanotubes, CARBON, 46, 833–840, 2008.

Deretzis, I. and La Magna, A.: Role of contact bonding on electronic transport in metal-carbon nanotube-metal systems, Nanotechnology, 17, 5063–5072, 2006.

Donato, N., Latino, M., and Neri, G.: Novel Carbon Nanotubes-Based Hybrid Composites for Sensing Applications, Carbon nanotubes – From Research to Applications, 14, 229–242, 2011.

Iijima, S., Ichihashi, T., and Ando, Y.: Pentagons, heptagons and negative curvature in graphite microtubule growth, Nature, 356, 776–778, 1992.

Neophytou, N., Ahmed, S., and Klimeck, G.: Influence of vacancies on metallic nanotube transport properties, Appl. Phys. Lett. 90, 182119–182121, 2007.

Robinson, J. A., Snow, E. S., Bădescu, Ş. C., Reinecke, T. L., and Perkins, F. K.: Role of Defects in Single-Walled Carbon Nanotube, Chemical Sensors. Nano Lett., 6, 1747–1751, 2006.

Scalese, S., Scuderi, V., Bagiante, S., Gibilisco, S., Faraci, G., and Privitera, V.: Order and disorder of carbon deposit produced by arc discharge in liquid nitrogen, J. Appl. Phys. 108, 064305–064309, 2010.

Scuderi, V., La Magna, A., Pistone, A., Donato, N., Neri, G., and Scalese, S.: Use of the electric fields for the manipulation of MWCNTs, Carbon-Based Low Dimensional Materials – Proceedings of the 2nd CARBOMAT Workshop ISBN 978-88-124-5, 86–89, 2012a.

Scuderi, V., Bongiorno, C., Faraci, G., and Scalese, S.: Effect of the liquid environment on the formation of carbon nanotubes and graphene layers by arcing processes, Carbon, 50, 2365–2369, 2012b.

Scuderi, V., Tripodi, L., Piluso, N., Bongiorno, C., Di Franco, S., and Scalese, S.: Current-induced defect formation in multi-walled carbon nanotubes, J. Nanopart. Res., 16, 2287–2292, 2014.

Zhang, T., Mubeen, S., Myung, N. V., and Deshusses M. A.: Recent progress in carbon nanotube-based gas sensors, Nanotech. 19, 332001–332014, 2008.

Zhou, O., Shimoda, H., Gao, B., Oh, S., Flaming, L., and Yue, G.: Materials science of carbon nanotubes: Fabrication, integration, and properties of macroscopic structures of carbon nanotubes, Accounts Chem. Res., 35, 1045–1053, 2002.

Investigation of dielectric properties of multilayer structures consisting of homogeneous plastics and liquid solutions at 75–110 GHz

M. Klenner[1], T. Abels[1], C. Zech[1], A. Hülsmann[1], M. Schlechtweg[1], and O. Ambacher[1,2]

[1]Fraunhofer Institute for Applied Solid State Physics (IAF), Freiburg im Breisgau, Germany
[2]Department of Microsystems Engineering (IMTEK), University of Freiburg, Freiburg im Breisgau, Germany

Correspondence to: M. Klenner (mathias.klenner@iaf.fraunhofer.de)

Abstract. In this paper, we demonstrate an active 3-D millimeter wave (mmW) imaging system used for characterization of the dielectric function of different plastic materials and liquid solutions. The method is based on reflection spectroscopy at frequencies between 75 and 110 GHz, denoted as W-band, and can be used to investigate homogeneous dielectric materials such as plastics or layered structures and liquid solutions. Precise measurement of their dielectric properties not only allows for characterization and classification of different fluids, but also for reliable detection and localization of small defects such as voids or delamination within multilayer structures built from plastic materials. The radio frequency (RF) signal generation is based on circuits that have been designed and fabricated at the Fraunhofer Institute for Applied Solid State Physics (IAF) using a 100 nm InGaAs mHEMT process (Tessmann et al., 2006; Weber et al., 2011).

1 Introduction

The development of weight- or stability-optimized materials such as plastics or laminates has become the basis of most modern technologies. Manufacturers must meet highest demands in terms of quality, reliability and cost efficiency. Consequently, there is an increasing interest in high-end product surveillance systems that allow for non-destructive material testing at the production stage as well as in operation. For most non-destructive testing methods, either ultrasonics (Blitz and Simpson, 1996) or electromagnetic radiation (Blitz, 1997) are used. The main disadvantage of ultrasonic methods is that a couplant between the transducer and the device under test (DUT) is required (Bourne, 2001), so that they are not suitable for the investigation of large areas. In contrast, remote operation can be realized by using electromagnetic waves. Accordingly, the size of the DUT is not limited, as it can either be moved through a fixed beam or scanned by mechanical or digital beam-forming (Wirth, 2001). Besides visible light, X-rays and terahertz frequencies, the millimeter wave spectrum that spans from 30 to 300 GHz has become well established for non-destructive material analysis.

At the Fraunhofer Institute for Applied Solid State Physics, an active 3-D imaging tomograph operating at frequencies between 75 and 110 GHz (W-band) has been successfully realized. The system can not only be used for detection of concealed objects (Zech et al., 2011), but also for visualization and localization of small defects within multilayer structures consisting of known dielectric materials (Klenner et al., 2013).

In this paper, we present a new experimental setup based on the imaging scanner and investigate the chances and challenges of non-destructive material characterization at W-band frequencies. The method is based on reflection spectroscopy and allows for measurements of the refractive indices of layered plastic samples and determination of the complex dielectric function of different liquid solutions using a Debye model.

2 Experimental setup

Figure 1 shows the mmW imaging scanner operating at frequencies between 75 and 110 GHz, denoted as W-band.

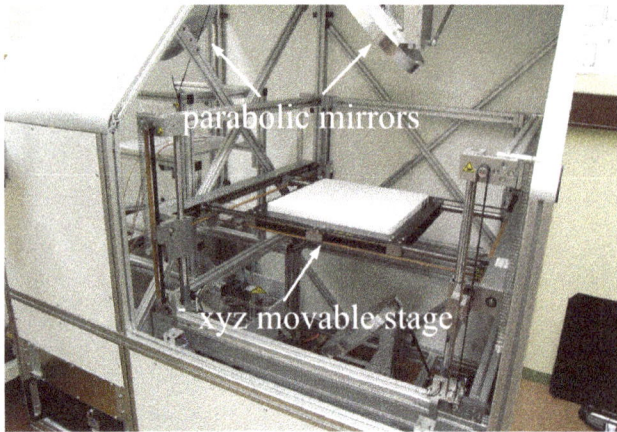

Figure 1. Photograph of the mmW focusing system.

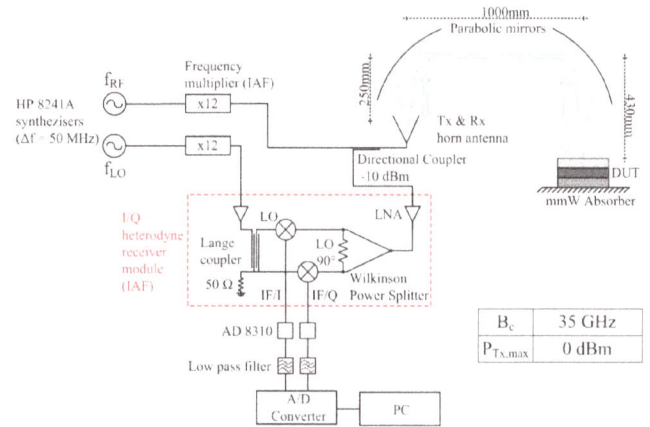

Figure 2. Block diagram of the millimeter wave imaging system used for reflectometric measurements.

Since the wavelength of W-band frequencies is small enough to approximately describe the signal by quasi-optical ray tracing models, parabolic mirrors have been designed to focus the signal on a movable sample stage. In order to perform reflection spectroscopy on different plastic samples, the system must be able to transmit and receive a signal at fixed frequencies between 75 and 110 GHz. This is realized by multiplying the output signals of two HP 8241A synthesizers using in-house $\times 12$ frequency multipliers as shown in the block diagram in Fig. 2. While one of the synthesizers is used to generate the transmit signal, the second one feeds the LO input of an in-house I/Q heterodyne receiver module. Thus, its output frequency is shifted by $\Delta f = 50$ MHz. A Tx & Rx horn antenna is used for transmission and reception of the mmW beam that is focused at the DUT on the sample stage using parabolic mirrors. A directional W-band coupler (-10 dBm isolation) is used to feed the RF input of the heterodyne receiver module such that RF and LO signals can be mixed to create in-phase (IF-I) and quadrature (IF-Q) output signals at a fixed intermediate frequency (IF) of 50 MHz. Both signals are then amplified and detected using two AD 8310 logarithmic power detectors with a 95 dB dynamic (DC to 440 MHz). It turned out that additional low-pass filters are needed to reduce the influence of unwanted system-related interference effects. The setup provides a total bandwidth of $B_c = 35$ GHz and an output power of $P_{Tx,max} \approx 0$ dBm.

A sketch of two exemplary DUT used for the characterization of different materials is shown in Fig. 3. While homogeneous plastics can be investigated by measuring the reflected intensity of arbitrary multilayer stacks built from several plastic discs, water-based liquid solutions need to be filled in a custom-designed cuvette acting as a Fabry–Perot interferometer.

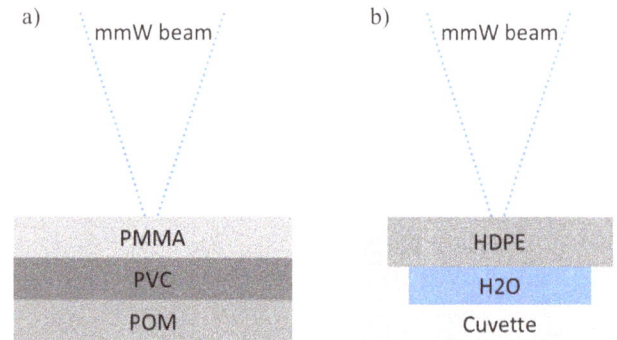

Figure 3. (a) Exemplary three-layer DUT consisting of discs made from POM, PVC and PMMA. (b) Custom-designed cuvette for investigation of liquid solutions such as H_2O. The HDPE layer acts as a Fabry–Perot interferometer.

3 Data processing

Since the absorption of millimeter waves in most homogeneous plastics is comparably small (Lamb, 1996), the complex part of the refractive index

$$n = n' - jn'' \tag{1}$$

can be neglected for investigation of the dispersion of multilayer structures built from homogeneous plastics. For this purpose, theoretical models based on calculation of **S** matrices (Kühlke, 2011) are used to fit the measured data in order to obtain the refractive indices of a multilayer sample. Since the experimental setup provides (approximately) perpendicular beam incidence on the sample stage, Fresnel's reflection and transmission coefficients corresponding to a layer i

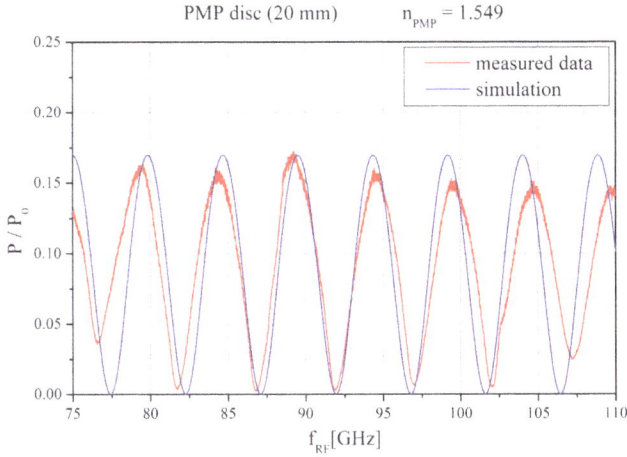

Figure 4. Simulated and measured power spectra corresponding to a 20 mm thick single-layer PMP sample.

within a multilayer structure are given by

$$r_{i-1,i} = \frac{n_{i-1} - n_i}{n_{i-1} + n_i}, \tag{2}$$

$$t_{i-1,i} = \frac{2n_{i-1}}{n_{i-1} + n_i}, $$

where n_i is the refractive index of the material in layer i. Furthermore, the phase shift of an electromagnetic wave passing through this layer can be calculated as

$$\delta_i = \frac{2\pi}{c_0} d_i n_i f, \tag{3}$$

where f is the frequency of the transmitted signal, d_i is the layer thickness and c_0 is the speed of light in vacuum. Based on these equations, an $\mathbf{S}$ matrix of the form

$$\mathbf{S} = \mathbf{S}(n_0, n_1, \ldots, n_M, d_1, \ldots, d_M, f) \tag{4}$$

and the corresponding reflectivity

$$\rho = \rho(n_0, n_1, \ldots, n_M, d_1, \ldots, d_M, f) \tag{5}$$

can be derived for arbitrary multilayer structures consisting of M layers. In order to determine the refractive indices of the layers within a sample, the refractive index of the medium surrounding the sample n_0 as well as the layer thickness $d_1, \ldots, d_M$ are fixed in the model, while the refractive indices of the layers within the DUT $n_1, \ldots, n_M$ can be optimized by a Levenberg–Marquardt algorithm (Levenberg, 1944). This model intrinsically assumes that there is no dispersion, as the Levenberg–Marquardt algorithm tries to find a single, optimal refractive index per layer for the full W-band. Therefore, dispersion will qualitatively become visible as a discrepancy between model and experimental data, as shown in Fig. 4.

In order to investigate material dispersion in a more quantitative way, it is necessary to select narrow frequency ranges rather than the full W-band to fit the model more accurately. The following method can be used.

1. Fit the experimental data for the full W-band to obtain a refractive index n_i^{Full} for every layer i.

2. Define upper and lower limits for possible values of the refractive indices based on n_i^{Full}. This will increase the convergence rates of the following fits.

3. Identify the first maximum or minimum and set a corresponding counter variable to $k = 0$.

4. Select a narrow frequency range around the kth extremum in the measured spectrum.

5. In consideration of the upper and lower limits defined in step 2, fit the measured data within the narrow range to obtain refractive indices n_i^k.

6. Slightly vary the values of n_i^k (manually) until the model does significantly deviate from the measured data. This allows for estimation of the errors in n_i^k.

7. Identify the center frequency f_m^k of the maximum or minimum and estimate its uncertainty s_f^k.

8. Save data point $\left(f_m^k, n_i^k\right)$ and corresponding error estimations.

9. Select the next extremum ($k = k + 1$).

10. Repeat steps 4 to 9 for every maximum and minimum in the spectrum.

The first two steps are mandatory for obtaining reasonable results from the Levenberg–Marquardt algorithm when only a narrow range of measurement data is used. In particular, there is no unique solution for multilayer DUT, and the algorithm does not converge properly without limiting the fitting parameters. In addition, it is useful to define a measure of inaccuracy for the fitted model such as the following sum of squares:

$$\sigma = \sqrt{\frac{1}{N} \sum_{j=1}^{N} \left(P_j - T(f_j)\right)^2}. \tag{6}$$

P_j is the jth of N measured values and corresponds to the frequency f_j. Accordingly, $T(f_j)$ is the value of the theoretical model at f_j. At the first step of the method described above, σ can be used for a first qualitative characterization of the strength of dispersion of a material over full W-band. It is, however, neither a quantitative measure of the relaxation strength nor a confidence limit on the refractive indices. Since there is only a slight dispersion in most materials that have been investigated, one could also use a model containing an additional distribution parameter for the refractive index instead of the method described above. In the case of a single-layer sample, the convergence rate of a model considering dispersion is still sufficient, while it is significantly decreasing when analyzing two- or three-layer DUT. Therefore, the algorithm described above is, in general, superior regarding convergence rates.

In principle, the analysis of liquid samples can be performed in a similar way. However, the assumption of negligible contribution of the complex part of the dielectric function no longer holds for water solutions, as they are significantly absorbing in the mmW regime (Peacock, 2009). Consequently, the algorithm presented above leads to significant uncertainties. In order to reliably fit the measured data using a model where dispersion is parameterized, the lid of the cuvette (cf. Fig. 3) must be made of a material with very weak dispersion such that the sample only consists of one layer where dispersion must be considered. This is realized by using PMMA and HDPE lids and a Debye-type model of the form

$$\epsilon(\omega) = \epsilon'(\omega) - j\epsilon''(\omega) = \epsilon_\infty + \frac{\epsilon_0 - \epsilon_\infty}{1 + j\omega\tau} \quad (7)$$

to parameterize the complex dielectric function of the liquid layer. Hereby, ϵ_0 and ϵ_∞ describe, respectively, the low- and high-frequency limits of the dielectric function. The parameter τ is the relaxation time of the dipole polarization process within the liquid solution caused by an incident electromagnetic field. Since a custom-designed cuvette is used for measurements of solutions (cf. Fig. 3), the layer structure of different DUTs is always identical and the sample can be regarded as a Fabry–Perot interferometer. Hence, instead of using an **S** matrix formalism, the reflected intensity from the cuvette can be described by the simpler relation

$$\frac{I_r(\omega)}{I_0(\omega)} = \left| r_{01} + t_{01} \, r_{12} \, t_{10} \, e^{-2i\delta} \sum_{j=0}^{\infty} \left(r_{10} \, r_{12} \, e^{-2i\delta} \right)^j \right|^2, \quad (8)$$

where

$$r_{12}(\omega) = \frac{n_1 - \sqrt{\epsilon_\infty + \frac{\epsilon_0 - \epsilon_\infty}{1 + i\omega\tau}}}{n_1 + \sqrt{\epsilon_\infty + \frac{\epsilon_0 - \epsilon_\infty}{1 + i\omega\tau}}} \quad (9)$$

is the reflection coefficient corresponding to the interface with the liquid solution, which is described by the Debye model, and δ is the phase shift of the electromagnetic field within the material on top of the liquid (cf. Eq. 3). By optimization of the Debye parameters ϵ_0, ϵ_∞ and τ, the model can be fitted to the measured data to obtain the evolution of the complex dielectric function of liquid solutions at W-band frequencies.

4 Dispersion of layered plastics

Table 1 shows the results of fitting theoretical models describing the reflectivity of single discs of POM, PVC,

Table 1. Average refractive indices of different plastics at W-band frequencies.

Material	n^{Full}	σ^{Full}
Polyoxymethylene (POM)	1.752	0.055
Polyvinyl chloride (PVC)	1.703	0.057
Polymethyl methacrylate (PMMA)	1.590	0.027
Polymethylpentene (PMP)	1.549	0.053
High-density polyethylene (HDPE)	1.512	0.021

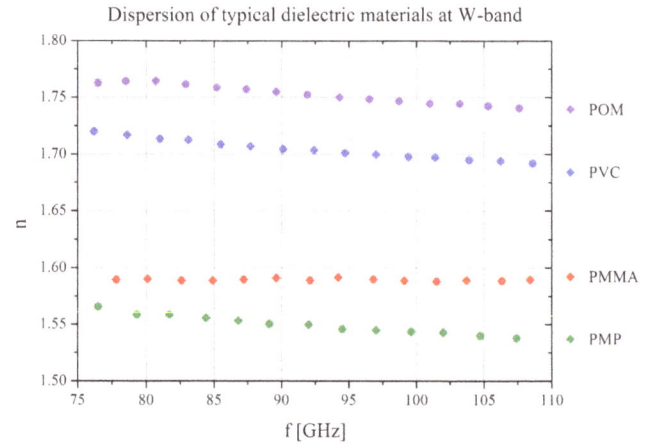

Figure 5. Dispersion of different plastics measured from single discs.

PMMA, PMP and HDPE (each $d = 20$ mm thick) to the measured data. The refractive indices n^{Full} have been found by optimization using a Levenberg–Marquardt algorithm. Additionally, the confidence limits σ^{Full} of the respective optimized model are shown for each refractive index. Note that this is not an explicit error of n^{Full} but a measure of how well the model, which assumes constant refractive indices, describes the corresponding experimental data that are in general influenced by dispersion. Assuming that the measurements have been performed under similar conditions, a relative difference in σ^{Full} between two materials is therefore essentially caused by their different strengths of dispersion.

By using the method described in Sect. 3, the refractive indices of narrow regions around the maxima and minima of the measured interference spectra (see Fig. 4) have been determined for each of the four plastic discs. The optimization algorithm did converge within all the selected regions such that a refractive index could be determined at every extremum. The resulting frequency-dependent refractive index is plotted for all the discs in Fig. 5. While the refractive indices of POM, PVC and PMP slightly decrease over W-band, PMMA shows nearly no dispersion. Thus, the dispersion plots are in agreement with the data shown in Table 1.

Multilayer samples have been investigated by building arbitrary stacks consisting of two, three or four discs that are either 10 or 20 mm thick. An exemplary dispersion plot cor-

(a)

Two-layer DUT
PVC (20 mm) on PMMA (10 mm)

(b)

Three-layer DUT
PVC (10 mm) on PMMA (10 mm) on PMP (10 mm)

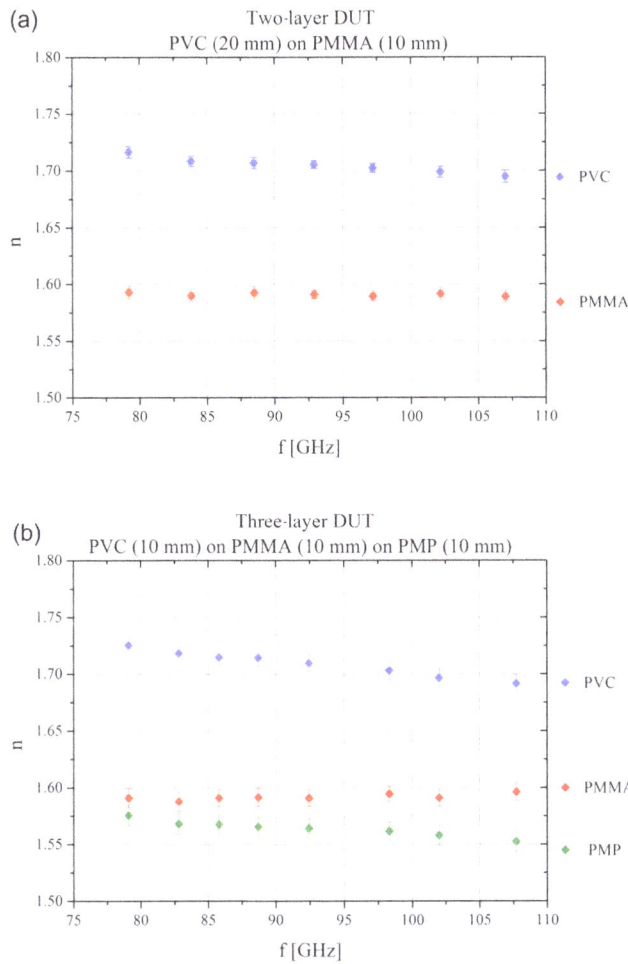

Figure 6. (a) Dispersion of different plastics measured from a double-layer structure. (b) The refractive index of PMP measured from a three-layer sample is overestimated compared to the single-disc measurements.

Table 2. Conductance and pH values of the different solutions.

DUT	Conductance (mS cm^{-1})	pH value
DI water	0.1	7.9
HCL solution	740.0	0.4
NaCl solution	171.2	8.2
NaOH solution	401.0	12.6

responding to a two-layer stack is shown in Fig. 6a. Since the power spectra become more irregular with the number of layers, frequency resolution of the dispersion plots is decreased compared to the single-disc measurements. However, both measurements are in agreement, and this is also true for any other combination of two discs.

Measurements of three- and four-layer samples have shown that, in several configurations, the refractive indices of deeper layers are overestimated with respect to the single-

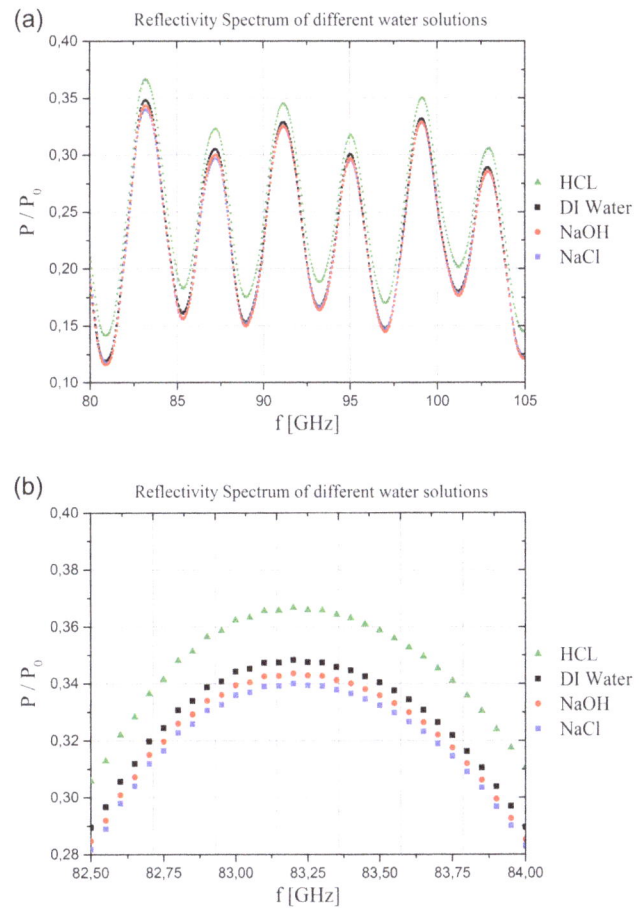

(a)

Reflectivity Spectrum of different water solutions

(b)

Reflectivity Spectrum of different water solutions

Figure 7. (a) Reflectivity spectrum of a HDPE cuvette filled with different solutions. (b) Error bars on the data points are much smaller than the markers used for visualization. Therefore, the different solutions are clearly distinguishable.

layer measurements. This behavior occurs if the refractive indices of upper layers are larger than in lower layers. Figure 6b shows the dispersion curves obtained by measuring the reflectivity of a three-layer sample where the lowest layer is a PMP disc and the corresponding refractive index is overestimated compared to the measurements shown in Fig. 5. The overestimation is most probably caused by the assumption of perpendicular incidence on the sample, which is not entirely true. In particular, beams get refracted away from the perpendicular in deeper layers if there are primarily transitions from a medium of higher refractive index to a medium of lower refractive index. Furthermore, parts of the signal get refracted out of the beam path of the system, causing a lower convergence rate of the Levenberg–Marquardt algorithm and therefore larger error bars compared to the single-disc measurements.

5 Complex dielectric function of water solutions

In order to investigate whether water solutions of different physical or chemical properties can be uniquely distinguished using reflection spectroscopy at W-band frequencies, a set of four sample solutions based on DI water has been prepared. By measuring their conductances and pH values, the solutions have been characterized as shown in Table 2. For the measurements, each solution is filled in a custom-designed cuvette with a 25 mm thick top layer made of HDPE. Accordingly, dispersion of the top layer can be neglected at W-band frequencies. Due to the finite isolation of the directional coupler used in the setup, a high-frequency interference pattern overlays the raw data of the spectra. Thus, a Savitzky–Golay algorithm (Savitzky and Golay, 1964) is used to smoothe the data sets. Furthermore, the reflectance of each DUT is measured ten times in succession to determine the reproducibility of the spectra by averaging and calculating the standard deviation. Figure 7 shows the smoothed and averaged reflection spectra of cuvettes filled with pure DI water as well as HCL, NaCl and NaOH solutions. The standard deviation of the curves is on the order of 10^{-5}. Error bars are therefore much smaller than the markers used for visualizing the data. Accordingly, reflection spectroscopy at W-band frequencies can be reliably used to detect or uniquely differentiate solutions of different conductances or pH values.

The complex dielectric function of these solutions can be calculated using the Debye model represented by Eq. (7). Thus, the parameters ϵ_0, ϵ_∞ and τ must be determined from the measured data sets. Therefore, the reflection coefficient corresponding to the liquid interface is parameterized by the Debye model (cf. Eq. 9) so that a Levenberg–Marquard algorithm can be used to fit Eq. (8) to the measured data by optimization of the Debye parameters.

However, simulations have revealed that all parameters influence the spectra in a similar way and that τ is by far the most sensitive parameter (Abels, 2014). These theoretical results are in accordance with other investigations showing that the presence of foreign molecules in water mainly result in the reduction of the potential barrier of the dipole reorientation process or an enhanced formation of hydrogen bonds (Kaatze et al., 2002). Both effects directly influence the dipole relaxation time and therefore the parameter τ. Accordingly, the convergence rate of the Levenberg–Marquardt algorithm is low without weighting or limits on the Debye parameters. In order to take this effect into consideration, the Debye parameters are first determined for a pure DI water solution. The results are in good accordance with other values found in the literature (Peacock, 2009), so that the weights of the optimization process can be calibrated with respect to the DI water sample. Assuming that the presence of HCL, NaOH or NaCl molecules in the water solution mainly influences the relaxation time τ, the dielectric function has been determined for each of the samples as shown in Fig. 8. While the dielectric function corresponding to the basic solutions is

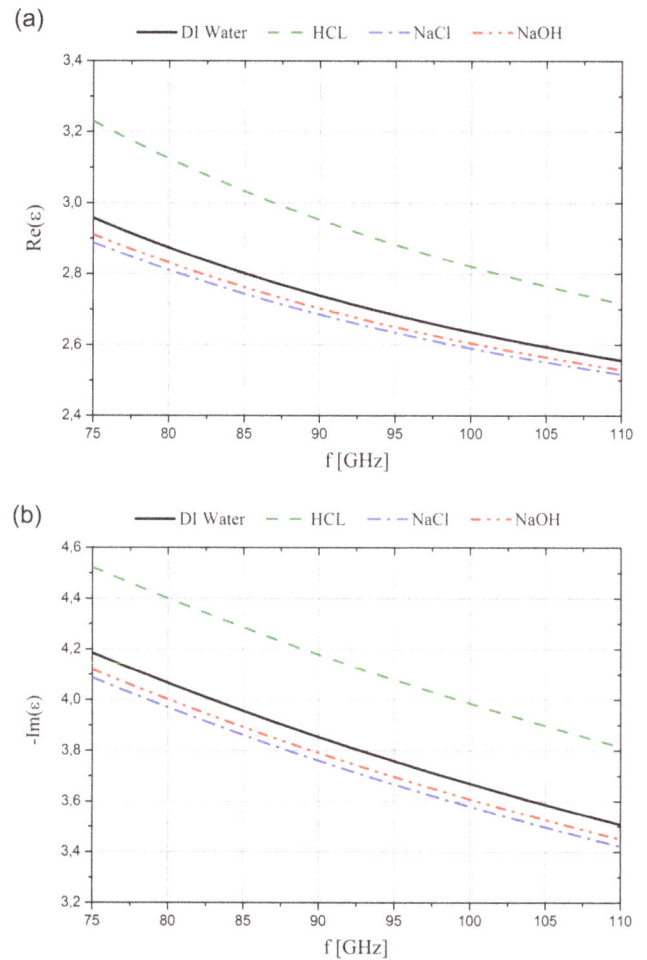

Figure 8. Real and imaginary parts of the dielectric function of different water solutions. Due to measurement inaccuracies, the dielectric function of the NaCl and NaOH solutions can not be uniquely distinguished.

lower compared to DI water, the permittivity spectrum of the acidic HCL solution is increased. These results are in contrast to prior investigations at frequencies of up to 40 GHz (Kaatze, 1983), where the real part of the dielectric function of hydrochloric acid solutions is lower compared to pure water.

A dielectric decrement can be attributed to the specific conductivity of the acidic solution. If this decrement is strong enough to significantly superimpose the relative increase of the dielectric function caused by its relaxation time, the above assumption no longer holds and a more detailed model must be considered to fit the measured data. Even though, to the authors' knowledge, there are no comparable measurements in the frequency range between 75 and 110 GHz, it appears likely that this is the case. However, due to the necessity of strong smoothing of the spectra, which influences both the amplitude and position of peaks, the use of more complex models is not feasible using the current experimental setup, and uncertainties in the parameters of the fitted Debye model are already significant. Hence, NaOH and NaCl solutions are not uniquely distinguishable.

6 Conclusions

We have demonstrated a reflectometric setup to measure the refractive indices and dispersion of different multilayer samples made from homogeneous plastics as well as the Debye parameters corresponding to a relaxation model describing the complex dielectric function of liquid solutions. The setup is included in a mmW imaging scanner so that the information on the dielectric behavior of different DUT can be directly used for mmW imaging.

Small defects in multilayer structures have been successfully visualized and localized based on the data obtained by reflectometry measurements (Klenner et al., 2013). In general, the presented method works well with stacks consisting of two or three layers. However, the refractive indices of lower layers in several configurations of three- and four-layer samples are overestimated if the signal gets significantly refracted away from the perpendicular. The performance of the system is strongly dependent on the convergence rate of the Levenberg–Marquardt algorithm and can be increased if some material parameters are already known. Accordingly, the method can for example be used to detect specific materials within unknown multilayer stacks, in particular if the dielectric properties of the target material adequately differ from its surroundings.

Furthermore, the reflectances of acid and basic solutions have been investigated and compared to pure DI water. While the reflectance spectra of all the samples are clearly distinguishable with respect to statistical uncertainties, the strong parasitic interference caused by the directional coupler used in the setup leads to significant inaccuracies regarding the model parameters when the data are fitted. As a consequence, more complex models could not be investigated in more detail. In addition, samples that only differ slightly in their dielectric behavior, such as the two basic solutions that have been investigated, can not be uniquely distinguished.

Acknowledgements. The authors would like to thank the workshop of the Fraunhofer IAF for precise mechanical prototype fabrication of the parabolic mirrors used in the imaging system and the dielectric samples.

References

Abels, T.: Reflexionsspektroskopie an verschiedenen Wasserproben im Millimeterwellenbereich, Bachelor Thesis, University of Freiburg im Breisgau, Germany, 44 pp., 2014.

Blitz, J.: Electrical and magnetic methods of non-destructive testing, Chapman & Hall, London, New York, USA, 1997.

Blitz, J. and Simpson, G.: Ultrasonic methods of non-destructive testing, Chapman & Hall, London, UK, 1996.

Bourne, S.: Novel hydrophilic polymer couplant for application in ultrasonic non destructive testing, PhD Thesis, Cranfield University, UK, 2001.

Kaatze, U.: Dielectric Effects in Aqueous Solutions of $1:1$, $2:1$, and $3:1$ valent Electrolytes: Kinetic Depolarization, Saturation, and Solvent Relaxation, Zeitschrift für Physikalische Chemie, 135, 51–57, 1983.

Kaatze, U., Behrends, R., and Pottel, R.: Hydrogen network fluctuations and dielectric spectrometry of liquids, J. Non-Cryst. Solids, 305, 19–28, 2002.

Klenner, M., Zech, C., Hülsmann, A., Tessmann, A., Leuther, A., Schlechtweg, M., Wagner, J., and Ambacher, O.: Multilayer material analysis using an active millimeter wave imaging system, in: 14th International Radar Symposium (IRS), Dresden, Germany, 19–21 June 2013, 1, 207–213, 2013.

Kühlke, D.: Optik – Grundlagen und Anwendungen, Europa-Lehrmittel, Edition Harri Deutsch, Frankfurt am Main, Germany, 407 pp., 2011.

Lamb, J. W.: Miscellaneous data on materials for millimetre and submillimetre optics, Int. J. Infrared Milli., 17, 1997–2034, 1996.

Levenberg, K.: A method for the solution of certain non-linear problems in least squares, Q. J. Appl. Math., II, 164–168, 1944.

Peacock, J. R.: Millimetre wave permittivity of water near $25\,°C$, J. Phys. D-Appl. Phys., 42, 205501, doi:10.1088/0022-3727/42/20/205501, 2009.

Savitzky, A. and Golay, M. J. E.: Smoothing and Differentiation of Data by Simplified Least Squares Procedures, Anal. Chem., 36, 1627–1639, 1964.

Tessmann, A., Kuri, M., Riessle, M., Massler, H., Zink, M., Reinert, W., Bronner, W., and Leuther, A.: A Compact W-Band Dual-Channel Receiver Module, in: Microwave Symposium Digest, 2006, IEEE MTT-S International, San Francisco, CA, USA, 11–16 June 2006, 85–88, 2006.

Weber, R., Lewark, U., Leuther, P., and Kallfass, I.: A W-Band 12 Multiplier MMIC With Excellent Spurious Suppression, IEEE Microw. Wirel. Co., 21, 212–214, 2011.

Wirth, W.: Radar Techniques Using Array Antennas, 2nd Edition, IET – The Institution of Engineering and Technology, 560 pp., 2001.

Zech, C., Hülsmann, A., Kallfass, I., Tessmann, A., Zink, M., Schlechtweg, M., Leuther, A., and Ambacher, O.: Active millimeter-wave imaging system for material analysis and object detection, in: Millimetre Wave and Terahertz Sensors and Technology IV, Proceedings of SPIE 8188, 81880D, 2011.

3

Lab-on-Spoon – a 3-D integrated hand-held multi-sensor system for low-cost food quality, safety, and processing monitoring in assisted-living systems

A. König and K. Thongpull

Institute of Integrated Sensor Systems, TU Kaiserslautern, 67663 Kaiserslautern, Germany

Correspondence to: A. König (koenig@eit.uni-kl.de)

Abstract. Distributed integrated sensory systems enjoy increasing impact leveraged by the surging advance of sensor, communication, and integration technology in, e.g., the Internet of Things, cyber-physical systems, Industry 4.0, and ambient intelligence/assisted-living applications. Smart kitchens and "white goods" in general have become an active field of R&D. The goal of our research is to provide assistance for unskilled or challenged consumers by efficient sensory feedback or context on ingredient quality and cooking step results, which explicitly includes decay and contamination detection. As one front end of such a culinary-assistance system, an integrated, multi-sensor, low-cost, autonomous, smart spoon device, denoted as Lab-on-Spoon (LoS), has been conceived. The first realized instance presented here features temperature, color, and impedance spectroscopy sensing in a 3-D-printed spoon package. Acquired LoS data are subject to sensor fusion and decision making on the host system. LoS was successfully applied to liquid ingredient recognition and quality assessment, including contamination detection, in several applications, e.g., for glycerol detection in wine. In future work, improvement to sensors, electronics, and algorithms will be pursued to achieve an even more robust, dependable and self-sufficient LoS system.

1 Introduction

The joint surging advance of sensors, communication, and integration technology allows the realization of more versatile and pervasive systems in nearly all domains of industry and daily life. Established and emerging application domains are, e.g., measurement, instrumentation, and automation, Industry 4.0, the Internet of Things, cyber-physical systems, and ambient intelligence/assisted living. Smart environments, in particular, in homes are a prominent example, where deeply embedded intelligent sensory systems add significant functionality unobtrusively merged into everyday life structures and devices. Miniaturization of such autonomous, potentially wireless, sensory systems for distributed measuring and observation can be found from sensate floors, over leading edge integrated data loggers to lifestyle and sportive gadgets, such as smart watches (Edwards, 2013) or activity trackers (Meyer and Boll, 2014).

Further intriguing research work is done in the field of lab-on-chip devices; see, e.g., Xu and Chakrabarty (2009), Spiller et al. (2006), Yang and Bashir (2008), and Bajwa et al. (2013). Zhao and Chakrabarty (2010) cover a wide field from medical to food applications. Point-of-care diagnostics are one thrilling field of lab-on-chip application systems. Advanced microelectromechanical systems (MEMS) and packaging technology are employed for potentially disposable system solutions with a high level of sophistication and potential price tags. Though this class of systems has numerous features in common with the Lab-on-Spoon (LoS) research presented in this paper, and has inspired the naming of the project, subtle differences, e.g., in cost, embodiment, reuse, mobility issues, autonomous system implementation, and system level integration can be identified. In perspective, a convergence of lab-on-chip technology and the Internet of Things, cyber-physical production systems, and Industry 4.0 application fields, e.g., for in-line and portable measurement, can be expected. The information obtained by such sensory systems can serve to achieve improved or novel assistance functionality in various domains of daily life, e.g., in the domain of nutrition, to assist the user in estimating achieved calorie burning in sports (Meyer and Boll, 2014) and the calorie contents of food, or even in providing better performance to unskilled users or supporting impaired persons in restoring lost sensing capability. In the kitchen environment, numerous product and research activities can be found to improve device performance or to achieve assistance system functionality. Relevant commercial and research work has been summarized, e.g., in König (2008), indicating the potential of sensing and sensory context to achieve a new class of assistance system for this domain, denoted as culinary-assistance systems (König, 2008). A second, related field of application, or better, concern, has emerged in the last years. Sources of unintentional or intentional contamination of soil and sea, e.g., by radiation, chemical, or biological pollution, have increased substantially and, correspondingly, contaminated food can enter the food chain from various sources and reach the consumer unnoticed. For instance, the omnipresent problem of product fraud or falsification, e.g., for frying oil (Qian and Xiaofang, 2014), has also become more noticeable in semi-processed and processed food supplies. The increasing need for food quality and safety monitoring adds further momentum to the outlined research to provide such support at a feasible cost in the consumer's home.

This challenge, in addition to the work presented here, triggered activities on assistive technologies in food safety and food analysis. One example is the portable Bio-Scout of SARAD GmbH (Streil, 2012), designed to discover radiation in food. Quite recently, e.g., the Vessyl (Vessyl, 2014), a metal cup claiming to be able to identify cup contents and give a reckoning of the calories, and the Thai e-tongue (Fuller, 2014), as a particular representative of the research on artificial degustation systems or e-tongues, reviewed in Tahara and Toko (2013), conceived to measure and assure the quality and authenticity of original Thai food, have emerged.

In our research, we aspire to contribute to advanced living-assistance systems for smart homes and, in particular, kitchen environments (see Fig. 1), that receive information from new smart autonomous devices, which are inspired by the technologies summarized above and embodied as common items of daily life, e.g., bowls, cups, forks, or spoons. The focus is on sensing principles and packaging technology that allow the achievement of low-cost, low-power, high-volume, multi-sensory integrated intelligent sensory systems and devices for both cooking assistance and food safety.

To achieve this goal, pioneering work by MIT Media Lab on smart or intelligent spoons by Selker (2013) is picked up and extended to wireless communication, advanced packaging, and low-cost multi-sensor capability, in particular impedance spectroscopy, which is a method of increasing impact and applicability, e.g., in Macdonald (1992), Spiller et al. (2006), and Yang and Bashir (2008). However, in the majority of applications, powerful but expensive and bulky

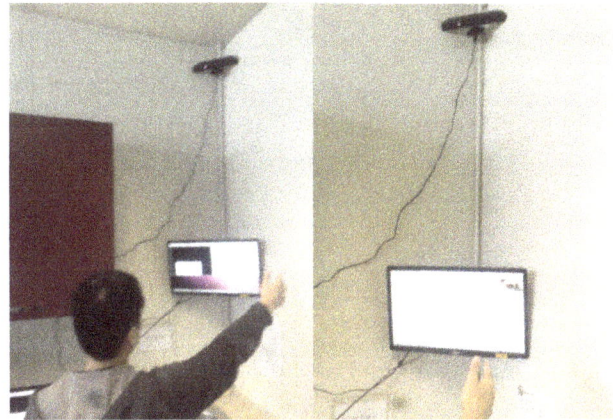

Figure 1. Gesture-controlled interactive cookbook for LoS sensor context acquisition before or after recipe food processing steps.

desktop equipment, e.g., an HP4195A network analyzer with an impedance measurement extension, Agilent 4294, LCZ meter model 4277A, or Xiton Hydra 4200, etc., are used. Applications are in the field of bio-impedance spectroscopy (Spiller et al., 2006) and electrochemical-impedance spectroscopy (Yang and Bashir, 2008), and medical tasks like skin cancer or wound healing monitoring (Schröter et al., 2013) or fish, liver, or meat freshness determination (Guermazi and Kanoun, 2013), tea quality (Xi-Ai et al., 2011) or general food monitoring in the food industry (Ghosh and Jayas, 2009), as well as water monitoring and detergent concentration determination (Gruden et al., 2013) in, e.g., dishwashers. The size of common instrumentation equipment hampers the system realization beyond discrete proof-of-principle prototypes. This has motivated various dedicated embedded designs based on off-the-shelf components and PCB integration. However, the existing commercial solutions, such as the AD5933 chip, cover only a small part of the interesting impedance, frequency range, and measurement quality for the different application domains, which stimulates ongoing dedicated chip design activities.

In this work, the concept and the first prototype of our Lab-on-Spoon will be presented. Section 2 will describe the concept and architecture of LoS, Sect. 3 will give details of the first LoS prototype, and Sect. 4 will describe selected applications and conducted experiments, including applied computational intelligence methods and tools. Concluding, the motivation for a custom CMOS chip and a reconfigurable, potentially MEMS-switch-based LoS will be discussed, and a preview of ongoing activities for an improved LoS version will be given.

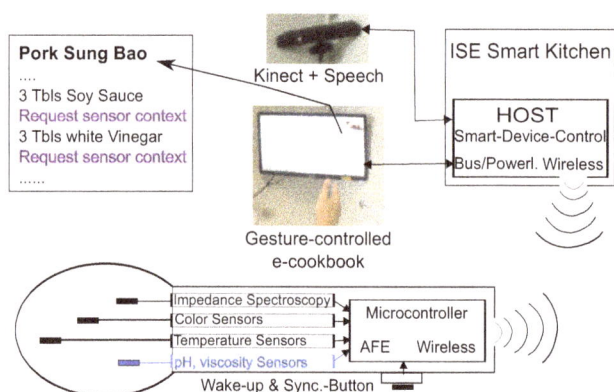

Figure 2. Block diagram of the aspired-to LoS system cooperation with the smart kitchen host for sensor context acquisition.

2   LoS concept and living-assistance system architecture

Figure 2 illustrates by a block diagram the concept of the proposed living-assistance system, and, in particular, the LoS as one possible autonomous sensory front end to it. The institute of integrated sensor systems (ISE) central smart kitchen host is designed to communicate with various smart devices for activity recognition and sensor context acquisition by wire, e.g., standard bus or power-line communication, or standard wireless communication, such as XBee. For instance, Sensitec current sensors are applied for activity recognition and loading assessment of electrical appliances, and the popular MS-Kinect sensor is applied for gesture control together with an emerging electronic or interactive cookbook (see Fig. 1), which is inspired by professional tools like ChefTec (see König, 2008). As exemplified in Fig. 2 for a few extracted recipe lines from a simple Chinese dish, for each step of ingredient inclusion and preparation, the e-cookbook is conceived to call on the sensory context, which will be provided here by the LoS, and by further resources emerging in our current research and development. In this step, ingredients can be checked by LoS for agreement with the current preparatory step, freshness, fraud, or contamination. The LoS is equipped with a microcontroller that runs the measurement control, the sensor readings, and the communication software to collaborate and exchange data with the smart kitchen host indicated in Fig. 2. Power awareness of the autonomous measurement system equipped with a rechargeable accumulator is achieved, e.g, by employing power-saving devices and sleep modes, etc. Conceptually, sensors for integrated temperature, color, infrared, impedance, pH, viscosity, weight, as well as radiation measurement, are aspired to.

Wireless communication with the smart kitchen host is realized by an RF module, e.g., by the XBee standard. The sensory context for each step in the recipe can be acquired and processed by computational intelligence methods on the host for food assessment. Thus, wrong or inadequate or even

dangerous ingredients potentially can be detected, indicated to the user, and excluded from further processing. In addition to ingredient monitoring, quantity determination for dosing, and the assessment of intermediate cooking step results are aspired to as part of the concept.

LoS data have to be processed by suitable means to allow the assessment according to common fuzzy textual statements in recipes on color, consistency, crispiness, etc., of the meal components. In addition to standard textual recipe information, LoS sensory contexts from successful (expert) meal preparations could be stored for each step to provide an even better basis for assessment of the current user activities.

It is anticipated that the spoon will be in sleep mode and get woken up by either a button press or, alternatively, a wake-up by radio from the host. The button is allocated in the spoon handle and serves also to synchronize the taking of the measurement. The system can visually, possibly complemented by standard speech output, prompt the user to enter the required ingredient into the spoon and confirm this by pressing the same button. That procedure helps to avoid the spurious analysis of LoS contents, e.g., of an empty spoon, of a spoon filled with residuals of prior activity, or of a spoon in cleaning. Measurement data will be acquired, sensory context data will be communicated to the host, and the LoS will go back to sleep again.

The potentially wide ranges of measurement quantities for different ingredient categories as well as tolerance and calibration issues require the on-the-fly reconfiguration capability for LoS. Some reconfiguration and self-x features are already included; more are under consideration for the implementation of the next LoS version. This will be discussed in more detail in Sect. 4.3.

The physical realization of the concept will exploit contemporary techniques of 3-D printing and the corresponding 3-D integration or packaging of sensors and electronics. 3-D printing allows the easy, low-cost, and rapid realization of arbitrary prototype shapes, including even sintered metals in the spectrum of materials. Suitable spoon shapes can be created in which sensors and electronics could be snapped into place. More advanced system-in-package approaches and technologies, e.g., the molded interconnect device or the active multi-layer technology (Hofmann GmbH, 2013), merge electronics and package design, and also seem very promising for LoS embodiment.

3   First multi-sensor LoS prototype

A subset of the outlined multi-sensor LoS concept and architecture outlined in the previous Sect. 2 has been realized in a hand-held, multi-sensor, and autonomous system implementation and embodied in spoon shape for the first time.

In Fig. 3, the block diagram of the LoS prototype, including three different sensors, is given. Their geometrical alignment in the spoon cavity is outlined in Fig. 4 accord-

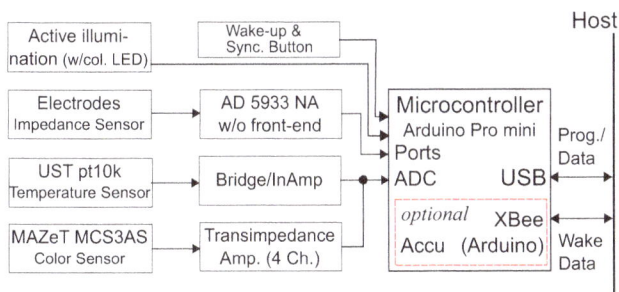

Figure 3. Block diagram of the current multi-sensor LoS prototype.

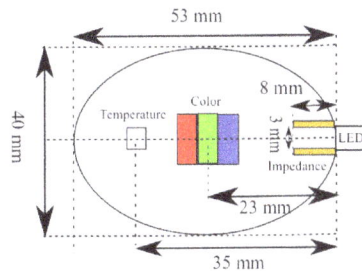

Figure 4. Sketch of the LoS cavity with geometrical information of cavity sizing and sensor placement.

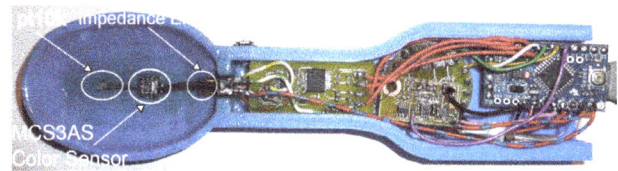

Figure 5. Multi-sensor electronics with USB interface for programming and host communication in a 3-D printed spoon of a LoS prototype before finalization and encapsulation.

ing to the LoS picture given in Fig. 5. The spoon cavity has a maximum depth of 7 mm. A ceramic substrate pt10k temperature sensor of UST GmbH, which comes with a custom calibrated PCB along with a corresponding third-order calibration polynomial, is placed in the center front of the cavity. The MCS3AS true color sensor with its corresponding MTI04QS transimpedance four-channel amplifier chip from MAZeT GmbH is placed in the center of the cavity with the objective to be always completely submerged in the liquid to be analyzed. The impedance spectroscopy measurement unit consists of the AD5933 (Analog Devices, 2011) network analyzer chip and gold-plated electrodes, which are placed at the center back of the cavity, just below the active illumination LED. By this placement, they will be completely immersed even for a low degree of spoon cavity filling. The placement of the sensors is mainly subject to the constraint to avoid variations in measurement due to uncertainty in filling level. For the impedance spectroscopy, the basic two-wire measurement approach is applied together with an optional simple analog front end for materials of very low impedance, e.g., ingredients of high salinity.

This complementing standard circuit reduces the output voltage and prevents the AD5933 from overloading by excessive output current in low-impedance measurements below the $1\,k\Omega$ range. For substances of higher impedance, e.g., oils, the AD5933 will be just used straight, bypassing the front end. Of course, the amplification of the AD5933 input stage has to be reconfigured too with regard to the aspired-to impedance measuring range and the use or bypass of the front end. These setting requirements and cali-

bration issues give rise to the reconfiguration concepts discussed in Sect. 4. Additional reconfiguration requirements come with the color sensor. Depending on the illumination intensity, the transimpedance of the sensor electronics can be digitally adapted or programmed in three stages in the MAZeT MTI04QS chip.

However, to provide reduced vulnerability to environmental illumination variations, LoS has been equipped with an active illumination of the spoon content by a white-light LED, which is activated during color measurement. Temperature and color values will be converted to digital by the 10 bit ADC of the microcontroller, while impedance values will be converted by the internal 12 bit AD5933 ADC.

In this context, a further feature for user interaction has been added, which gives a haptic feedback on the spoon contents' temperature. Inspired by previous activities of, e.g., MIT's smart sink (Selker, 2013), where tap water was illuminated by a color coding the temperature, from cold (blue) to very hot (red), to warn the user and avoid injury due to scalding, the LoS was extended. In wake-up state, without host request on data, the spoon in this mode continuously measures the temperature of the spoon contents and illuminates the spoon contents in a corresponding color by a full color LED. This is illustrated in Fig. 6. In addition to that warning function, thresholds and color assignments can be reconfigured by the host, e.g., in tasks where a liquid has to be in an arbitrary interval or even meet an exact value. The water temperature for yeast bacteria cultivation in bread making is one example, which should best be about 30 °C.

The microcontroller and system of choice for LoS, after considering and testing alternatives, in particular the low-power EnergyMicro Cortex M3 EFM32G890 F128 microcontroller, was the Arduino system family, due to the well-known properties of flexibility, easy and rapid prototyping, and a large portfolio of modules and accessories. The Arduino system family with wireless extensions has evolved to be very popular for Internet of Things realizations. For reasons of constrained space, the Arduino Pro Mini board, 5V supply version, equipped with the ATmega 328 microcontroller had been chosen, which provides time-multiplexed ADC inputs for temperature and color sensor reading, I2C bus support for the AD5933 communication, and general purpose digital I/O, including a PWM option for button reading, as well as white light and color LED control.

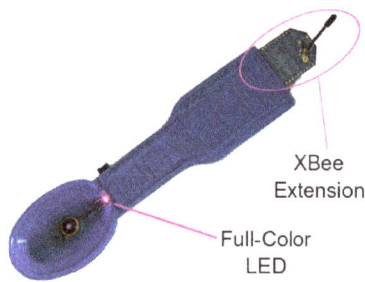

Figure 6. Autonomous LoS prototype with XBee extension board and haptic color feedback on spoon content temperature.

A USB module has been included in the system design for (re)programming of the microcontroller, as well as wired supply, host communication and data transfer for basic LoS operation in the development phase. Figure 5 shows an early prototype of our LoS system without accumulator/XBee extension and before finalization and package sealing. For autonomous operation, the system is completed by a lithium polymer accumulator and a standard XBee communication module. This option is pointed out in Fig. 3 by the only dashed box, labeled Accu/XBee(Arduino), and the corresponding physical realization is illustrated in the top right corner of Fig. 6 as an plug-on board extension to the spoon handle. The spoon package itself has been shaped in a first-cut design employing the Blender software and the Makerbot Replicator I 3-D printer. The currently employed thermoplastic spoon package serves only for the first step of evolution and has to be replaced by a material more robust to the full range of common cooking temperatures and food safety regulations.

The LoS embedded system software is developed in C in the standard Arduino development environment 1.5.6. On the host side, a Python-based interface communicates with the LoS via the serial interface and reads the data in for the application software. Depending on the connection, directly by USB during development or by wireless in the actual application, a second Arduino/XBee module is required as a gateway on the host side.

4 Experiments and results

In the following, front-to-back application of the LoS from sensory registration to data analysis and recognition system design and employment will be investigated for selected liquid food samples. The proprietary ISE QuickCog tool (König, 2008; König et al., 1999) offers the required facilities to analyze, fuse, and recognize the acquired multi-sensor data, but the integration of LoS spoon control and a data acquisition interface seemed to be more promising on a multiplatform flexible and open system, which is provided by the Python-based Orange system (Demšar et al., 2013). So, in this work, Orange was extended on both the Windows and Linux platforms by the LoS interface and a QuickCog interface to immediately exploit features and efficient methods from the field of computational intelligence still unavailable in Orange. Moving these features and methods to Orange and public availability is one of our research goals in the context of LoS research.

Plots for interactive data visualization and analysis can be generated by, e.g., the choice of two particularly relevant variables from the measurement, or by dimensionality reducing mapping techniques, e.g., by Sammon's nonlinear mapping and related fast techniques (König, 2001; König et al., 1999). The first option is very transparent and the axes of such a scatterplot have a clear physical notion. However, in particular, in multi-sensory systems, the required salient information rarely is provided by only two or three measurement inputs or variables. The latter option can deal with measurement data of arbitrary dimensionality and reduces the data dimensionality to a 2- or 3-D plot under the constraint of, e.g., preserving the distances of the data points as undistorted as possible. These distances represent the similarities of data points and their underlying physical measurement data. In such a plot, commonly denoted as a feature map or a feature space projection, the plot axes have no assignment to a particular physical notion.

In gas sensing, commonly linear discriminant analysis is also employed for the same visualization purpose. As this is a supervised method, which includes the labeling or class affiliation of the measurement in the dimensionality reducing mapping, we prefer the unsupervised Sammon nonlinear mapping and employ it for LoS data presentation and assessment in the following.

The maps thus generated can help in understanding and optimizing both data acquisition and processing for ingredient recognition/identification or grading in new applications (see Figs. 10–16). The feature map or feature space projection can be complemented by labeling the data points with class information as well as temperature context information, which could stem from the pt10k sensor in the spoon volume or the AD5933 or other chips' internal temperature sensors. Furthermore, QuickCog provides a set of automated feature selection (AFS) options, which assist in efficient feature-level fusion from the different sensor channels to obtain well-discriminating but lean intelligent systems. For the generation of the following feature maps and the final results table, the q_{oi} overlap measure with k neighbor parameter $k = 5$ and the simple sequential-forward selection (SFS) scheme (König et al., 1999) has been employed. AFS leads to more lean, in some cases better performing decision systems, but the particular advantage in impedance spectroscopy is that only a few case-specific spectral components have to be measured, and the measurement time can be reduced significantly. The full sweep will only be needed in the analysis of new tasks and related measurement data.

In the following, two different kinds of experiments were conducted. In the first group, the discrimination capability

```
#Lab-on-Spoon waiting for Button press
#Number of data lines: 516
#Number of measurements: 30
#Temperature:22.70
#Color-B: 1.6396
#Color-G: 1.7476
#Color-R: 1.7932
#Data:
Time,  Mag,  Phas,   Freq
3.205,  37.33,  181.18, 10.000
3.267,  36.47,  176.09, 10.176
3.330,  36.49,  179.43, 10.352
3.391,  37.44,  180.45, 10.528
 .
 .
 .
34.494,56.76,187.05, 99.936
#End of Measurement Cycle: 1
```

Figure 7. LoS output of temperature, color, and 4 of 512 lines of raw uncalibrated impedance data sent to the host for further processing.

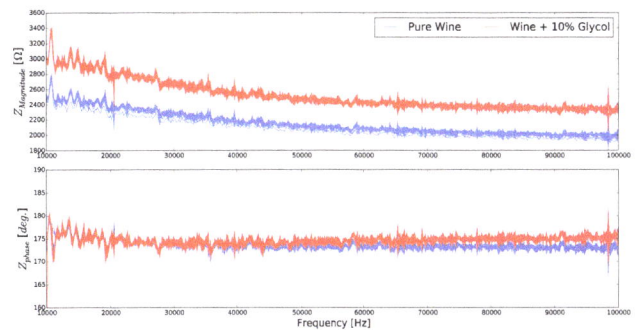

Figure 8. Plot of impedance magnitude and phase spectra for the ingredient data given in Fig. 16. The ordinate values of the magnitude plot have been computed by multiplying the raw data from Fig. 7 by the calibration factor 54.89 of the employed LoS prototype.

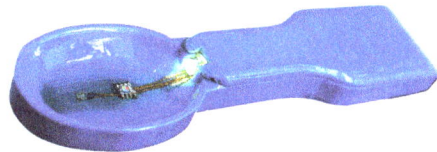

Figure 9. Completed LoS prototype with active white LED illumination for the color measurement phase.

of the LoS for various common cooking ingredients will be investigated and demonstrated.

In the second group of experiments, the LoS grading capability with regard to loss of freshness or the presence of contamination will be tentatively investigated and demonstrated.

For all experiments, the following settings for sensor signal conditioning have been applied. The transimpedance setting of the LoS color sensor is set to 500 k. The frequency sweep range of the AD5933 chip is 10–100 kHz, with a frequency increment of about 175 Hz. The temperature sensor module is calibrated, has no setting options, and returns a temperature value in degrees Celsius. LoS gives, for each measurement, 1 temperature value, 3 values of RGB channels for color registration, and 512 complex impedance values, given as 1024 magnitude and phase values. This amounts to the data vector of 1028 entries exemplified in Fig. 7.

In addition, Fig. 8 illustrates the plot of the impedance magnitude and phase spectra from LoS measurement data for several examples of the contamination recognition and data given in the following Fig. 16. Each point in Fig. 16 corresponds to one measurement, i.e., one of the spectra given in Fig. 8.

For the measurement of substances with a low impedance or high impedance range, two different LoS prototypes with an employed or bypassed analog front end, as indicated in Fig. 3, were employed in the work described in the following.

In the following, 30 measurement repetitions have been adopted as the standard for each substance or ingredient.

The measurement currently proceeds in three phases measuring the temperature first, followed by accumulative color registration of 50 samples of each RGB channel with active white-LED illumination switched on (see Fig. 9), and con-

cluding with the impedance spectroscopy measurement in the given frequency sweep range. After the scheduled number of measurements, which were triggered by the push button and/or by wireless, the LoS goes back to sleep.

4.1 Ingredients recognition

In the first experiment, the discrimination of basic liquid cooking ingredients has been examined, e.g., by filling or immersing the spoon in plain tap water, salted water, soy sauce, or white wine vinegar.

Figure 10 shows the resulting feature map with an AFS result of four features, i.e, three color values and and feature 2 from impedance magnitude, which groups the four ingredients clearly with regard to their basic conductivity.

The next application tried to distinguish four brands of beer. The result along with the brand names is given in Fig. 11. Color was not as helpful as in previous cases, as two of the Pils-type beers nearly feature the same color. The list of the 118 features selected from impedance magnitude data can be found in Appendix A1.

Though the basic discrimination is possible by LoS, the margins between the classes with regard to intra-class scatter are not as favorable as in the previous case. This demands the improvements in the electronics as outlined in the discussion in Sect. 4.3.

The LoS tasting ability has been further challenged by the task to distinguish up to seven kinds of wine. The result along with the wine kind names and origin is given in Fig. 12. The

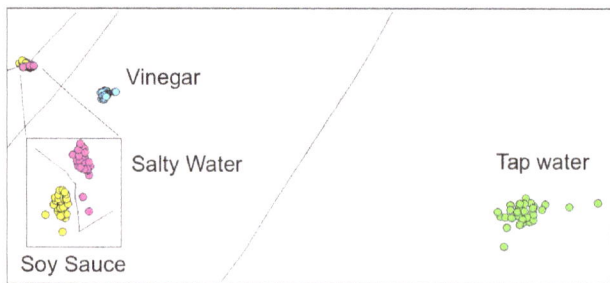

Figure 10. LoS measurement results for the recognition of plain tap water, salted water, soy sauce, and white wine vinegar.

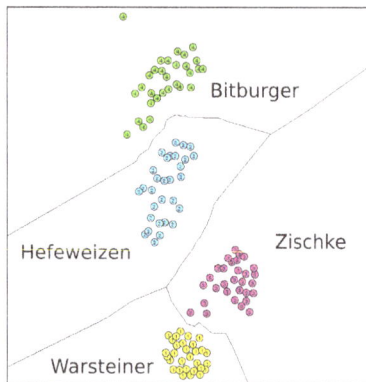

Figure 11. LoS measurement results for the recognition of four kinds of beers.

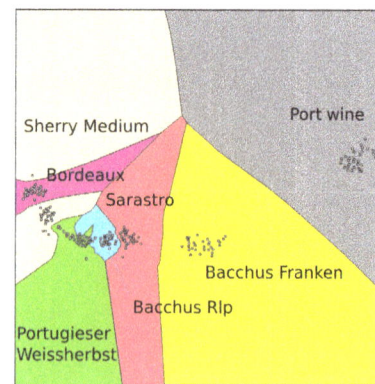

Figure 12. LoS measurement results for the discrimination of seven kinds of wine.

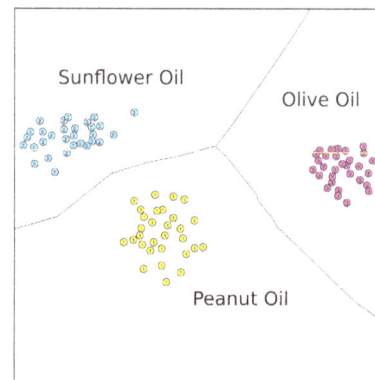

Figure 13. LoS measurement results for the recognition of three plant oils.

list of the 33 features selected from impedance magnitude data can be found in Appendix A2.

This example shows the basic distinguishing capability of the current LoS prototype, but also the need for improvement for more comprehensive and robust operation.

In contrast to the first experiments with ingredients of rather low impedance, oils feature high to very high impedance values. The last experiment of the first group shows the high-impedance LoS capability to distinguish three different kinds of common cooking oils, i.e., sunflower, peanut, and olive oil.

Figure 13 shows the resulting feature map based on six features, i.e., the three color values and the impedance magnitude values 385, 387, and 437, with clear discrimination capability.

4.2   Ingredient grading

In addition to the correctness or appropriateness of the ingredient with regard to the current preparation step, the state or the quality of the ingredient needs assessment. This could be accomplished with the goals of determining the freshness or potential rottenness of the food or the presence of contaminations in the otherwise fine ingredient.

The degradation of oil is a common issue, which will be regarded first in the following. The plant oil of a home frying machine, e.g., for french fries preparation, was investigated with regard to wear-out and the need for oil exchange. Fresh oil was compared to used oil, which according to human impressions of look, smell, and cycles of use, was due for exchange. Data were acquired as in the experiments before and grading pursued in a crisp form, distinguishing just fresh and worn-out oil.

Figure 14 shows the corresponding feature map with two distinct, well-separated clusters for fresh (left) and heavily used (right) frying oil. Here, the color information with the three color values as features itself is already very meaningful. The impedance magnitude delivers similar information with the selections 370, 443, 445, and 491 as features. In the following classification, feature-level fusion of these groups will be applied. It is acknowledged that infrared spectroscopy is a common and successful method for oil and lubricant quality sensors, but in the food domain, a low-cost solution would be welcome.

Also, the spoon embodiment is not required for the frying machine or related machinery. The frying oil classification or grading opens the door to a separate application system, i.e., a stationary oil sensor for indication of wear-out determined

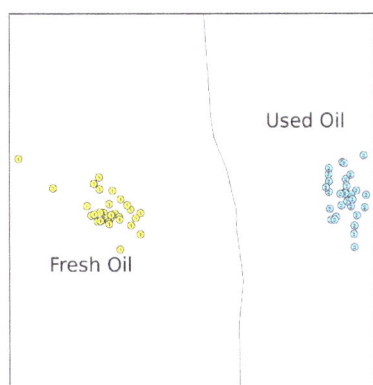

Figure 14. Feature space projection of the frying oil data set.

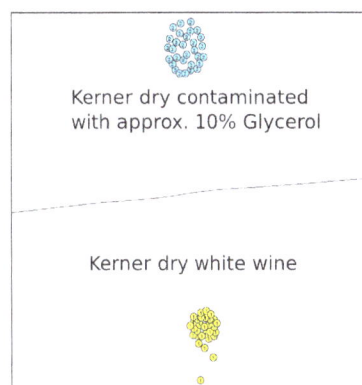

Figure 15. Feature space projection of the milk decay data set.

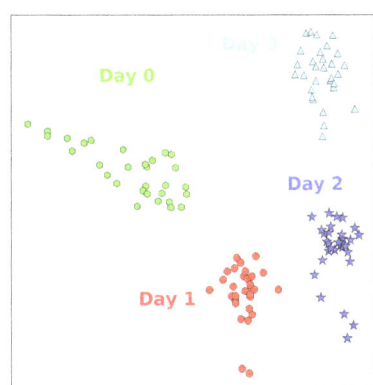

Figure 16. Feature space projection of pure and contaminated white wine.

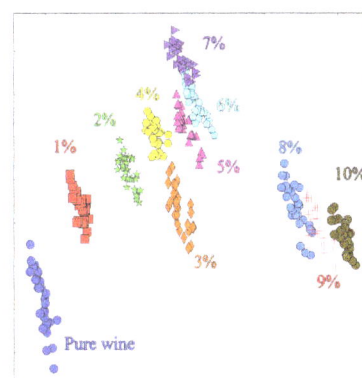

Figure 17. Feature space projection of pure and ten different glycerol concentrations from 1 to 10 % of contaminated white wine.

oil exchange. In the next example, the detection of decay or rotting in milk by LoS, as a common food of daily use, is investigated tentatively. The first sample is taken of fresh milk with 3.5 % fat content. A quantity of about 150 mL of the same milk is left outside the refrigerator and uncovered in a glass and over a period of 4 days; 1 time a day, LoS is filled with about 10 mL taken from the glass and 30 samples of the spoon contents are measured.

Figure 15 shows the feature space projection obtained from 4 × 30 samples. AFS chose the red color value and the values 4, 18, and 337 from the impedance magnitude as features.

After the fourth day, the milk started to be clearly degraded from visual and olfactory appearance. Due to the variety of milk, involved bacteria, and other potential influences on the rotting process, this experiment is clearly indicated as tentative, but it nevertheless shows the LoS basic ability to give a warning about potential degradation of the ingredient milk and to help the assistance system to dissuade the user from further use.

The last example deals with the issue of detecting food contamination. As an example, we were inspired by a real occurrence about two decades ago, where wine was sweetened by the addition of about 5 % of a chemical substance

(diethylen glycol) usually serving as an anti-freezing agent. Here, we use the less poisonous chemical glycerol and add it in a ratio of about 10 % to the dry Kerner white wine in a first experiment, which has been the basis for numerous LoS life demonstrations. Figure 16 shows the result from one of the conducted measurements for selected features 465 and 497 of the impedance spectroscopy magnitude. The obtained data were employed to train and validate an SVM classifier (Muller et al., 2001) as given in Fig. 18. This example was successfully used for life classification in LoS demonstrations, as intended in the actual practical use of the spoon.

In addition to this demonstration, a more detailed investigation of LoS sensitivity has been carried out by acquiring a new data set, measuring pure wine and ten contaminated samples with an increase in glycerol concentration of 1 % from 1 to 10 %. In Fig. 17, the result of this measurement series with 11 × 30 samples is shown. Clearly, all glycerol concentrations can be well separated from the pure wine data. This has been confirmed by the ensuing classification experiments given in Table 1, both for the full feature set and the AFS reduced set with the impedance magnitude features 1, 3, and 4. In addition to the standard experiments summarized

Data Sampling

Data Sample → Data

Save Train data set

Data

Remaining Data → Data

Recorded data file   Data Sampler

Save Test data set

Training Flow

Selected Data → Data   Classifier

Data

Save Classifier

SVM

LoS Feature Selector
train

Load Train data set   Support Vectors → Data

Data

Selected Data → Data

Selected Data → Separate Test Data

Load Test data set

Validation   Info

LoS Feature Selector
test

Live Classification

Received Data → Data   Selected Data → Data

Selected Data → Predictions

LoS Serial Port
Interface

Received Data → Data   Selected Data → Data

Predictions

Predictions → Data

Load Classifier

Save fresh data   LoS NIF file
converter   Classification
results

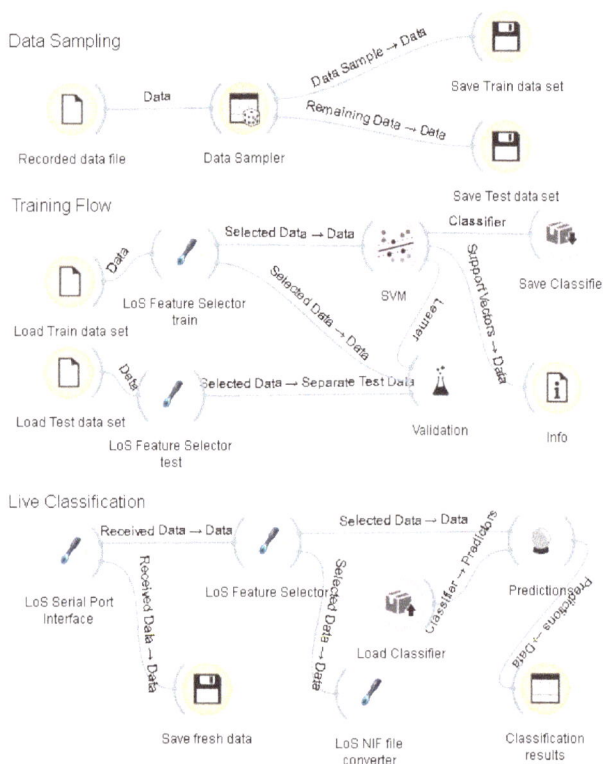

Figure 18. SVM-based classification system for wine contamination detection in Orange running on the smart kitchen server as a subroutine of the e-cookbook.

below, for the complete feature set, data have been sampled with only 20 % training and 80 % testing data. Even then, the same results as in Table 1, i.e., 100 % classification accuracy, have been obtained for all classes, which means perfect generalization. This result implies that LoS could also be employed to predict the concentration, at least in steps of 1 %.

The extension of LoS from the presented multi-class recognition to continuous grading of food properties by function approximation, e.g., RBF networks (Haykin, 1994) and/or support-vector regression (SVR), for oil, milk, wine, and related liquid ingredient assessment, is currently in progress, with promising perspectives.

It is assumed that for each task in ingredient or processing result inspection, trained decision making units are archived and are modularly available in the e-cookbook along with the recipes.

All the described experiments have been subject to classification investigations to also numerically assess LoS discrimination abilities. Table 1 shows the results of a hold-out approach with random data splitting in 50 % training and 50 % testing employing an SVM RBF-kernel classifier. Complete and selected data have been used, based on the detailed feature selection lists given in the text above or in Appendices A1 and A2 for each example. The generalization capability, but for two data sets, was perfect.

Table 1. LoS results for ingredient recognition and grading.

Experiment	Sel. features (col.; mag.)	SVM par.		No. of SV	CA (%)
		C	γ		
Soy	All	2	0.03125	31	100
	(3; 1)	512	0.03125	15	100
Beer	All	128	0.125	60	98.33
	(0; 118)	512	0	36	98.33
Wine	All	32	0.125	94	99.05
	(0; 33)	512	0.03125	15	98.10
Oil	All	512	0.125	45	100
	(3; 3)	8	0	17	100
Used oil	All	512	0.5	41	100
	(3; 4)	128	2	7	100
Milk decay	All	128	0	43	100
	(1; 3)	2	0.5	14	100
Glycol in wine	All	512	2	20	100
	(0; 2)	512	0.5	2	100
1–10 % glycol in wine	All	512	0	164	100
	(0; 3)	512	8	165	100

4.3 Discussion

The analysis of the current LoS implementation showed useful sensitivity (see Table 1) and the basic capability to fulfill the raised goals, but also several needs for improvements to sensory data quality and the potential for a significant performance increase. The measurements above show that the intra-class scatter is quite high and, in some cases, approaches the inter-class distances. This can be reduced to about 50 % by the use of an external precision clock, a higher-quality 3V supply, and a front end without DC excitation voltage output as in our related DeCaDrive project (Li et al., 2013) for the AD5933. The DC voltage output degrades the measurement in general, e.g., due to polarization effects. Color measurement currently is impaired by the limited ADC on the microcontroller; i.e., RGB values are sequentially obtained and at mediocre bit resolution. Therefore, the three-channel RGB color sensor will be replaced by the multi-spectral MAZeT MMCS6CS color sensor and related improved signal conditioning and MCDC04 conversion chips with synchronous high resolving measurement of all channels. These changes will significantly improve LoS selectivity and stability. Furthermore, additional digital reconfiguration and self-x features (see, e.g., Johar et al., 2011), based on suitable MOS- or MEMS-switching resources, are considered (Johar and König, 2013). This is required for, e.g., switching from high to low impedance ranges, changing the feedback resistor in the I/O amplifier, and alternating between calibration elements and actual measurement impedance, e.g., immersed electrodes.

The extension of LoS implementation with integrated pH-value and viscosity sensing, the improvement in the 3-D spoon shape and electronics packaging, e.g., with 3-D printing technologies with higher resolution, higher temperature tolerance, and food safety regulation compliance, as well as the advance of related sensor fusion and intelligent system design techniques, are the next steps on the agenda.

A key issue remains employing the range-limited AD5933 chip for impedance spectroscopy realization. Employing an Agilent network analyzer for the same and related tasks showed that, even for a sweep range of up to 2–4 MHz, previously regarded tasks could be solved with higher accuracy, and a wider scope of applications and substances can be distinguished. Thus, the design of a more able dedicated CMOS chip for impedance spectroscopy, which employs differential current stimulation and the four-wire measurement approach, and which is conceived to be applicable for the needs of a wider range of integrated/embedded impedance spectroscopy applications (König, 2008; Li et al., 2013; Guermazi and Kanoun, 2013; Gruden et al., 2013; Schröter et al., 2013), is currently being pursued at ISE.

5 Conclusions

This paper presented the concept and the first embodiment of the Lab-on-Spoon autonomous, hand-held, multi-sensory system as a low-cost front end of an intelligent assisted-living system with a distributed sensor network for home-based smart kitchen applications (König, 2008). The LoS is designed to provide sensory context to the ingredient and preparatory step results monitoring of recipes in an interactive cookbook and, thus, support both unskilled or challenged persons by improving or partially restoring perceptive and assessment ability. Furthermore, LoS is conceived to detect decay and/or contamination in food, which might be imperceptible to humans. With regard to the severe and increasing pollution of soil and water worldwide, contaminations could reach the consumer undetected, and strongly advocate the creation of capable, yet affordable, local sensing capability at the consumer's end of the food chain.

The first LoS prototype was implemented in our work on the favorable Arduino platform with temperature, color and impedance sensing, and integrated into the Orange system for capable multi-sensor signal processing and recognition and life or online classification, e.g., on CeBIT 2014 (König, 2014). With regard to our goals, it showed encouraging capabilities and sensitivity for a challenging application spectrum from food classification to grading. Improvements in electronics, packaging, sensor palettes, and algorithms, as outlined in the discussion above, are on the way. Self-x features for dependable and self-sufficient operation, as needed for related cyber-physical systems, the Internet of Things, or Industry 4.0 domains, are under investigation for LoS. Further inspirations are expected from the thriving lab-on-chip research field.

In addition to advancement of the scientific LoS development and improvement, including the abstraction to alternative devices, e.g., a lab-on-fork, lab-in-bowl, etc., to probe non-liquid food and materials, such as meat or cheese, etc., commercialization with industrial partners is aspired to, with a potential mass market in mind. Recent competitive approaches underpin this view, e.g., Vessyl (2014) and Fuller (2014). From the results on frying oil, and preliminary experiments on combustion engine oil, an extension of the project to simple and low-cost motor and gear oil sensors seems to be promising and straightforward.

Appendix A: Details on feature selection

In the following, the larger selection lists of Sect. 4 will be given. These lists, though extensive, are relevant for the potential data analysis, repetition of experiment, and possible reduction of measurement time in future applications.

A1   Selection list for Beer experiment

The selected 118 magnitude features are 8, 9, 23, 43, 48, 52, 62, 65, 74, 77, 84, 86, 87, 89, 94, 96, 103, 105, 107, 116, 124, 126, 134, 141, 144, 148, 149, 152, 155, 157, 161, 181, 191, 196, 198, 203, 205, 206, 208, 210, 224, 226, 228, 234, 242, 250, 261, 263, 265, 269, 270, 272, 277, 282, 286, 296, 298, 303, 307, 309, 310, 314, 318, 322, 325, 330, 332, 333, 338, 342, 348, 350, 352, 361, 366, 370, 372, 378, 380, 382, 383, 384, 385, 387, 391, 395, 397, 398, 401, 407, 411, 412, 421, 427, 428, 429, 430, 431, 434, 436, 441, 442, 444, 449, 450, 453, 454, 457, 461, 462, 466, 472, 484, 491, 494, 495, 496, and 501.

A2   Selection list for wine experiment

The following 33 magnitude features have been selected: 1, 35, 46, 71, 99, 111, 142, 148, 152, 161, 229, 265, 270, 275, 280, 286, 294, 335, 352, 370, 372, 373, 382, 385, 398, 403, 435, 438, 444, 462, 484, 490, and 494.

Author contributions. Andreas König conceived the Lab-on-Spoon idea, initiated, and guides the research project as principal investigator. He designed and assembled the Arduino-based LoS prototypes shown in this paper, including the embedded software, designed the experiments, did most of the described measurements described in Sect. 4, and evaluated the data and developed recognition systems based on QuickCog. He wrote the majority of this paper. Kittikhun Thongpull developed the interface for LoS communication and data acquisition based on Python and Orange and an interface to QuickCog, did the milk decay and 1–10 % glycol-in-wine contamination experiments, developed and applied a complete trainable SVM-based recognition system, added life LoS data acquisition capability for demonstration purposes to Orange, and conceived a robustly working life demo for contamination detection in wine.

Acknowledgements. This work follows up and exploits a previous funding by the German Federal Ministry of Education and Research (BMBF) in the mst-AVS program, project PAC4PT-ROSIG grant no. 16SV3604. The work on color sensor modules of Thomas Gräf from 2004 in the Ambient Intelligence of Rhineland-Palatina priority program, the code for AD5933 programming from the master project of Thomas Bölke in our DeCaDrive project (Li et al., 2013), the analog AD5933 front-end board and the 3-D spoon prints from David Los Arcos, and the temperature measurement circuit of UST from the ROSIG project have been employed in adapted form for the reported work. These contributions and those of Abhay C. Kammara to the smart kitchen research and the support of Dennis Groben in software and electronics issues are gratefully acknowledged. In particular, the sponsorship of MAZeT GmbH of the LoS project is gratefully acknowledged.

References

Analog Devices: AD5933 datasheet, Analog Devices, Inc., 2011.

Bajwa, A., Tan, S. T., Parameswaran, A. M., and Bahreyni, B.: Automated rapid detection of foodborne pathogens, in: Solid-State Sensors, Actuators and Microsystems (TRANSDUCERS EUROSENSORS XXVII), 2013 Transducers Eurosensors XXVII: The 17th International Conference, 337–340, June, 2013.

Demšar, J., Curk, T., Erjavec, A., Gorup, Č., Hočevar, T., Milutinovič, M., Možina, M., Polajnar, M., Toplak, M., Starič, A., Štajdohar, M., Umek, L., Žagar, L., Žbontar, J., Žitnik, M., and Zupan, B.: Orange: Data mining toolbox in python, J. Mach. Learn. Res., 14, 2349–2353, 2013.

Edwards, C.: Watch clever, Engineering & Technology, 8, 30–35, 2013.

Fuller, T.: You call this thai food? the robotic taster will be the judge, available at: http://www.nytimes.com/2014/09/29/world/asia/bad-thai-food-enter-a-robot-taster.html (last access: 14 January 2015), 2014.

Ghosh, P. and Jayas, D.: Use of spectroscopic data for automation in food processing industry, Sensing and Instrumentation for Food Quality and Safety, 3, 3–11, 2009.

Gruden, R., Köbele, W., Tran, D., and Kanoun, O.: Online Detection of the Critical Micelle Concentration of Commercial Detergents by Impedance Spectroscopy, in: Abstract Book, Int. Workshop on Impedance Spectroscopy IWIS 2013, Poster, 56–57, 25–27 September, Chemnitz, 2013.

Guermazi, M. and Kanoun, O.: Feature Extraction for Meat Characterization, In: Abstract Book, Int. Workshop on Impedance Spectroscopy IWIS 2013, Poster, 95 pp., 25–27 September, Chemnitz, 2013.

Haykin, S.: Neural Networks, A Comprehensive Foundation, 2nd Edn., Pearson Education Inc., 1994.

Hofmann GmbH: AML-Technology, available at: http://www.hofmann.de, last access: 2 February 2013.

Johar, M. A. and König, A.: Advanced Sensory Electronics and Systems with Self-x Capabilities by MEMS Switch Integration, Proc. of Int. Conf. on SENSORS 2013, 346–351, AMA, Nürnberg, 2013.

Johar, M. A., Freier, R., and König, A.: Adding Self-x Capabilities to AMR Sensors as a First Step Towards Dependable Embedded System, in: Proc. of the 9th WISES 2011, 41–46, 7–8 July, Regensburg, 2011.

König, A.: Dimensionality Reduction Techniques for Interactive Visualisation, Exploratory Data Analysis, and Classification, in: Pattern Recognition in Soft Computing Paradigm, World Scientific, FLSI Soft Computing Series, Vol. 2, edited by: Pal, N. R., ISBN 981-02-4491-6, 1–37, January, 2001.

König, A.: Automated and Holistic Design of Intelligent and Distributed Integrated Sensor Systems with Self-x Properties for Applications in Vision, Robotics, Smart Environments, and Culinary Assistance Systems, Invited Talk, Int. Conf. On Neural Information Processing of the Asia-Pacific Neural Network Assembly (ICONIP'08), Book of Abstracts, 69–70, 25–28 November, Auckland, New Zealand, 2008.

König, A.: Lab-on-Spoon – Multi-Sensorial 3D-integrated Measurement System for Smart-Kitchen and AAL Applications, available at: http://www.eit.uni-kl.de/koenig/gemeinsame_seiten/projects/LabonSpoon.html (last access: 14 January 2015), 2014.

König, A., Eberhardt, M., and Wenzel, R.: Quickcog self-learning recognition system – exploiting machine learning techniques for transparent and fast industrial recognition system design, Image Processing Europe, Vol. Sept./Oct, 10–19, 1999.

Li, L., Bölke, T., and König, A.: Can Impedance Spectroscopy Serve in an Embedded Multi-Sensor System to Improve Driver Drowsiness Detection, in: Abstract Book, Int. Workshop on Impedance Spectroscopy IWIS 2013, 48–49, 25–27 September, Chemnitz, 2013.

Macdonald, J. R.: Impedance Spectroscopy, Ann. Biomed. Eng., 20, 289–305, 1992.

Meyer, J. and Boll, S.: Digital health devices for everyone!, Pervasive Computing, IEEE, 13, 10–13, 2014.

Muller, K., Mika, S., Ratsch, G., Tsuda, K., and Scholkopf, B.: An introduction to kernel-based learning algorithms, IEEE Transactions on Neural Networks, 12, 181–201, 2001.

Qian, Y. and Xiaofang, P.: Methods for differentiating recycled cooking oil needed in china, available at: http://www.aocs.org/Membership/FreeCover.cfm?itemnumber=18028 (last access: 14 January 2015), 2014.

Schröter, A., Gerlach, G., Rösen-Wolff, A., Wendler, J., Nocke, A., and Cherif, C.: Miniaturized Wound Sensors for Chromatin Detection, in: Abstract Book, Int. Workshop on Imp. Spectr. IWIS 2013, Poster, 95 pp., 25–27 September, Chemnitz, 2013.

Selker, T.: Counter-Intelligence Project, MIT, available at: http://www.media.mit.edu/ci/, last access: 2 February 2013.

Spiller, E., Schöll, A., Alexy, R., Kämmerer, K., and Urban, G. A.: A microsystem for growth inhibition test of Enterococcus faecalis based on impedance measurement, Sensor. Actuat. B-Chem., 118, 182–191, doi:10.1016/j.snb.2006.04.016, 2006.

Streil, T.: Bio-Scout, Product Information, SARAD GmbH, available at: http://sarad.de/cms/media/docs/datenblatt/Bio-Scout_Infoblatt_20-12-12.pdf (last access: 14 Januar 2015), 2012.

Tahara, Y. and Toko, K.: Electronic tongues – a review, IEEE Sens. J., 13, 3001–3011, 2013.

Vessyl: available at: www.myvessyl.com (last access: 14 January 2015), 2014.

Xi-Ai, C., Guang-Xin, Z., Ping-Jie, H., Di-Bo, H., Xu-Sheng, K., and Ze-Kui, Z.: Classification of the green tea varieties based on support vector machines using terahertz spectroscopy, in: Instrumentation and Measurement Technology Conf. (I2MTC), 2011 IEEE, 1–5, May, 2011.

Xu, T. and Chakrabarty, K.: Design-for-testability for digital microfluidic biochips, in: VLSI Test Symposium, VTS'09. 27th IEEE, 309–314, May, 2009.

Yang, L. and Bashir, R.: Electrical/electrochemical impedance for rapid detection of foodborne pathogenic bacteria, Biotechnol. Adv., 26, 135–150, 2008.

Zhao, Y. and Chakrabarty, K.: Digital microfluidic logic gates and their application to built-in self-test of lab-on-chip, IEEE Transactions on Biomedical Circuits and Systems, 4, 250–262, 2010.

4

Thin film sensors for measuring small forces

F. Schmaljohann, D. Hagedorn, and F. Löffler

Physikalisch-Technische Bundesanstalt, Bundesallee 100, 38116 Braunschweig, Germany

Correspondence to: F. Schmaljohann (frank.schmaljohann@ptb.de)

Abstract. Especially in the case of measuring small forces, the use of conventional foil strain gauges is limited. The measurement uncertainty rises by force shunts and is due to the polymer foils used, as they are susceptible to moisture. Strain gauges in thin film technology present a potential solution to overcome these effects because of their direct and atomic contact with the measuring body, omitting an adhesive layer and the polymer foil.

For force measurements up to 1 N, a suitable deformation element was developed by finite element (FE) analysis. This element is designed for an approximate strain of $1000 \, \mu\text{m} \, \text{m}^{-1}$ at the designated nominal load. The thin film system was applied by magnetron sputtering. The strain gauge structure is fabricated by distinct photolithographic steps.

The developed sensors were tested with different load increments. The functional capability of the single resistance strain gauges could be proven. Moreover, a developed sensor in a full bridge circuit showed a linear characteristic with low deviation and good stability.

1 Introduction

For many years, the use of resistance strain gauges (RSGs) for deformation measurements of parts has been state of the art. As a consequence thereof, the measurement of force is feasible with force transducers. The force is applied to a geometrically defined deformation element. Due to the deformation, the resistance of the applied RSG changes and, with the knowledge of the material parameters, the applied force, can be determined with high accuracy.

Most common are foil strain gauges, where the strain-sensitive pattern is applied to a flexible polymer foil or comparable backing material. These gauges are then fixed to the designated areas of deformation with a special glue.

Nevertheless, the measurement of small forces with foil gauges is limited, especially in the case of metrological intention. Sources of error that have an influence on the measurement uncertainty are located in force shunts with the RSG and with the susceptibility to moisture of the polymer foil. Additionally, the thickness of the backing material has to be mentioned, as this creates a distance between the actual deformation of the part and the strain-sensitive pattern. Furthermore, both the foils and the glue have a limited durability at high or very low temperatures.

Promising alternatives to foil gauges are thin film resistance strain gauges. The coating method and the subsequent structuring of the film allow an application directly onto the deformation element, without any kind of backing material except, at need, a thin insulation.

In 1992, Karaus and Paul (1992) presented a simple strain gauge based on thin film technology. Furthermore, the potential of this sensor technology for metrology was tested by a more sophisticated strain gauge applied to a 10 kN reference block. Thereby, a consistently higher sensitivity compared to conventional foil strain gauges was determined (Buß et al., 2008).

A technology has been developed at PTB to achieve thin film sensors applied onto metallic materials by sputtering, utilizing a thinner electrical insulation layer. In combination with a very flexible structuring technique, the fabrication is possible also for three-dimensional workpiece geometries (Hagedorn et al., 2007; Schmaljohann et al., 2012).

The high sensitivity and low creep behavior as stated in Buß et al. (2008) are due to the direct and atomic contact achieved by sputtering. In addition, the technology allows further miniaturization of the sensor and, hence, of the force transducer itself.

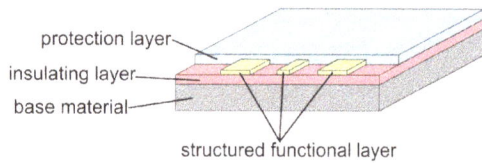

Figure 1. Film system used for the thin film sensors.

This paper presents the design of a compact deformation element for a nominal load of 1 N suited for the application of thin film strain gauges. In addition to the application of the sensors, different tests were carried out. Weights applied to the force transducer facilitate the functional support and its characteristic values. The developed prototype is connected in a full bridge circuit and demonstrates good stability and a linear behavior during strain measurement.

2 Thin film sensor technology

The thin film sensors introduced in this paper are initially based on an electrically insulating layer completely covering the body. By this layer, an electrical separation of the metallic and therefore electrically conducting deformation element from the following metallic sensor layer is achieved. If needed, a top layer can be applied to protect the sensor against environmental influences. The complete film system is depicted in Fig. 1.

To achieve the conductive paths and, thereby, the geometry of the sensor pattern, the second layer (i.e., the sensor layer) is structured by photolithography.

Consequently, there is no need for an adhesive layer and backing material, as is the case when using foil gauges. This makes the thin film sensor thinner by about 3 orders of magnitude.

2.1 Film system and application

All layers were deposited in magnetron sputter system "LS320S" by von Ardenne.

On the substrate, in this case the force transducer, an electrically insulating layer is deposited. This silicon oxide layer has a thickness of less than $5 \mu m$.

Afterwards, an electrically conducting layer is applied, which serves as the sensor later on. Depending on the intended use, a suitable material can be chosen. In this work, a Cu–Ni alloy was chosen to ensure a low temperature dependency of the sensor. It is possible to match the film thickness to the desired sensor characteristics and, in this case, it is only a few $100 \, nm$ thick.

Subsequent to the structuring of the sensor layer, a protection layer can be sputtered on top. Again, this layer is made of an electrically insulating material such as aluminum oxide or silicon oxide.

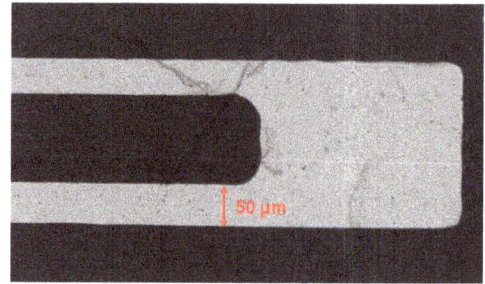

Figure 2. Detail of a structured sensor layer.

Thus, the film system has a thickness of less than $10 \mu m$ in total.

2.2 Structuring

The sensor layer is structured by a self-developed photolithographic process. Details of this process are covered in Schmaljohann et al. (2011).

The light-sensitive photoresist is sprayed uniformly onto the sensor layer. Afterwards, the paths of the sensor structure have to be exposed to light.

For this purpose, a UV light exposure system was developed and manufactured at PTB. The system is equipped with a blue-ray laser diode and specially adapted optics, so that the photoresist can be exposed with a precision in the micrometer range (Schmaljohann et al., 2013). In contrast to the commonly used exposure masks, the layout of the sensor can be changed easily. The demand for diverse requirements or dimensions of the sensors, which is especially the case in prototype construction, can be fulfilled by this very flexible technique.

After exposure, the photoresist is developed. The areas that were not exposed to light are removed and only the later pattern of the thin film sensor is protected by the resist for the next process step. The following wet-chemical etching process removes the uncovered part of the sensor layer. Finally, the remaining photoresist is stripped, leaving only the actual sensor pattern.

The thin film strain gauges presented have a structure size of about $50 \mu m$. Smaller structure sizes can also be achieved by this process.

A detailed view of such a sensor structure is depicted in Fig. 2.

3 Force transducer layout

The goal in the engineering and design of the deformation element was to make it compact in size and suitable for force measurements up to 1 N. Furthermore, it should be possible to connect four RSGs in a full bridge circuit.

The computer-aided design of the force transducer was adjusted and optimized by the finite element (FE) analysis

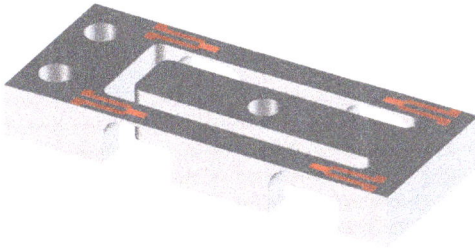

Figure 3. Deformation element as CAD model, utilized with RSGs in the areas of maximum strain.

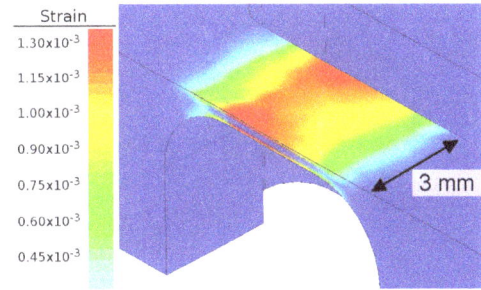

Figure 4. FE analysis of a single elastic deflection area of the deformation element.

Figure 5. Overview of the structured thin film RSG.

method. After this step, the electrical properties of the RSG were calculated and the sensor design was fitted to the areas of maximum strain.

3.1 Development of the deformation element

By finite element analysis, a deformation element had to be found, which shows an elongation at the surface of 1 ‰ at a nominal load of 1 N. The material used for the design was EN AW-2024, a typical aluminum alloy for force transducers. Nevertheless, the prototypes were manufactured from the EN AW-2007 alloy because of its good machinability, while having a comparable material parameter.

The design of the deformation element has four areas of a defined and reduced cross section. The location for the application of force was selected such that under load, two deflection areas are elongated, while the other two are compressed in an equal manner. Therefore, the requirement for the installation of a full bridge is met.

Furthermore, the geometry of the deformation element allows the application of the full bridge on the planar upper side of the part. The limitation on only one side helps to simplify the surface coating and structuring process considerably.

The deflection areas had to be reduced to a width of 3 mm at a thickness of only 0.3 mm to achieve the desired strain of $1000\,\mu m\,m^{-1}$ at a load of 1 N. In contrast to foil strain gauges, the comparably small area is no problem for the application due to the miniaturization capabilities of the thin film RSGs.

The computer-aided design (CAD) model of the deformation element with the four areas for the strain gauges is depicted in Fig. 3.

Figure 4 shows the result of the finite element analysis under a nominal load of 1 N. The orange-colored areas mark the desired strain of about $1000\,\mu m\,m^{-1}$. At the same time, a higher stress is located in the boundary areas, illustrated by the red coloration. One potential reason for this occurrence are insufficient boundary conditions. However, as a stress step-up in this area is also quite possible, it will be considered for the sensor design.

3.2 Sensor layout

The layout of the sensor pattern was matched to the width of the deflection area and the result of the finite element analysis.

To get a sufficient but limited averaging of the strain measurement, a length of 1.5 mm for the strain pattern was chosen. With a structure width of $50\,\mu m$ and a distance between the conducting paths of $100\,\mu m$, six parallel conducting paths are possible in total. Therefore, the total width of the strain pattern is only 0.8 mm. The influence on the force measurement by stress step-up in the boundary areas stated above is minimized. Two conducting paths with a width of 0.5 mm lead to the contact pads with a size of 1 mm^2 each.

Despite the small-sized sensor pattern, the layout stipulates a sensor resistance of up to 1000 Ω. This high resistance value, in comparison to other RSGs, should have a positive effect on the sensor's sensitivity, and it minimizes the impact of contact resistances and measuring lines. The high resistance value can be realized by a very thin sensor layer of about 100 nm, but, nevertheless, it can be changed easily to meet modified requirements by changing the layer thickness or by the line width of the structure itself.

Figure 5 shows a thin film RSG on a deformation element after the structuring process. A full view of the sputtered and structured deformation element is depicted in Fig. 6. For the following test procedures, thin wires were soldered to the contact pads of each RSG.

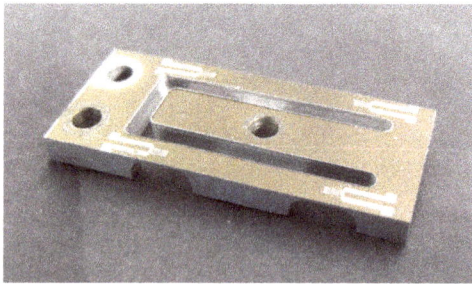

Figure 6. Deformation element with four sputtered RSGs in the deflection areas.

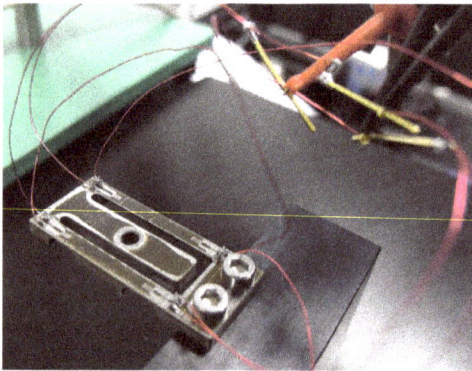

Figure 7. Measurement setup with a deformation element connected as a full bridge circuit.

4 Test procedure and measurement results

Initially, the base resistance of the RSG and the analysis of possible defects in the sensor structure, i.e., a good electrical insulation of the substrate, is of primary concern. Therefore, the electrical conductance of the sensor layer to the deformation element was checked with a lab multimeter.

Subsequently, for the functional tests, the sensors were connected to measurement amplifier "MX440A" by HBM GmbH. The force transducer is fastened at the two screw holes on one side. The measurement setup is depicted in Fig. 7.

In the first instance, only a quarter bridge, i.e., one single RSG and its shift of the resistance value, was measured. Hence, no compensation of the measurement bridge was needed. Instead, direct conclusions of the single sensors and their function were made.

The measurement procedure is based on the DIN EN ISO 376 international standard. The force was applied manually by calibrated single weights directly to the deformation element. During the measurements, the room temperature had a steady value of $22\,°C$.

After the connection of the sensor to the measurement amplifier, the measurements were not carried out until after an idle time of over 30 min. At the beginning of the tests, the resistance value and variation were measured at no load for

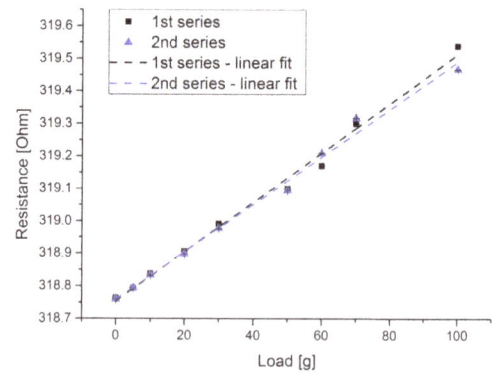

Figure 8. Diagram of the measured resistance of one single RSG and the applied load in grams.

30 s. Afterwards, the load was increased in eight steps up to 100 g, which is about 1.0 N, pausing for 30 s between each step. The respective resistance measurement values were logged for 30 s at an interval of 1 Hz. At the end of the first series of measurements, a second one was started after an idle time of 3 min.

For each step, the average value and standard deviation based on the 30 measurands were calculated. The results including the linear fits of a typical thin film strain gauge are shown in the diagram of Fig. 8.

The standard deviations of the values are not visible in the diagram as they are in the range of only $0.4\,m\Omega$. During the tests, the sensors showed a very quick response and good damping behavior. Besides basic operation of the strain gauge, the test reveals the desired and almost linear characteristic. Nevertheless, differences in the two measurement series can be observed, possibly because of slight temperature changes that cannot be compensated for by this setup; this is one of the drawbacks in a quarter bridge setup, besides its limited resolution. Likewise, this might be the explanation for the slight deviation to linearity, and the occurrence of unwanted side forces is possible.

In addition to the measurement of one single RSG, four RSGs were connected in a full bridge configuration on another force transducer. The measured single resistances were in the range from 420 to $470\,\Omega$. Therefore, the bridge was compensated for by additional resistors up to $470\,\Omega$ each. The rather small differences in the resistance values of about 10 % are due to slight variations in the width of the sensor pattern because of the wet etching step and, to a smaller degree, because of film thickness deviation.

The measurement procedure for this force transducer is the same as stated above. The results in the diagram of Fig. 9 reveal a very linear behavior over the full measurement range, much better compared to the quarter bridge setup. The calculated value of a strain of $1000\,\mu m\,m^{-1}$ at the nominal load of about 1 N is reached. Moreover, the standard deviation of

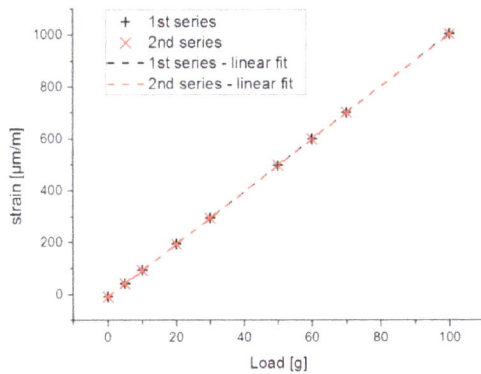

Figure 9. Diagram of the measured strain and the applied load in grams of a full bridge circuit

the single measurands is in the range of only $0.05\,\mu\text{m}\,\text{m}^{-1}$ (average of $0.03\,\mu\text{m}\,\text{m}^{-1}$).

A valid analysis of the measurement uncertainty was not conducted at the time of this writing, since this would imply better knowledge of the influences, for example, gravitational acceleration, and the accuracy of the weights. Nevertheless, besides the general functionality of the sensors in thin film technology, the results show the favored, linear dependence of the strain in the deflection areas and the change in resistance of the thin film sensor. Regarding the prototype status of the transducer assembly, the measured deviations of about 5×10^{-5} are very promising results.

5 Summary and outlook

The present paper shows the development of a force transducer for small forces up to 1 N in the metrological field. After optimization of the deformation element for the application of thin film sensors and small forces, the sensor layout was adapted to the deflection areas.

The coating and structuring technologies are applicable to the EN AW-2007 aluminum alloy and can be transferred to the EN AW-2024 typically used for force transducers.

The results show the general functionality and correct dimensioning of the deformation element and, in particular, the thin film strain gauges. The measurements based on international standards showed a very sensitive but stable behavior, with small deviations for a force transducer at this stage of development. In summary, the force transducer equipped with thin film sensors is very well eligible for measuring small forces up to 1 N. As a matter of course, the measurement uncertainty has to be determined to ensure its qualification.

The manufacturing procedures will be further improved to achieve full bridge compensation. In addition to the tests presented here, a comparison of the measured values with a deadweight force standard machine is planned, as well as long-term testing of the creep behavior. Likewise, the force transducer is to be tested for its adequacy for precision measurements in a special force standard machine for small forces (Schlegel et al., 2010). Because of the observed short response times and good damping behavior, dynamic testing of the sensor is needed.

Acknowledgements. The authors would like to thank Mr. Philip Heinisch for his support in the development of the light exposure system and his assistance in conducting the measurements.

References

Buß, A., Illemann, J., Kumme, R., Hagedorn, D., Meeß, R., and Löffler, F.: Entwicklung und Untersuchung von Kraftaufnehmern zur hochgenauen Messung statischer und dynamischer Kräfte, in: Sensoren und Messsysteme 2008, no. 2011 in VDI-Berichte, 2008.

Hagedorn, D., Meeß, R., and Löffler, F.: Fabrication of Sputtered Resistance Strain Gauges on Curved Surfaces, in: Proceedings of the 7th euspen International Conference, May 2007, Bremen, 2007.

Karaus, A. and Paul, H.: Load cells with small nominal load based on strain gauges using thin-film techniques, Measurement, 10, 133–139, doi:10.1016/0263-2241(92)90009-s, 1992.

Schlegel, C., Slanina, O., Haucke, G., and Kumme, R.: Construction of a Standard Force Machine for the Range of $100\,\mu\text{N}$–200 mN, in: IMEKO 2010 TC3, TC5 and TC22 Conferences, 2010.

Schmaljohann, F., Hagedorn, D., Buß, A., Kumme, R., and Löffler, F.: Entwicklung von Dünnschichtsensoren mit kleiner Strukturbreite auf dünnen isolierenden Schichten, in: MikroSystemTechnik Kongress 2011: Proceedings, 764–767, VDE-Verlag, 2011.

Schmaljohann, F., Hagedorn, D., Buß, A., Kumme, R., and Löffler, F.: Thin-film sensors with small structure size on flat and curved surfaces, Measurement Science and Technology, 23, 074019, doi:10.1088/0957-0233/23/7/074019, 2012.

Schmaljohann, F., Hagedorn, D., Meeß, R., and Löffler, F.: Blu-ray laser system for light exposure of photoresists on three dimensional surfaces, in: Proceedings of the 13th euspen International Conference – Berlin, 344–347, 2013.

5

Silicon micro-levers and a multilayer graphene membrane studied via laser photoacoustic detection

Z. Zelinger[1], P. Janda[1], J. Suchánek[1,2], M. Dostál[1,2], P. Kubát[1], V. Nevrlý[2], P. Bitala[2], and S. Civiš[1]

[1]J. Heyrovský Institute of Physical Chemistry AS CR, Prague, Czech Republic
[2]Faculty of Safety Engineering, VŠB – Technical University of Ostrava, Ostrava, Czech Republic

Correspondence to: Z. Zelinger (zelinger@jh-inst.cas.cz)

Abstract. Laser photoacoustic spectroscopy (PAS) is a method that utilizes the sensing of the pressure waves that emerge upon the absorption of radiation by absorbing species. The use of the conventional electret microphone as a pressure sensor has already reached its limit, and a new type of microphone – an optical microphone – has been suggested to increase the sensitivity of this method. The movement of a micro-lever or a membrane is sensed via a reflected beam of light, which falls onto a position-sensing detector. The use of one micro-lever as a pressure sensor in the form of a silicon cantilever has already enhanced the sensitivity of laser PAS.

Herein, we test two types of home-made sensing elements – four coupled silicon micro-levers and a multilayer graphene membrane – which have the potential to enhance this sensitivity further. Graphene sheets possess outstanding electromechanical properties and demonstrate impressive sensitivity as mass detectors. Their mechanical properties make them suitable for use as micro-/nano-levers or membranes, which could function as extremely sensitive pressure sensors.

Graphene sheets were prepared from multilayer graphene through the micromechanical cleavage of basal plane highly ordered pyrolytic graphite. Multilayer graphene sheets (thickness $\sim 10^2$ nm) were then mounted on an additional glass window in a cuvette for PAS. The movements of the sheets induced by acoustic waves were measured using an He–Ne laser beam reflected from the sheets onto a quadrant detector. A discretely tunable CO_2 laser was used as the source of radiation energy for the laser PAS experiments. Sensitivity testing of the investigated sensing elements was performed with the aid of concentration standards and a mixing arrangement in a flow regime. The combination of sensitive microphones and micromechanical/nanomechanical elements with laser techniques offers a method for the study and development of new, reliable and highly sensitive chemical sensing systems. To our knowledge, we have produced the first demonstration of the feasibility of using four coupled silicon micro-levers and graphene membranes in an optical microphone for PAS. Although the sensitivity thus far remains inferior to that of the commercial electret microphone (with an S/N ratio that is 5 times lower), further improvement is expected to be achieved by adjusting the micro-levers and membrane elements, the photoacoustic system and the position detector.

1 Introduction

The objective of this paper lies in the context of new sensing technologies based on micromechanical sensing elements, including functional materials for gas sensing. These elements could be employed in the laser photoacoustic spectroscopy (PAS) method as part of a sensitive optical microphone. This technique offers several advantages compared with conventional spectroscopy; some of its major benefits include the fact that the detected signal is directly proportional to the laser intensity, the elimination of false absorption resulting from scattered light and the lack of a need for photodetectors, which perform poorly in the mid-infrared region.

PAS is a spectroscopic method that differs from other absorption spectroscopic techniques in the manner in which the absorbed radiation is detected. The absorbed light is converted into heat, which leads to gas expansion. If the excitation light is modulated or chopped, then the resulting pressure waves can be sensed by a microphone. Although conventional condenser microphones have reached their limits of sensitivity, the development of new pressure sensors offers an opportunity to increase the sensitivity of this technique.

It has been proposed that a cantilever-type pressure sensor be used in PAS in place of microphones to achieve optimal sensitivity (Kauppinen et al., 2004; Wilcken and Kauppinen, 2003; Kuusela and Kauppinen, 2007; Koskinen et al., 2008, 2006). The primary benefits of a cantilever are the very low string constant and the extremely wide dynamical range that can be achieved in the cantilever movement. The string constant can be 2 or 3 orders of magnitude smaller than that of the membrane of a condenser microphone, and the movement of the cantilever can span tens of micrometres without suffering any non-linear or restricting effects. A non-contact (deflection, interferometric) measurement of cantilever movement is required to avoid any damping caused by the probe and to maintain the wide dynamic range (Li et al., 2012).

Micromechanical sensors represent a new branch of chemical sensing that utilizes a microfabricated spring (cantilever), as originally applied in atomic force microscopes (AFMs), for the recognition of interfacial mass- and charge-transfer processes with very high sensitivity. A detection system based on cantilever micromechanical behaviour already exists and is identical to that employed in AFMs (Jalili and Laxminarayana, 2004). This system has allowed the detection of analytes in concentrations down to the picomolar and sub-picomolar levels to become feasible in a variety of processes, including deposition/dissolution, adsorption/desorption (Ji et al., 2001), solution pH changes and surface-confined charge-transfer reactions (Tabard-Cossa et al., 2005). For detection, both bending- and frequency-readout micro-lever sensors can be employed (Battiston et al., 2001). Further development will be focused on attempts to construct cantilever arrays (Lang et al., 2005) modified by various receptors to increase the selectivity of the sensor response (Grogan et al., 2002) and to gain the ability to perform multicomponent analysis in a single step. The major advantages of microcantilever array sensors are their micrometre-scale size, high sensitivity and short response time (Berger et al., 1997).

A good approximation to the ideal cantilever – i.e. a one-dimensional single nanocrystal – can be found in multi-/single-layer graphite crystals – i.e. multi-/single-layer graphene (MLG, SLG) (Novoselov et al., 2004). Graphene sheets possess outstanding electromechanical properties (Geim and Novoselov, 2007; Lee et al., 2008; Castro Neto et al., 2009) and demonstrate impressive sensitivity as mass detectors (Chen et al., 2009; Avdoshenko et al., 2012). Because of its highly uniform (if defect-free) structure, relatively high chemical stability, and outstanding mechanical properties such as a high Young's modulus and a low specific weight, which allow it to reach a high differential mass

ratio and hence a high sensor sensitivity, graphene represents almost the ideal material for nanomechanical sensors. Nanomechanical vibrations have been investigated for potential application in nanothermometers (Rahmat et al., 2010).

In previous studies (to improve the physical modelling of urban air pollution), we have developed and employed laser photoacoustic spectrometry (Zelinger et al., 2004, 2006, 2009). The objective of this paper is to present the design and fabrication of several home-made sensing elements of the cantilever-and-membrane type and to test and characterize their mechanical properties with the aid of the developed PAS method. Transducers composed of various materials (silicon, carbon) were employed in the design of new gas-sensing elements.

The purpose of experiments that involve microphones, graphene-based membranes and silicon cantilevers is to examine the differences among various sensors to optimize the range of their utilization. Unlike microphones, cantilevers allow for detection in both frequency and resonance modes in addition to the deflection mode. In the case of a multilayer graphene membrane detector, we have now proven the feasibility of its use. This detector represents a fusion of a microphone membrane and a cantilever, the deflection of which is probed by a reflected laser beam. Its prospective advantage lies in the possibility of tuning its sensitivity by decreasing its stiffness. This can be accomplished by decreasing the number of graphene layers, which will also cause a shift in the resonant frequency. Most importantly, however, our approach includes an examination of a multicantilever/multimode set-up, which offers variability in both the mechanical properties and detection modes of the cantilevers and thus the choice of the detector with the most suitable response for a given application. There is also the possibility of receiving signals from a multidetector system and thus of improving the signal-to-noise ratio and minimizing detector failure. Both features are important for prospective utilization in detector units and for increasing automatic sensor-network coverage.

2 Experimental

A discretely tunable home-made CO_2 laser emitting at rotation–vibration transitions of CO_2 in the bands $\Sigma_u^+(00^01)$ – $\Sigma_g^+(10^00)$ and $\Sigma_u^+(00^01)$ – $\Sigma_g^+(02^00)$ in a spectral range of 9–$11\,\mu$m was used as the source of radiation energy for laser photoacoustic spectroscopy (PAS). The photoacoustic (PA) cell was designed as a cylindrical cell of 300 mm in length and 6 mm in diameter equipped with a microphone as well as either a graphene membrane or a cantilever sensor, which could be exchanged with each other or for a different membrane or cantilever. All sensor units were placed in the middle of the longitudinal dimension of the cell. The total internal volume of the PA cell was $19.5\,$cm^3. To obtain the largest possible PA signal, it is necessary that the interior space of the PA cell, which is not exposed to laser radiation, is minimized. This means that in the case of a cylindrical PA cell, the inner diameter of the PA cell must be minimized with respect to the laser beam. The diameter of the laser beam that was used was approximately 9–10 mm, and the laser beam was focused by a lens with a 25 cm focus onto the centre of the PA cell; this set-up motivated the geometry of the PA cell described above.

The sensitivity levels of the sensing elements were tested using concentration standards based on the permeation method (Okeeffe and Ortman, 1966; Barratt, 1981). The concentration standards were prepared in the form of closed tubes composed of permeable material (silicon, polyethylene, Teflon), and these standards were placed in a thermostatted chamber at 35 °C for long-term weighing and storage. For measurements, each concentration standard was placed in a chamber through which carrier air was flowing. Using such a mixing arrangement, we were able to prepare a given concentration level in a flow regime (flow rate $1 \, \text{cm}^3 \, \text{s}^{-1}$) and under atmospheric pressure, with methanol vapour as the testing gas. The CO_2 laser was tuned to the 9 P(34) CO_2 laser line (1033.488 cm^{-1}), which corresponds to the fundamental CO stretching band of methanol, where the maximal absorption cross section is $72 \times 10^{-20} \, \text{cm}^2$ (Loper et al., 1980).

We studied several AFM-based silicon cantilevers and graphene sheets via laser PAS. The membranes for the photoacoustic detector – graphene sheets – were prepared from MLG through micromechanical cleavage of basal plane **H**ighly **O**rdered **P**yrolytic **G**raphite (HOPG, ZYH grade, Bruker, USA) (Novoselov et al., 2004). Graphene sheets in the form of circular membranes (Fig. 1a) and AFM-based silicon cantilevers in a square arrangement (Fig. 1b) were tested. The MLG sheets (thickness $\approx 10^2$ nm) were mounted on an additional glass window in the cuvette for PAS (Fig. 1a). The movements of the graphene sheets or the cantilever ends that were induced by acoustic waves were measured by an He–Ne laser beam that was reflected from the sheets or the ends onto a quadrant detector (a red-enhanced quad-cell silicon photodiode, SD 085-23-21-021, Laser Components). The signals from both the microphone and the quadrant detector were processed by an oscilloscope (LeCroy 9361).

Two types of cantilevers and one type of membrane were used: commercial (Bruker) silicon cantilevers of the OTESPA (metallized) type or the NP (metallized) type, with an arm length of 200 μm and with resonant frequencies of 300 kHz in ambient gas and 10 kHz in liquid, and home-made multilayer graphene membranes. The primary benefits of using layered graphene as a membrane lie in its high elasticity (YM ~ 1 TPa, defect free), the possibility of modifying its stiffness by simply removing graphene layers (down to the level of a single monolayer) and thus increasing its bending sensitivity, and, finally, its reflectivity, which allows the bending of the membrane to be probed using a reflected laser

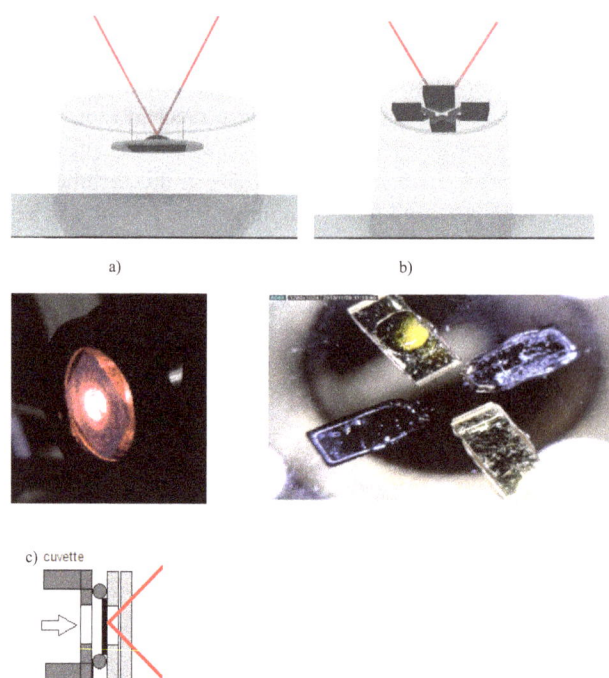

Figure 1. Schematic illustrations of the construction of the sensing elements and their implementations: (**a**) graphene-based sensing elements, (**b**) AFM-cantilever-based sensing elements, and (**c**) schematic illustration of the graphene membrane detector assembly on the cuvette window sealed with a silicone O-ring, where the red lines represent the laser beam and the arrow indicates the pressure wave.

beam. Compared with microphone membranes, cantilevers and graphene membranes exhibit lower damping and lower stiffness, and, therefore, higher sensitivity in general can be achieved. The graphene was of grade ZYH (Bruker, USA). The thickness of the multilayer graphene (MLG) could be simply determined through an AFM profile measurement. The thickness of the graphene that was used in these experiments was typically in the range of 10^2 nm, although this thickness could be further decreased.

The graphene membrane was attached to the glass window by an epoxy glue film and was pressed by an O-ring over the circular opening through which the pressure wave from the cuvette was transferred to the membrane. The membrane deflection was measured using a laser beam reflected from the membrane onto the quadrant detector. Details regarding the membrane mounting are provided in Fig. 1c.

The micromechanical sensing elements were installed in the cuvette for PAS together with a sensitive microphone (electret microphone EK 23024, Farnell) for comparison. The experimental set-up is depicted in Fig. 2. Quantitative measurements were performed using a phase-sensitive lock-in amplifier (Stanford Research Systems SR530 lock-in amplifier). The CO_2 laser beam was modulated by a chopper, and the photoacoustic signals from the microphone and the

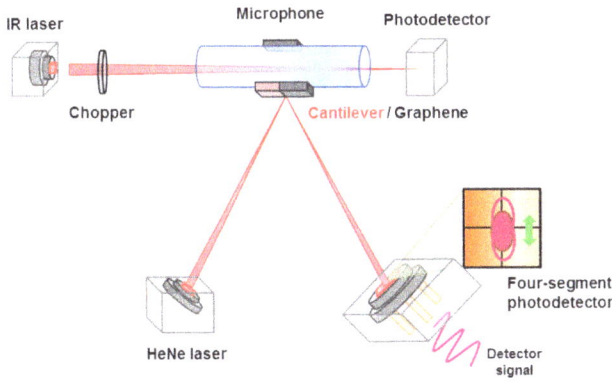

Figure 2. Experimental set-up for the tests of the sensing elements.

graphene were demodulated by the lock-in amplifier and subsequently processed using a PC. The signals from the microphone and from the cantilevers or graphene membrane were passed through preamplifiers to lock-in amplifiers. The reference signal from the chopper, i.e. the frequency of interruptions of the laser beam, served as a reference signal for the lock-in amplifiers. The input signal amplitude for the lock-in amplifiers was 3 mV for the microphone and 5 mV for the cantilever and graphene sensors.

3   Results and discussion

AFM-based silicon cantilevers and MLG sheets were tested as photoacoustic detectors in these studies. Figure 1 shows a schematic diagram of the set-up for the testing of the MLG-based and silicon–AFM–cantilever-based sensing elements (henceforth referred to as graphene sensors and cantilever sensors, respectively, for brevity). In the case of a classical capacitive microphone, there is a space both in front of and behind the membrane that is connected by a balancing channel to eliminate any damping of the membrane's movement during its return. The same function is also performed in our sensing elements. The free ends of the cantilevers in the cantilever sensors can move freely in space. In the case of the graphene sensors, an additional glass slide is used as a specialized holder for the MLG-sheet-based membrane (Fig. 1a). The elimination of damping effects is proven by the experimental data presented in Fig. 3a. Figure 3a presents the motion of the MLG sheet as indicated by the He–Ne laser beam reflected onto the quadrant detector, which translates the movement into a voltage amplitude proportional to the pressure changes inside the chamber. The low-frequency (4 Hz) modulation of the laser by the chopper causes symmetrical movements of the graphene membrane.

The experimental data presented in Fig. 3a demonstrate the detection of the motion of the graphene sensor in the form of rising and falling exponentials. These observations are the result of the absorption of radiation, which heats the gas in the chamber. By virtue of the fixed volume, the time-varying thermal fluctuations generate pressure variations (sound waves), which are detected by the graphene sensor. The following conditions correspond to the typical case and are assumed to apply: the source is modulated such that the time profile approximates a square wave of period T; the molecular radiative and collisional relaxation times are small and can be neglected in comparison with the period of the chopper modulation T; and the detector chamber is a cylinder of length l and radius a, where $l \gg a$. Under these conditions, it can be theoretically derived that the time dependence of the pressure response $p(\mathrm{t})$ is given in the following forms:

$$p(t) = c_1 + c_2(1 - e^{-t/\tau}) \qquad 0 \le t \le T/2,$$

$$p(t) = c_2 e^{T/2\tau} e^{-t/\tau} \qquad T/2 \le t \le T,$$

where τ is the thermal relaxation time to the temperature of the chamber walls and

$$c_1 = \Delta p \frac{(e^{-T/2\tau} - 1)}{(e^{-T/2\tau} - e^{T/2\tau})}, \qquad c_2 = \Delta p \frac{(1 - e^{-T/2\tau})}{(e^{-T/2\tau} - e^{T/2\tau})},$$

where Δp corresponds to the fully developed pressure amplitude ($T \to \infty$).

A gradual change from an exponential waveform to a sine waveform occurs as the modulation frequency is increased, at a frequency of approximately 23 Hz (Fig. 3b). Such waveforms are suitable for processing using a phase-sensitive amplifier (lock-in amplifier), thus allowing us to compare the sensitivities of the microphone and the sensing elements. The detection chamber was equipped with both the microphone and the tested sensing element (graphene sheet or silicon cantilevers). Both signals – from the microphone and from the quadrant detector (representing the detection of the motion of the sensing element: the graphene sensor or the cantilever sensor) – were processed using lock-in amplifiers. As a testing gas, we used methanol vapour. The detection chamber was connected to the flow system for precise monitoring of the flow rate of the carrier air using mass flow meters and controllers (FMA Series, OMEGA). A sampling chamber, into which the concentration standards were inserted, was placed in the flow system just in front of the detection chamber. These concentration standards provided a calibration of the measurements based on the permeation method (Okeeffe and Ortman, 1966; Stellmack and Street, 1983; Zelinger et al., 1988) and the long-term weighing of the standards.

Equivalent measurements were performed for the cantilever sensors (Fig. 1b). Silicon cantilevers offer the advantage of better optical reflection compared with the diffuse reflection from the MLG-sheet-based sensing elements (Fig. 1a). These experiments directly followed the research work previously performed by Kaupinnen et al. (Kauppinen et al., 2004; Koskinen et al., 2006, 2008; Kuusela and Kauppinen, 2007).

Waveforms were simultaneously recorded to monitor the time dependence of the intensity of the laser radiation, the

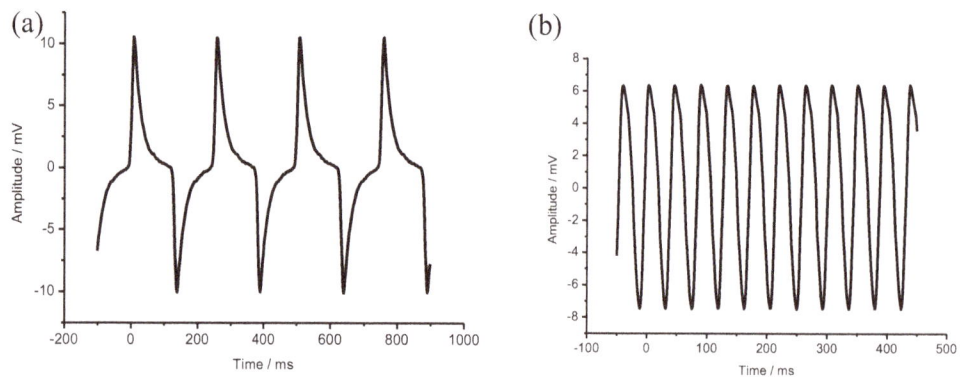

Figure 3. Experimental signals recorded from the graphene-membrane-based sensing elements at low frequency: (**a**) 4 Hz and (**b**) 23 Hz.

time dependence of the response of the microphone as processed by the lock-in amplifier, and the time dependence of the response of the graphene sensor as processed by the lock-in amplifier; the results are depicted in Fig. 4. The concentration of the methanol vapour that was introduced into the detection chamber was ~10 ppm. The following steps were performed during each 500 s scan: the laser was initially off (OFF) – the laser was turned on (ON) – the laser was turned off (OFF) – the laser was turned on (ON) – the laser was turned off (OFF). Both the microphone and the graphene sensor responded to the turning on and off of the laser (Fig. 4). A rise and fall in the signals from both sensors can be observed upon the powering on and off of the laser, respectively, in Fig. 4; the magnitude of the difference in the signal is determined predominantly by the concentration of the absorbing gas, i.e. 10 ppm of methanol. The background signal level (the difference between the signals recorded in the ON and OFF states of the laser at zero concentration of the absorbing gas) is almost negligible; in the case of the microphone, it is approximately twice as high as the noise level, and in the case of the cantilever, it is approximately half of the noise level.

In addition to the ON–OFF mode of the laser, we also tested a similar ON–OFF mode for the concentration standard (Fig. 5). When we inserted a concentration standard (~2 ppm) into the sample chamber (concentration ON), both the signal from the microphone and that from the graphene sensor (as processed by the lock-in amplifier) increased; see Fig. 5a. The same measurement was performed using the microphone and cantilever system; see Fig. 5b. Then, we removed the concentration standard from the flow system (concentration OFF) and monitored the resulting decrease in the signal (Fig. 5c, d). The responses of the microphone and graphene sensor are shown in Fig. 5c, and the responses of the microphone and cantilever system are shown in Fig. 5d. The perturbations observed at the beginning of the increase in the concentration signal (Fig. 5a, b) and at the beginning of the decrease in the concentration signal (Fig. 5c, d) were caused by pressure disturbances created in the sample cham-

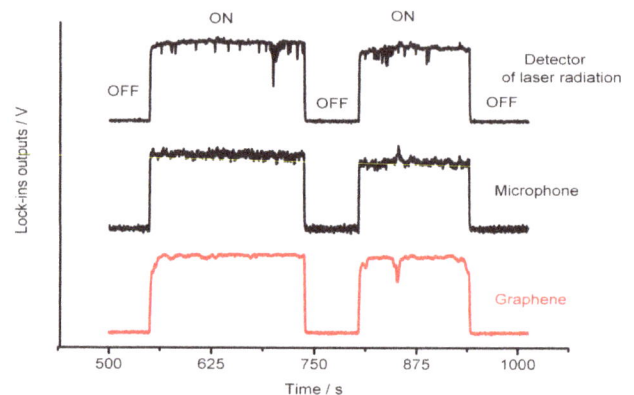

Figure 4. Experimental data for comparison of the signals detected by the microphone and by the graphene-based membranes at the same gas concentration (methanol, ~ 10 ppm), as processed by the lock-in amplifiers.

ber when the concentration standard was inserted and removed.

A comparison of the noise levels indicated by the experimental data presented in Fig. 5 yields the corresponding signal-to-noise ratios (S/N). The graphene sensors and the cantilever systems have different connections to the PA cell. The connection is in the middle of the cell and thus exerts a strong influence on the acoustic properties of the interior of the cell. In the connection of the graphene sensors or the cantilever systems to the PA cell, different regions are created that affect the acoustic signal gain, giving rise to a difference in the resulting sensitivities of the two systems. In the case of the microphone and the graphene sensor, the signal-to-noise ratio for the microphone was $S/N \approx 25$ and that for the graphene sensor was $S/N \approx 5$. In the case of the microphone and the cantilever system, the signal-to-noise ratio for the microphone was $S/N \approx 50$ and that for the cantilever system was $S/N \approx 36$. It is observed that the signal detected by the graphene sensor suffered from a higher noise level than did the signal from the cantilever. Possible reasons for this find-

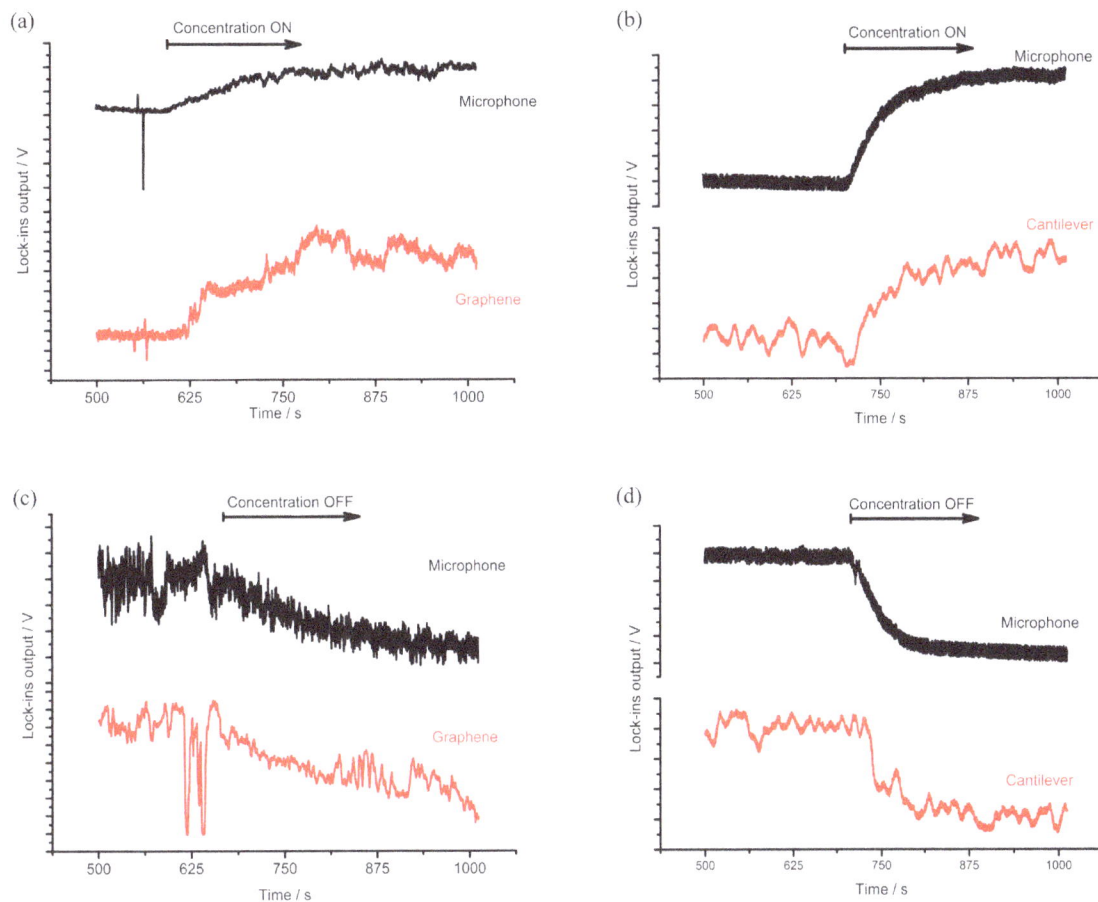

Figure 5. Experimental signals recorded from the lock-in amplifiers for a generated concentration of methanol of ~ 2 ppm: **(a)** concentration ON for the microphone and graphene sensor, **(b)** concentration ON for the microphone and cantilever system, **(c)** concentration OFF for the microphone and graphene sensor, and **(d)** concentration OFF for the microphone and cantilever system.

ing could lie in the adsorption properties of graphene with respect to those of methanol vapour; this phenomenon will be the subject of further studies. The cantilever system is not as strongly affected by the adsorption properties of the gas, and, therefore, its sensitivity approaches that of the microphone.

Thus far, the sensitivity of these elements has been inferior to that of a commercial microphone. We expect that further enhancement of the sensitivity can be achieved by adjusting the elements in combination with the entire photoacoustic system. To determine the optimal responses of the membranes, graphene sheets of different thicknesses will be prepared, and the dimensions of the cuvette (and thus the resonant frequency) will be varied. Reduction of the diffusion of the reflected laser beam can be achieved by depositing a thin layer of metal onto the membranes. Future work will focus on few-layer graphene cantilevers/membranes and single-layer graphene, which will be prepared using the chemical vapor deposition (CVD) technique and then transferred to supports, thus allowing for free-standing mounting.

4 Conclusions

The first experimental results of analytical testing of MLG-sheet-based sensing elements as well as results obtained for AFM-based silicon cantilever systems are presented. A method for the comparison of the responses of various sensing systems, including graphene sensors, cantilever systems and microphones, has been developed. We obtained the first quantitative data regarding the sensitivity of the tested graphene sensors and cantilever systems by applying concentration standards to generate concentrations of methanol in a well-defined way. The combination of sensitive microphones and micromechanical elements with advanced laser techniques offers possible opportunities for the development of new, reliable and highly sensitive chemical sensing systems.

Acknowledgements. The authors are grateful for the financial support provided via project no. LD14022 within the framework of COST action TD 1105 funded by the Ministry of Education, Youth and Sports of the Czech Republic, via project no. 14-14696S funded by the Grant Agency of the Czech Republic and via project no. R200401401 in the framework of the regional cooperation of the Academy of Sciences of the Czech Republic.

References

Avdoshenko, S. M., Gomes Da Rocha, C., and Cuniberti, G.: Nanoscale ear drum: Graphene based nanoscale sensors, Nanoscale, 4, 3168–3174, doi:10.1039/c2nr30097d, 2012.

Barratt, R. S.: The preparation of standard gas mixtures a review, The Analyst, 106, 817–849, 1981.

Battiston, F. M., Ramseyer, J. P., Lang, H. P., Baller, M. K., Gerber, C., Gimzewski, J. K., Meyer, E., and Güntherodt, H. J.: A chemical sensor based on a microfabricated cantilever array with simultaneous resonance-frequency and bending readout, Sens Actuators, B. Chem., 77, 122–131, doi:10.1016/s0925-4005(01)00683-9, 2001.

Berger, R., Gerber, C., Lang, H. P., and Gimzewski, J. K.: Micromechanics: A toolbox for femtoscale science: "Towards a laboratory on a tip", Microelectron. Eng., 35, 373–379, 1997.

Castro Neto, A. H., Guinea, F., Peres, N. M. R., Novoselov, K. S., and Geim, A. K.: The electronic properties of graphene, Rev. Mod. Phys., 81, 109–162, doi:10.1103/RevModPhys.81.109, 2009.

Chen, C., Rosenblatt, S., Bolotin, K. I., Kalb, W., Kim, P., Kymissis, I., Sormer, H. L., Heinz, T. F., and Hone, J.: Performance of monolayer graphene nanomechanical resonators with electrical readout, Nat. Nanotechnol., 4, 861–867, doi:10.1038/nnano.2009.267, 2009.

Geim, A. K. and Novoselov, K. S.: The rise of graphene, Nat. Mater., 6, 183–191, doi:10.1038/nmat1849, 2007

Grogan, C., Raiteri, R., O'Connor, G. M., Glynn, T. J., Cunningham, V., Kane, M., Charlton, M., and Leech, D.: Characterisation of an antibody coated microcantilever as a potential immuno-based biosensor, Biosens. Bioelectron., 17, 201–207, doi:10.1016/s0956-5663(01)00276-7, 2002.

Jalili, N. and Laxminarayana, K.: A review of atomic force microscopy imaging systems: Application to molecular metrology and biological sciences, Mechatronics, 14, 907–945, doi:10.1016/j.mechatronics.2004.04.005, 2004.

Ji, H. F., Hansen, K. M., Hu, Z., and Thundat, T.: Detection of pH variation using modified microcantilever sensors, Sens Actuators, B. Chem., 72, 233–238, doi:10.1016/s0925-4005(00)00678-x, 2001.

Kauppinen, J., Wilcken, K., Kauppinen, I., and Koskinen, V.: High sensitivity in gas analysis with photoacoustic detection, Microchem. J., 76, 151–159, doi:10.1016/j.microc.2003.11.007, 2004.

Koskinen, V., Fonsen, J., Kauppinen, J., and Kauppinen, I.: Extremely sensitive trace gas analysis with modern photoacoustic spectroscopy, Vib. Spectrosc., 42, 239–242, doi:10.1016/j.vibspec.2006.05.018, 2006.

Koskinen, V., Fonsen, J., Roth, K., and Kauppinen, J.: Progress in cantilever enhanced photoacoustic spectroscopy, Vib. Spectrosc., 48, 16–21, doi:10.1016/j.vibspec.2008.01.013, 2008.

Kuusela, T. and Kauppinen, J.: Photoacoustic gas analysis using interferometric cantilever microphone, Appl. Spectrosc. Rev., 42, 443–474, doi:10.1080/00102200701421755, 2007.

Lang, H. P., Hegner, M., and Gerber, C.: Cantilever array sensors, Mater. Today, 8, 30–36, doi:10.1016/s1369-7021(05)00792-3, 2005.

Lee, C., Wei, X., Kysar, J. W., and Hone, J.: Measurement of the elastic properties and intrinsic strength of monolayer graphene, Science, 321, 385–388, doi:10.1126/science.1157996, 2008.

Li, P., You, Z., and Cui, T.: Graphene cantilever beams for nano switches, Appl. Phys. Lett., 101, 093111, doi:10.1063/1.4738891, 2012.

Loper, G. L., Calloway, A. R., Stamps, M. A., and Gelbwachs, J. A.: Carbon dioxide laser absorption spectra and low ppb photoacoustic detection of hydrazine fuels, Appl. Opt., 19, 2726–2734, 1980.

Novoselov, K. S., Geim, A. K., Morozov, S. V., Jiang, D., Zhang, Y., Dubonos, S. V., Grigorieva, I. V., and Firsov, A. A.: Electric field in atomically thin carbon films, Science, 306, 666–669, doi:10.1126/science.1102896, 2004.

Okeeffe, A. E. and Ortman, G. C.: Primary standards for trace gas analysis, Anal. Chem., 38, 760–763, doi:10.1021/ac60238a022, 1966.

Rahmat, F., Thamwattana, N., and Hill, J. M.: Carbon nanotube oscillators for applications as nanothermometers, J. Phys. Math. Theor., 43, 405209–405209 doi:10.1088/1751-8113/43/40/405209, 2010.

Stellmack, M. L. and Street, K. W.: Permeation devices for high-pressure gases, Anal. Lett. Pt. A, 16, 77–100, 1983.

Tabard-Cossa, V., Godin, M., Beaulieu, L. Y., and Grütter, P.: A differential microcantilever-based system for measuring surface stress changes induced by electrochemical reactions, Sens Actuators, B. Chem., 107, 233–241, doi:10.1016/j.snb.2004.10.007, 2005.

Wilcken, K. and Kauppinen, J.: Optimization of a Microphone for Photoacoustic Spectroscopy, Appl. Spectrosc., 57, 1087–1092, doi:10.1366/00037020360695946, 2003.

Zelinger, Z., Papouskova, Z., Jakoubkova, M., and Engst, P.: Determination of trace quantities of freon by laser optoacoustic detection and classical infrared-spectroscopy, Collect. Czech. Chem. Commun., 53, 749–755, 1988.

Zelinger, Z., Střižík, M., Kubát, P., Jaňour, Z., Berger, P., Černý, A., and Engst, P.: Laser remote sensing and photoacoustic spectrometry applied in air pollution investigation, Opt. Lasers Eng., 42, 403–412, doi:10.1016/j.optlaseng.2004.03.005, 2004.

Zelinger, Z., Střižík, M., Kubát, P., Lang, K., Bezpalcová, K., and Jaňour, Z.: Model and real pollutant dispersion: Concentration studies by conventional analytics and by laser spectrometry, Int. J. Environ. An. Ch., 86, 889–903, 2006.

Zelinger, Z., Střižík, M., Kubát, P., Civis, S., Grigorová, E., Janečková, R., Zavila, O., Nevrlý, V., Herecova, L., Bailleux, S., Horká, V., Ferus, M., Skřínský, J., Kozubková, M., Drábková, S., and Jaňour, Z.: Dispersion of light and heavy pollutants in urban scale models: co 2 laser photoacoustic studies, Appl. Spectrosc., 63, 430–436, doi:10.1366/000370209787944226, 2009.

6

Impedance spectroscopy characterization of an interdigital structure for continuous particle measurements in wood-driven heating systems

A. Weiss[1], M. Bauer[1], S. Eichenauer[2], E. A. Stadlbauer[2], and C.-D. Kohl[1]

[1]Institute of Applied Physics, JLU Giessen, Giessen, Germany
[2]Competence Centre for Energy and Environmental Engineering, University of Applied Science THM,
Campus Giessen, Giessen, Germany

Correspondence to: A. Weiss (alexander.weiss@ap.physik.uni-giessen.de)

Abstract. In the course of the climate change and increased focus on CO_2-neutral energy sources, the use of wood-driven small heating systems (SHS) becomes more important. But, their contribution to air pollution, especially particulate matter, is about as high as the emissions from car engines. The specific formation of harmful substances in wood fires and possible countermeasures by continuously operating sensor and control systems are covered.

Impedance spectra of interdigital electrode (IDE) structures are taken before and after mounting in wood-driven SHS to get information about the particles in the exhaust stream. It appears that the capacitive parts of the impedance spectra at a fixed frequency are appropriate for a fast signal evaluation. The good correlation with established offline measuring methods is discussed and the capability of thermal regeneration is demonstrated. The offline measurements of this work shall give the experimental basis for the development of online measurements in order to control the particle emissions of wood-driven SHS.

1 Introduction

The utilization of biomass for the generation of electricity and heat is becoming more important due to an increased focus on CO_2-neutral energy sources. Particularly the sector of small heating systems (SHS) has a rising demand for wood as an alternative fuel. The denotation "small heating system" is often interpreted in different ways. In this paper, it refers to boilers, stoves and furnaces with a nominal power output of less than 50 kW that are typically used in residential and small industry or business buildings. It also explicitly includes the sector of single-room heating systems, e.g., closed fire places. In Germany, roughly half of the SHS are already wood-driven (approximately 15 million). Their number is growing steadily due to increasing prices of fossil fuels. Especially in rural areas with easy access to the regional forestry, wood is established as a cost-efficient fuel.

1.1 Types of emissions

This growing number of wood-driven SHS is accompanied by an additional load of emitted air pollutants. One of them that is worth focusing on is the fraction of particulate matter (PM), because it is considerably higher than that of comparable oil or gas burners. According to the German Federal Environmental Agency, it already exceeds the PM emission caused by engines (cars, trucks, motorcycles, etc.) in 2003 (GFEA, 2007). The importance of that problem is also underlined by the constitution of the promoting cluster ("emission reduction in biomass-driven SHS") by the German Federal Environment Foundation (GFEF, 2014).

Fine dust is categorized by its aerodynamic particle diameter in the classes PM_{10} ($< 10 \mu m$), $PM_{2.5}$ ($< 2.5 \mu m$) and ultrafine dusts ($< 0.1 \mu m$). Besides this classification, which mainly provides information about the possibility of inhaling, the chemical nature of particles is also important for their effect on human health. Thereby, the primary aerosols are differentiated in soot, which mainly consists of carbon (black carbon, BC), mineral salts and condensable organic compounds (COC), also known as tars.

In addition, volatile organic compounds (VOC), ammonia, carbon monoxide and oxides of nitrogen build up during combustion. Besides their direct toxic impact, they also contribute to the secondary organic aerosols (SOA) as well as to the inorganic fraction by conversion processes in the atmosphere (Nussbaumer, 2010).

1.2   Health effects

The health risk of fine dust is primarily caused by its small particle sizes. While the human body has defense mechanisms against larger particles, fine dust is able to penetrate the lungs via the respiratory system. PM_{10} is named "inhalable", whereas $PM_{2.5}$ is already classified as "respirable". Ultrafine dust is able to enter the alveoles (alveolar) and is hardly or even not at all removable. At this point, even in other respects, nontoxic substances can cause health damage like coughing, inflammation of the respiratory system, bronchitis or asthma.

Due to the large surface area of the pulmonary tissue, particles can also migrate to the blood and lead to serious diseases of the cardiovascular system (e.g., heart attack), especially for long time exposures. Long-term studies show that there is a correlation between elevated mortality rate, particularly by respiratory system diseases or lung cancer, and the fine dust pollution (LUA, 2005).

Klippel and Nussbaumer (2007a) showed that health effects of fine dust depend not only on its particle sizes but are also mainly influenced by its chemical composition. Accordingly, the weakest health effects are caused by dust that mostly consists of mineral salts, such as those occurring in well-adjusted combustion systems. Compared to that, the cell toxicity of organic dusts from incomplete combustion is approximately 100 times higher. Besides the soot fraction (BC), they include mostly condensable organic compounds (COC). Additionally, a large number of different polycyclic aromatic hydrocarbons (PAH) can be found that are well known for their high carcinogenic impact. An overview of the health effects of fine dust is given by Nussbaumer (2012).

1.3   Legal framework

The German 1st Federal Immission Protection Directive (1st BImSchV) regulates the operation and inspection of small- and medium-sized heating systems (GFRG, 2010). The need for a reduction in pollutants is reflected in stricter emission limits in the latest version of 22 March 2010. Table 1 shows the two-stage lowering of the maximum concentrations of CO and dust allowed. There are some interim arrangements for existing systems. If single-room heating systems do not fulfill the requirements of maximum allowed emission levels of $0.15 \, \mathrm{g \, m^{-3}}$ dust and $4 \, \mathrm{g \, m^{-3}}$ CO in an in-

Table 1. Emission limits for small heating systems according to the 1st BImSchV (GFRG, 2010).

	From 22 Mar 2010		From 31 Dec 2014 (31 Dec 2016)[a]	
	CO ($\mathrm{g \, m^{-3}}$)	Dust ($\mathrm{mg \, m^{-3}}$)	CO ($\mathrm{g \, m^{-3}}$)	Dust ($\mathrm{mg \, m^{-3}}$)
Single-room heating system	2.0 0.4[b]	75 30–50[b]	1.25 0.25[b]	40 20–30[b]
SHS > 4 kW driven with natural wood	0.8[a] 0.5[b]	100[a] 60[b]	0.5	20

[a] Split logs, wood chips; [b] pellets.

spection, they have to be upgraded or put out of operation between 2014 and 2024, depending on the year of installation.

Currently, the 1st BImSchV does not define limits for the emission of nitrogen oxides (NO_x) from wood-driven SHS. NO_x regulations only exist for oil- and gas-driven SHS as well as for fuels with high nitrogen content like straw and grain. However, the DINplus seal of quality already limits the NO_x emission of wood-driven SHS to $200 \, \mathrm{mg \, Nm^{-3}}$, underlining their increasing relevance.

2   Characteristics of wood combustion

2.1   Properties of wood

Wood consists for the most part of cellulose and hemicellulose (60–80 % of the dry mass). These polysaccharides give tensile strength to the wood by its fibrillae. The phenolic macro-molecule lignin contributes about 30 % of the dry mass. It exhibits molecule masses in the range of 5000 to 10 000 u and, in contrast to the weak parts of the plant, it gives compression strength to the wood. In addition, there are some extractives like resins or minerals. The exact composition differs depending on the type of wood and individual growing conditions.

Besides these constituents, forming the dry mass, a considerable amount of water is present in wood. The wood moisture, which is defined as the ratio of water mass to the wood dry mass, can measure up to 150 % for fresh chopped trunks. During drying it is reduced, depending on the ambient conditions. Wood moisture is an important parameter for combustion, because it reduces the lower heating value (LHV) by its evaporation during combustion.

2.2   Combustion

According to Nussbaumer (2010), the combustion of wood can be generally divided into three phases. In the first phase, the wood is dried, causing the residual wood moisture to evaporate. Afterwards, gasification and pyrolysis begin, at which point volatile hydrocarbons are released into the gas phase and carbonaceous organic compounds (cellulose, hemicellulose and lignin) start to decompose. Coke, volatile compounds (CO, H_2 and VOC) and condensible organic compounds (COC) originate from this decomposition process. Under optimal conditions, these products are oxidized to CO_2 and H_2O in the third phase (Nussbaumer, 2010). Concerning the whole fuel volume, it is important to notice that all three phases take place at the same time. While oxidation takes place at one point of the wood, it influences drying and pyrolysis at another point of it.

2.3　Formation of pollutants

2.3.1　Dust

During combustion with a restricted oxygen supply, particles form from incompletely burned components. At high temperatures ($> 800\,°C$), the organic compounds are decomposed to a degree at which mainly carbonic soot (black carbon, BC) is generated. On the other hand, if the temperature of the fire is low, mainly COC arise, because the already formed tars cannot react any further and condense to particles in the atmosphere. Low fire temperatures can occur for different reasons, e.g., energy consumed by evaporation of high moisture content, retarded exothermal oxidation due to oxygen deficiency, or due to cooling by an excessive inlet air stream (Nussbaumer, 2010). In fact, VOC also contribute to the organic aerosols, but their conversion to particles takes place by secondary reaction in the atmosphere (secondary organic aerosols, SOA). In this process, the mass of the particles can exceed the original mass of VOC by a factor of 2, due to the uptake of further elements (Nussbaumer, 2010).

In contrast to the organic fraction, the emission of mineral components can also occur in complete combustion. They form either by evaporation and subsequent condensation in the cooling phase or are carried away from the ashes by the exhaust stream. They mainly consist of chlorides, sulfates, carbonites and oxides of the alkaline metal and alkaline earth metals (Oser et al., 2003).

2.3.2　Oxides of nitrogen

In contrast to other kinds of combustion systems, the temperatures in wood-driven SHS are low enough that the formation of NO_x from the atmospheric nitrogen (thermal NO_x) can be neglected. In this case, the nitrogen originates from the wood itself (fuel NO_x), where its content is significantly higher than in fossil fuels. However, the effectively emitted amount of NO_x can be considerably reduced by low-NO_x methods like fuel or air staging. The latter is characterized by the spatial separation of supply air into primary and secondary air. Thereby, the primary air area has to be highly substoichiometric to generate an oxygen-deficient reduction zone. Nitrogen-containing compounds can be reduced to molecular nitrogen at high temperatures (approximately $1100\,°C$) combined with a dwell time of at least $0.5\,s$ in this zone. The complete oxidation of the organic parts is carried out by the subsequent addition of secondary air (Hasler et al., 2000). A distinctive reduction zone ($0.2 < \lambda < 0.3$) also favors lower emissions of particulate matter.

2.4　Mode of operation and sensor-controlled operation

A consequence of varying fuel properties (wood moisture, loading of the combustion chamber, etc.) and the complex relationships between the operating parameters is a discontinuous combustion with partly high emissions. Hence, the accepted opinion in research and in the chimney sweeper trade is that the typical mode of operation of wood-driven SHS is far away from the optimum. Thus, in general, the actual emissions significantly exceed the values measured under norm conditions. This supports the assumption that the fraction of wood-fired heating as a source of fine dust is considerably underestimated (Nussbaumer, 2010). Studies in a German residential area show that wood-fired heating is responsible for up to 57 % of the ambient PM_{10} pollution under certain weather conditions (Bari et al., 2011).

Therefore, it is reasonable to adjust the available variables like primary air, secondary air and fuel supply to the current combustion conditions to achieve minimum emissions. To evaluate these conditions, a control system needs comprehensive sensor technology for the direct measurement of all relevant emission quantities. Similar approaches for small-scale biomass furnaces were recently investigated by Bischof et al. (2013).

3　Sensors

3.1　Systems currently in use

In contrast to large combustion systems or to the control of combustion engines (lambda probe), continuously working sensor systems are not common in wood-driven SHS. If continuous systems are present anyway, it is mostly about temperature-sensitive elements for power management. These can either mechanically control (bimetal, thermostat) the inlet air, causing bad combustion conditions, or regulate the fuel amount, e.g., in modern pellet combustion systems. Besides a measurement of the boiler temperature in water-bearing systems, there is the possibility of placing thermo-elements directly in the combustion chamber. In some rare cases, lambda probes are used to gauge the needed amount of inlet air. In contrast to combustion engines, which usually work in a narrow range around the stoichiometric mixture ($0.9 < \lambda < 1.1$), there are lower values in the reduction zone ($0.2 < \lambda < 0.7$) and higher ones in the oxidation zone ($1.4 < \lambda$).

The sensor systems currently in use with regard to SHS refer to the measurements observing emission limits that are regulated by law. In contrast to continuously working sensor systems, these kinds of snap sample measurements are not applicable for electronic combustion control. They

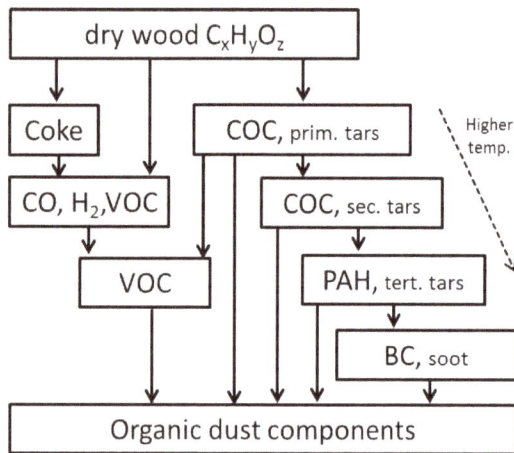

Figure 1. Ways of conversion of the hydrocarbons to organic dust components according to Nussbaumer (2010) (Weiss et al., 2014).

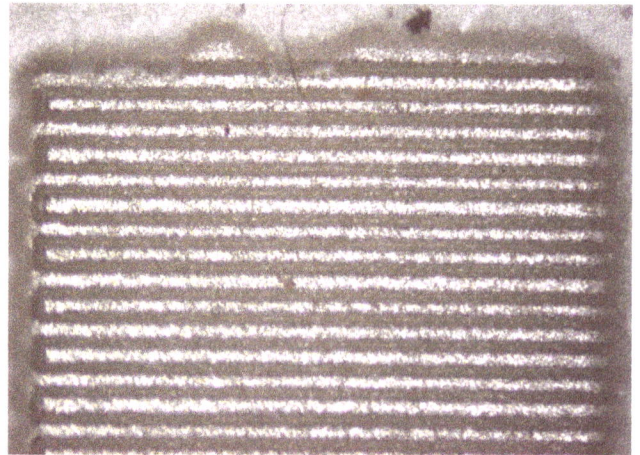

Figure 2. Used IDE structure consisting of a ceramic substrate (Al_2O_3) and platinum interdigital electrodes. The electrode width and distance is $27.5\,\mu m$, the length $1400\,\mu m$ and the number of legs 34 (Weiss et al., 2014).

are only suitable for the detection of leakages or maladjustments. These 15 min measurements are performed by chimney sweepers and serve for observation of the limits of dust and CO. Additionally, the temperature, oxygen content and pressure difference is taken to benchmark the operating status.

Measurement methods regarding PM include mainly gravimetric ones, but optical (scattered light, photoemission) and electrostatic methods are in use, too. The gravimetric methods feature, e.g., filter cartridges that are weighed after sampling or are part of a resonance circuit. The gaseous measurands (CO, O_2 and NO_x) are taken by electrochemical cells in some instruments. The temperature can be measured by thermocouples (e.g., by type K).

The instruments that are compliant for measurements according to 1st BImSchV exhibit a measurement uncertainty of 30 to 50 % concerning the dust content. Furthermore, the costs of EUR 5000 to EUR 20 000 per instrument are a high investment for the user.

3.2   Sensors for continuous measurements

To use sensors for the control of SHS operating parameters, the measurement method has to work continuously and must withstand the rough conditions in combustion systems by its technical implementation. Low costs per unit are essential for an economical realization.

In the field of gas sensing, metal oxide gas sensors (MOX) are suitable for this challenge. For example, tin oxide sensors for the measurement of CO have already been developed for fire detection in lignite-fired power plants. Despite the rough ambient conditions (particularly dust), they achieve lifetimes of up to 7 years and have been applied successfully in many power plants (Kohl et al., 2001). Also, the NO_x measurement can be carried out by MOX. Benner et al. (2002) showed that the used MOX withstand the combustion chamber conditions

of a power plant at up to 650 °C. Electrochemical sensors cannot be operated under these conditions. They are only suitable in bypass measurements at lower temperatures after exhaust stream cooling. Their lifetime is typically on the order of 1 to 3 years and thereby shorter than that of MOX.

Direct measurement of the dust emission is crucial for controlled operation, because it cannot be determined just by the knowledge of other measurands like the CO concentration due to the fact that there is no clear correlation (Klippel and Nussbaumer, 2007b). For this task, common quartz bulk oscillators (QCM) are not suitable, because the limited temperature stability of quartz does not allow thermal regeneration. But, other piezoelectric materials like langasite may also be candidates for high temperature application (Richter et al., 2008; Richter and Fritze, 2014). Surface acoustic wave devices are commonly used as sensors in various applications, especially for the detection of mass deposits on its surface. But, under rough and fluctuating conditions – like those present in SHS – they are hard to operate. Because temperature deviations of only 1 K can be enough to interfere with the pure sensor response, temperature compensation, e.g., by a dual-port SAW, is necessary (Mujahid and Dickert, 2014). Beyond that, the elastic and conductance properties can spoil the result by a factor of up to 2. Our own approach for a capacitive dust measurement procedure will be detailed in the following section.

4   Capacitive dust measurement

An approach for dust measurements in wood-driven SHS was investigated by the authors in a joint research project with partners from industry and a chimney sweep academy. The basis of the measurement is a ceramic substrate (Al_2O_3) with platinum interdigital electrodes (IDE); cf. Fig. 2. Comparable structures as resistive-type dust sensors are also in use for the monitoring of particulate filters in automotive applications (Bosch, Continental) and are currently under investigation (Bartscherer and Moos, 2013). Recently, they have been suggested for the detection of conductive smoke deposits in electrical installations (Cleary, 2014). Resistance measurements of such structures require a minimum level of coverage at which the percolation threshold is reached and a conducting path builds up. The disadvantages are that low levels of coverage cannot be determined and that only conducting particles contribute to the signal.

In contrast to the detection of filter damages in cars, where the relevant timescale is on the order of hours or days, a sensor for the control of wood combustion has to operate on scales of less than a minute. Quick reaction and the detection of small amounts of dust are crucial for the application. Therefore, a capacitive approach is deployed. Dust deposited on and between the electrodes gives rise to the capacitance of the device by its permeability even when no conduction path develops.

4.1 Technical realization

To design simple electronics for a capacitive online measurement, it is first necessary to analyze the IDE structure and dust properties in detail. Therefore, in contrast to the future application, in this proof of principle, all measurements were conducted in the lab of the university, not online in SHS. Thus, the fresh IDE structures were first investigated by impedance spectroscopy (IS), subsequently loaded with dust in SHS and afterwards characterized by IS again. Impedance spectra were taken using a Solartron SI1260 in combination with a Chelsea dielectric interface that facilitates the measurement of small currents and therefore extends the measuring range to small capacitances even at low frequencies. The complex capacitance C^* is calculated from the complex impedance Z^* according to Eq. (1).

$$C^* = \frac{Y^*}{i \cdot \omega} = \frac{1}{i \cdot \omega \cdot Z^*} \qquad (1)$$

$$\mathrm{Re}(C) = \frac{-\mathrm{Im}(Z)}{\omega (\mathrm{Re}(Z)^2 + \mathrm{Im}(Z)^2)}. \qquad (2)$$

The real part of C^* is expressed by Eq. (2) and describes the capacitance of a capacitor in an equivalent circuit model that is frequency independent for an ideal capacitor-like system. The IDE exhibits only minor frequency dependency in the range from 1 to 10^5 Hz ($< 0.5\%$). Thus, the sensor response C/C_0 for low (no conducting) coverage can easily be calculated by using one frequency (16.3 kHz), which is important for future device implementations.

The measured base capacitance C_0 (without dust load, at room temperature) of the IDEs is in the range from 2.5 to 2.9 pF due to minor production fluctuations. A theoretical description of such a system is given by Endres and

Figure 3. Socket mounted IDE structures. Left: fresh; middle: light loaded (response $\sim 3\%$); right: heavily loaded (response 419 %, percolation threshold exceeded) (Weiss et al., 2014).

Drost (1991) and Endres and Bock (2006), and results in Eq. (3):

$$C = (n-1) \cdot L \cdot \varepsilon_0 \cdot \left(\left(\frac{\varepsilon_L + \varepsilon_S}{2} \right) \cdot \frac{K \cdot (\sqrt{1 - (\frac{a}{a+b})^2})}{K \cdot (\frac{a}{a+b})^2} + \varepsilon_L \frac{h}{a} \right), \qquad (3)$$

where a, b, h, L and n are electrode distance, width, height, length and number, respectively, ε_L and ε_S are the relative permittivities of air and substrate material and $K(x)$ is the elliptic integral of the first kind (Endres and Drost, 1991; Endres and Bock, 2006). Taking average values of the used IDE ($a = 27.5\,\mu m, b = 27.5\,\mu m, h = 5\,\mu m, L = 1400\,\mu m, n = 34$, $\varepsilon_L = 1.000059$, $\varepsilon_S = 9.5$), Eq. (3) yields a capacitance of 2.9 pF, which is in good agreement with our results. In principle, electrode distances will not affect the results in resistive measurements (assuming a homogeneous coverage with sufficiently small particles). In a capacitive approach, it will influence the base capacitance (C_0) as well as the sensitivity. Therefore, an adjustment of the electrode structure can be utilized as an optimization parameter in further investigations.

The capacitance shows only small fluctuations between measurements from day to day (0.04 %) in the laboratory setup. Signal changes of 0.1 % can be clearly separated from the noise. The cross-sensitivity to humidity is quite low. The capacitance of the fresh IDE structure at 16.3 kHz changes by only 0.03 %, when the relative humidity is varied from 0 to 30 % at room temperature. Figure 3 shows examples of three IDE structures with different levels of soot loading from a wood-driven SHS. As can be seen, even hardly visible loadings lead to clearly measurable results (middle). The response will reach very high values for loads in excess of the percolation threshold (right).

4.2 Thermal regeneration

Besides IDE, the substrate possesses a platinum resistance wire (PT10) that is arranged in a U shape around the IDE in a sublayer and that serves as a resistance thermometer and heating. It allows an exact temperature control as well as thermal regeneration at temperatures up to 800 °C. The organic part of the dust deposits can be burned directly on the IDE structure. Optical investigations reveal that 650 °C

Figure 4. Single soot particle left between the platinum electrodes after a first regeneration step. The particle was removed by a second regeneration (electrode width 27.5 μm).

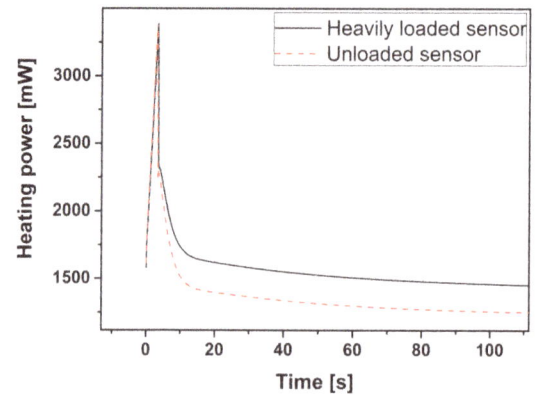

Figure 5. Dissipated heating power during thermal regeneration of a heavily loaded IDE structure (line) and a fresh IDE structure (dashed line).

are sufficient to relieve the surface of all black deposits. The required time depends on the amount of dust and varies between a few seconds for light loading and up to 5 min for heavy loading. Higher temperatures are able to fasten this procedure. In a series of ten regeneration tests, only one sensor showed a single residual soot particle after a first regeneration step (see Fig. 4). A second regeneration was sufficient to remove the particle. The complete cleaning of the sensor surface is verified by further IS measurements. It is shown that the capacitance returned to its former basic value of C_0.

The dissipated heating power during temperature-controlled regeneration can serve as a further measurand. As indicated in Fig. 5, higher dust loadings require an elevated amount of heat. This optional analysis step may deliver additional information about the dust loading and will be the object of further investigations, as well as the detailed temperature dependence of the IDE structure.

4.3   Results from the test grid

Series of measurements were performed with IDE structures loaded in different SHS (Buderus Blueline stove, Buderus Logano S231-25 wood gasifier) and with different kinds of wood-based fuels. In each case, the IDE structures are placed in the exhaust tube directly behind the SHS with the surface facing upstream in the direction of the combustion chamber. Because it is not possible to generate completely reproducible combustion situations (see Sect. 2), reference systems are necessary. The only feasible ones are the instruments according to 1st BImSchV and in accordance with VDI guideline 2066-1 from which two with different measurement principles are used. Due to the fact that they are not designed for continuous operation, they enforce 15 min measurements with corresponding high loading levels for the IDE sensor.

First results indicated that there is an acceptable correlation between the IDE sensor and references despite the fact that the capacitance signal is not volume weighted like the references (Weiss et al., 2014).

Further results from a series with four different wood-based fuels using the Buderus Blueline stove are shown in Fig. 6. The measurement numbers indicate eight separate fires. To evaluate the relative accuracy of the IDE sensors, each measurement is carried out using two of them in parallel. As can be seen, the relative deviations are small, with only 6 % of the response on average and some points so close to one another that they can hardly be differentiated on this scale. A distinguishable correlation between IDE sensor response and the reference values is achieved over all measurements, despite various fuels. The remarkable deviations of the two reference instruments (Fig. 6: nos. 5 to 8) may be caused by a saturation effect of reference instrument 1 due to the high dust concentrations ($> 150 \, \mathrm{mg \, m^{-3}}$).

Measurement no. 8 exhibits an overestimation of the dust content by the IDE sensor. Detailed inspection of the trends of temperature, dust, CO and O_2 reveals that the fire has come to the final combustion phase during the measurement. This has a strong influence on the pressure inside the systems and therefore on the exhaust velocity. In contrast to the IDE sensor, the volume-weighted values of the reference instruments compensate for that effect, which may explain the deviation.

5   Conclusions

The continuous in situ measurement of particles under firing conditions is an important task to reduce wood fire emissions, especially during the initial and ceasing states. This paper presents a first step in that direction, showing that cheap interdigital structures generate useful signals for an economical device. Further on, it could be shown that these devices can be used repeatedly in a cyclic manner. That opens the

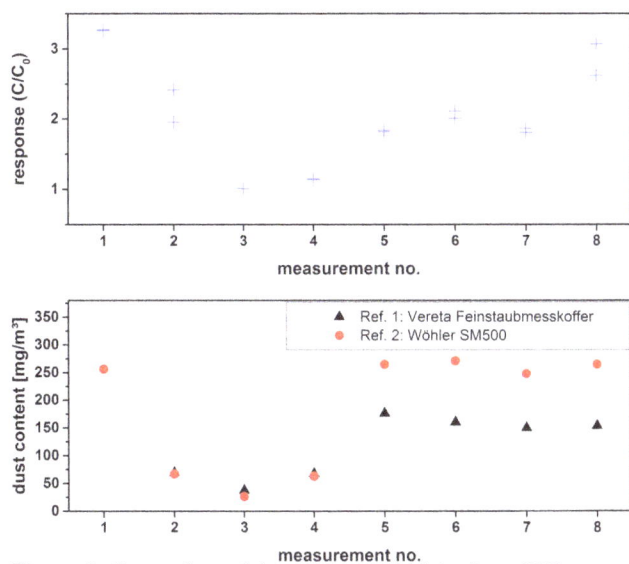

Figure 6. Comparison of the response (C/C_0) of two IDE sensors at a frequency of 16.3 kHz with values from two reference instruments. The response of both IDE sensors overlap at several points due to their small deviation. The fuels are HTC char (no. 1, no. 2), beech briquette (no. 3, no. 4), Pini&Kay briquette beech (no. 5, no. 6), Pini&Kay briquette spruce (no. 7, no. 8). The deposition time is 15 min according to the 1st BImSchV requirements of the reference instruments.

Acknowledgements. The authors thank Bernd Vollmer and Hans-Eberhard Kopp of the North Rhine-Westphalia chimney sweep academy (Schornsteinfegerakademie Dülmen) as well as Ulrich Strohal and Jochen Fey of the Strohal enterprise (Strohal Anlagenbau, Staufenberg) for their technical support and preparation of the test combustion systems. Further gratitude goes to the German Federal Environment Foundation for supporting the project.

possibility of applying them in an online control system for wood-driven SHS.

At this early stage of research, the sensors already show a good correlation with reference instruments in offline measurements, although the data analysis of the IDE sensor presented here is not explicitly dedicated to dust loadings beyond the percolation threshold. But, such high loading levels were reached in the measurements due to the 15 min time interval according to 1st BImSchV, which is necessary for the reference instruments. Furthermore, a large optimization potential can be utilized by taking the actual exhaust stream velocity into consideration by measuring or controlling it in ways of geometrical flow dynamic manipulation. Also, the electrode structure can be further adapted to the main particle properties.

The influence of humidity on the capacitance of loaded sensors should be part of comprehensive studies. The varying chemical composition and morphology of the particles may lead to different effects due to the environmental dependency of the water permittivity. It can drop by a factor of approximately 8 in close vicinity to surfaces (Paul and Paddison, 2001). Comparative investigations with other methods like differential thermal analysis are expedient.

First tests of capacitive sensor readout with simpler electronics are conducted under lab conditions with promising results. As mentioned above, the dissipated heating power as another measurand as well as the thermal dependence of the permittivity will be part of further investigation.

References

Bari, M. A., Baumbach, G., Kuch, B., and Scheffknecht, G.: Air pollution in residential areas from wood-fired heating, Aerosol Air Qual. Res., 11, 749–757, 2011.

Bartscherer, P. and Moos, R.: Improvement of the sensitivity of a conductometric soot sensor by adding a conductive cover layer, J. Sens. Sens. Syst., 2, 95–102, doi:10.5194/jsss-2-95-2013, 2013.

Benner, N., Eberheim, A., and Kohl, D.: Halbleiter-Gassensoren in der Rauchgasanalyse, 11. ITG/GMA Fachtagung Sensoren und Messsysteme, Ludwigsburg, 2002.

Bischof, J., Baumbach, G., and Struschka, M.: A new approach for CO-based control of small-scale biomass furnaces, Proceedings of the 21st European Biomass Conference and Exhibtion, Copenhagen, Denmark, 3–7 June 2013.

Cleary, T.: Effects of Soot Deposition on Current Leakage in Electronic Circuitry, Proceedings of 14th International Conference on Automatic Fire Detection, 14–16 October 2014.

Endres, H.-E. and Bock, K.: Impedanzspektroskopie und multivariante Signalverarbeitung für planare chemische Sensoren, Technische Mitteilungen 1/2 06, ISSN 0040-1439, Essen, 2006.

Endres, H.-E. and Drost, S.: Optimization of the geometry of gassensitive interdigital capacitors, Sens. Act., B4, 95–98, 1991.

GFEA, German Federal Environmental Agency: Die Nebenwirkungen der Behaglichkeit: Feinstaub aus Kamin und Holzofen, Berlin 2007.

GFEF, German Federal Environment Foundation: Kleine Holzfeuerungsanlagen: Klimafreundlich heizen und Feinstaubbelastung senken, DBU-aktuell No. 1, 2014.

GFRG, Government of the Ferderal Republic Germany: Erste Verordnung zur Durchführung des Bundes-Immissionsschutzgesetzes (Verordnung über kleine und mittlere Feuerungsanlagen – 1st BImSchV), Berlin, 2010.

Hasler, P., Nussbaumer, T., and Jenni, A.: Praxiserhebung über Stickoxid- und Partikelemissionen automatischer Holzfeuerungen, Bundesamt für Energie BFE, Zürich, 2000.

Klippel, N. and Nussbaumer, T.: Health relevance of particles from wood combustion in comparison to diesel soot, 15th European Biomass Conference and Exhibition, Berlin, 7–11 Mai 2007a.

Klippel, N. and Nussbaumer, T.: Einfluss der Betriebsweise auf die Partikelemission von Holzöfen, Bundesamt für Energie BFE, ISBN 3-908705-15-0, Zürich, 2007b.

Kohl, D., Kelleter, J., and Petig, H.: Detection of Fires by Gas Sensors, Sensors Update, 9, 161–223, 2001.

LUA NRW, Landesumweltamt Nordrhein-Westfahlen, Fachberichte 7/2005, ISSN 1613-0715, Essen, 2005.

Mujahid, A. and Dickert, F. L.: SAW and Functional Polymers, in Kohl, C.-D. and Wagner, T., Gas Sensing Fundamentals, Springer Series on Chemical Sensors and Biosensors, 15, 213–246, doi:10.1007/5346_2013_55, 2014.

Nussbaumer, T.: Emissionsfaktoren von Holzfeuerungen und Klimaeffekt von Aerosolen aus der Bio-masse-Verbrennung, 11. Holzenergie-Symposium, Zürich, 17 September 2010.

Nussbaumer, T.: Gesundheitsauswirkungen von Feinstaub aus Holzfeuerungen, 12. Holzenergie-Symposium, Zürich, 14 September 2012.

Oser, M., Nussbaumer, T., Mueller, P., Mohr, M., and Figi, R.: Final report of the research program 26688, Grundlagen der Aerosolbildung in Holzfeuerungen, Bundesamt für Energie BFE, ISBN 3-908705-02-9, doi:10.1109/ICSENS.2008.4716740, 2003.

Paul, R. and Paddison, S. J.: A statistical mechanical model for the calculation of the permittivity of water in hydrated polymer electrolyte membrane pores, J. Chem. Phys., 115, 7762–7771, 2001.

Richter, D., Sauerwald, J., Fritze, H., Ansorge, E., and Schmidt, B.: Miniaturized resonant gas sensor for high temperature applications, Proceedings of the 7th IEEE Conference on Sensors, Lecce, 1536–1539, 26–29 October 2008, doi:10.1109/ICSENS.2008.4716740, 2008.

Richter, D. and Fritze, H.: High temperature gas sensors, in: Gas Sensing Fundamentals, edited by: Kohl, C.-D. and Wagner, T., Springer Series on Chemical Sensors and Biosensors, 15, 1–16, doi:10.1007/5346_2013_56, 2014.

Weiss, A., Eichenauer, S., Stadlbauerm E. A., and Kohl, C.-D.: Sensor systems for the reduction of particulate matter in wood driven fire places, Proceedings of the 17th ITG/GMA symposium on sensors and measurement systems, Nuernberg, 3–4 June 2014.

7

High-temperature piezoresistive C/SiOC sensors

F. Roth[1], C. Schmerbauch[2], E. Ionescu[1], N. Nicoloso[1], O. Guillon[2], and R. Riedel[1]

[1]Technical University Darmstadt, Institute of Material Science, Jovanka-Bontschits-Strasse 2,
64287 Darmstadt, Germany
[2]Forschungszentrum Jülich, Institute of Energy and Climate Research, Wilhelm-Johnen-Strasse,
52425 Jülich, Germany

Correspondence to: F. Roth (roth@materials.tu-darmstadt.de)

Abstract. Here we report on the high-temperature piezoresistivity of carbon-containing silicon oxycarbide nanocomposites (C/SiOC). Samples containing 13.5 vol% segregated carbon have been prepared from a polysilsesquioxane via thermal cross-linking, pyrolysis and subsequent hot-pressing. Their electrical resistance was assessed as a function of the mechanical load (1–10 MPa) and temperature (1000–1200 °C). The piezoresistive behavior of the C/SiOC nanocomposites relies on the presence of dispersed nanocrystalline graphite with a lateral size ≤ 2 nm and non-crystalline carbon domains, as revealed by Raman spectroscopy. In comparison to highly ordered carbon (graphene, HOPG), C/SiOC exhibits strongly enhanced k factor values, even upon operation at temperatures beyond 1000 °C. The measured k values of about 80 ± 20 at the highest temperature reading ($T = 1200$ °C) reveal that C/SiOC is a primary candidate for high-temperature piezoresistive sensors with high sensitivity.

1 Introduction

The improvement of combustion processes relies on the exact control of the compression–combustion–exhaust cycles and thus there is a stringent need for pressure sensors with high sensitivity, low response time, high bandwidth of response and outstanding stability at high temperatures and aggressive environments. However, commercially available piezoresistive sensors, which are usually based on semiconductors or polymer composites, are limited by their low thermal stability in air (Kanda and Suzuki, 1991). Recently, polymer-derived ceramics (PDCs) such as silicon oxycarbides (C/SiOC) or silicon carbo(oxy)nitrides (C/SiCN, C/SiOCN) have been shown to combine piezoresistivity (Riedel et al., 2010; Zhang et al., 2008, Terauds et al., 2010) with outstanding temperature and oxidation stability (Riedel et al., 1995, 1996). Hence, they are promising candidates for future high-temperature pressure sensors. Concerning their structural features, PDCs can be described as amorphous and intrinsically nanoheterogeneous materials. The microstructure of C/SiOC with high carbon content has been described as a interpenetrating network of silica and carbon (Papendorf et al., 2013).

In the following, the results of temperature-dependent investigations of the k factor and of Raman spectroscopic studies are presented for a C/SiOC nanocomposite (13.5 vol% C) as clear experimental evidence for the intimate relationship between the carbon microstructure and the piezoresistive behavior of C/SiOC.

2 Experimental procedure

The polymeric precursor (poly(methylsilsesquioxane), PMS MK, Wacker AG, Munich, Germany) was cross-linked at 250 °C for 2 h, pyrolyzed at 900 °C for 2 h under flowing argon and subsequently ball-milled and sieved to a particle size $< 100 \mu$m. The sieved powder was hot-pressed at 1500 °C (30 MPa, Ar atmosphere, dwell 30 min) to obtain dense C/SiOC monoliths. Raman spectra were recorded with a Horiba HR800 micro-Raman spectrometer (Horiba JobinYvon, Bensheim, Germany) equipped with an Ar laser (514.5 nm). The measurements were performed by using a grating of 600 g mm^{-1} and a confocal microscope (magnification $100 \times$ NA 0.9 – numerical aperture) with a 100μm aperture, giving a resolution of 2–4 μm. The laser power (20 mW) was attenuated by using neutral density filters; thus,

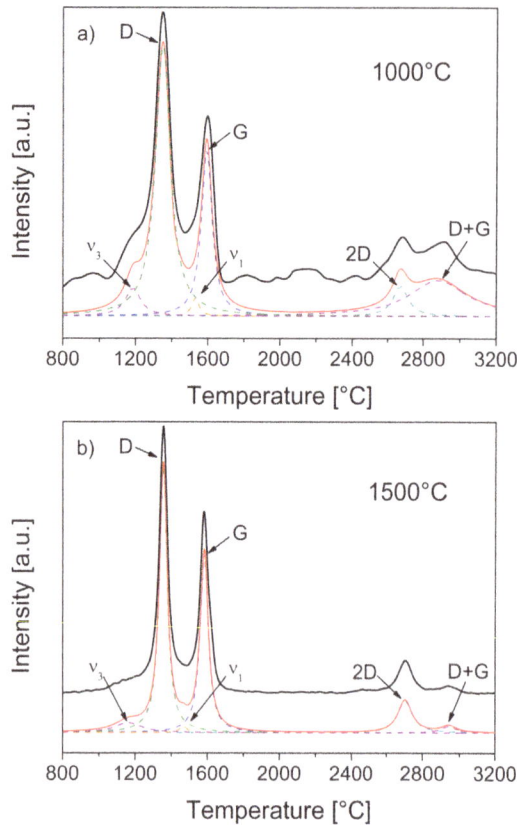

Figure 1. Raman spectra of C / SiOC (13.5 vol% C) at 1000 (**a**) and 1500 °C (**b**). Dashed lines represent the deconvoluted spectra.

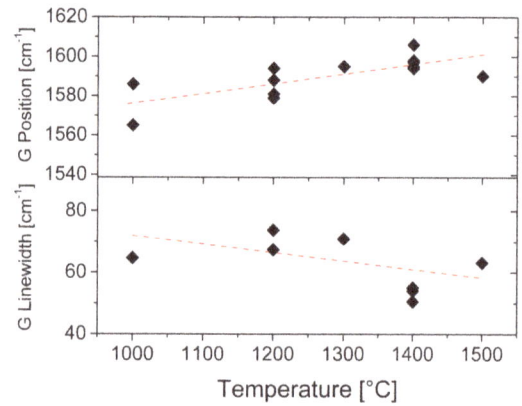

Figure 2. G position and linewidth as a function of sample preparation temperature.

the power on the sample was in the range from $6\,\mu$W to $2\,$mW. C / SiOC samples were placed in a cylindrical furnace allowing for resistivity measurements of up to 1500 °C. After achieving the desired temperature, the uniaxial load was applied by a mechanical testing machine (Model 5565, Instron Corp., Canton, MA, USA) using alumina rods as extensions. The resistivities of the loaded and unloaded samples, respectively, were calculated from the observed voltage changes and the applied currents.

3   Results and discussion

Piezoresistive materials are commonly classified by their k factor (gauge factor), which is defined as the change of the sample resistivity with applied stress:

$$k = \frac{Y}{R_0} \cdot \frac{\Delta R}{\Delta \sigma}, \tag{1}$$

with Y being the Young modulus ($Y = 85\,$GPa for C / SiOC; Papendorf et al., 2013), R_0 the resistivity of the stress-free sample, $\Delta \sigma$ the applied mechanical load and ΔR the change in the resistivity upon applying the mechanical load. As shown below, C / SiOC containing 13.5 vol% of dispersed carbon exhibits k values of $\approx 10^2$ in the temperature range

from 1000 to 1200 °C. The k factor decreases with increasing temperature, indicating a direct correlation with activated electronic transport. The Arrhenius plot provides an activation energy of $\geq 0.3\,$eV for k. A similar behavior has been observed for C / SiOCN nanocomposites containing 8.5 vol% segregated carbon (Terauds et al., 2010). However, the two composites differ with respect to the magnitude of k. Rather high k values ($k \approx 10^3$) have been reported for C / SiOCN in the temperature range $700 < T < 1000$ °C. We note that the C / SiOCN composite has a lower carbon content than our sample and, accordingly, a higher resistivity and a higher k. In the following we present Raman data of the carbon phase of C / SiOC and combine them with the piezoresistivity results to provide evidence that the piezoresistive effect is linked to the microstructure of the carbon phase, notably its disordered non-crystalline part.

3.1   Raman spectroscopy

Raman spectroscopy is a powerful method to characterize the various types of carbon, providing information about the degree of ordering of the carbon atoms, ranging from perfectly ordered sp^2-bonded carbon to less ordered noncrystalline carbon (Ferrari and Robertson, 2000; Dresselhaus et al., 2008; Pimenta et al., 2007). In graphitic materials, the Raman spectrum is dominated by two strong features: the G mode at $1581\,$cm^{-1} and the D mode at $1350\,$cm^{-1}. The G mode involves in-plane bond-stretching of sp^2 carbon (E_{2g}-symmetry) and is, together with its second harmonic mode (2-D), the only mode with significant intensity in graphene (two-dimensional single layer of graphite). Disordered carbon materials contain additional bands in their first-order Raman spectrum whose origin is still debated in literature. For example, the D band has been recently attributed to double resonant Raman scattering, questioning the assumption that it is strictly related to aromatic rings (Thomsen and Reich, 2000; Saito et al., 2001).

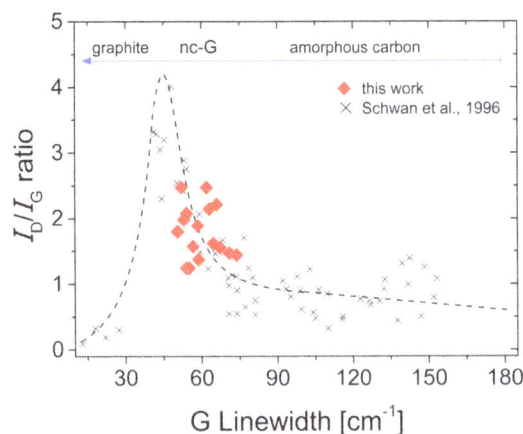

Figure 3. Ratio I_D / I_G vs. linewidth of G. The dashed line is a guideline for the eye taken from literature (Schwan et al., 1996). Increasing graphitization from amorphous carbon to nanocrystalline graphite (nc-G) and graphite is indicated by the arrow at the top.

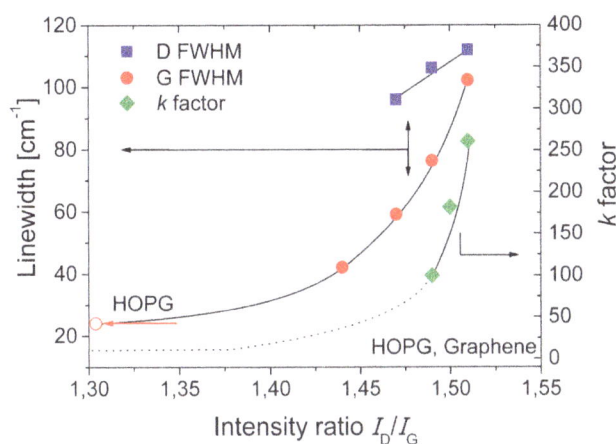

Figure 4. Correlation of Raman and piezoresistivity data for C/ SiOC (13.5 vol% C). The dashed area represents literature values for k and Raman linewidth of highly ordered carbon (HOPG, graphene).

An often used method to determine the degree of order in graphitic-like materials is to assess the intensity ratio I_D / I_G of the D and G bands. The Raman spectra of C/ SiOC samples prepared at 1000 and 1500 °C and rapidly cooled down to freeze the microstructure are shown in Fig. 1. The spectrum for C/ SiOC at 1000 °C reveals high-intensity D and G bands, less intense 2-D and D + G bands (the second harmonics), and weak ν_3 and ν_1 bands at 1180 and 1520 cm^{-1}. We assign the latter peaks to polyolefinic chains following Ferrari and Robertson (2004). With increasing temperature, the ν_1 peak is shifted to lower wave lengths, induced through chain growth. A comparison of the spectra recorded for the samples prepared at 1000 and 1500 °C indicates a clear tendency of ordering of the segregated carbon with increasing temperature, as I_D / I_G decreases. In line with this result, the

linewidth (full width at half maximum, FWHM) of G narrows by ≈ 15 cm^{-1} and the G position shifts by ≈ 25 cm^{-1} (Fig. 2). The Raman data suggests that sp^2 domains associated with the G band are progressively aligned yielding nanocrystalline graphite. Similarly, the polyolefinic structures represented by the ν_3 and ν_1 bands may be integrated into the aromatic domains by, e.g., a Diels–Alder reaction, explaining the observed decrease of the ν_3 and ν_1 bands (see Fig. 1). However, a non-crystalline part still remains, since we do not see a substantial change in the intensity of the D line. Thus, carbon appears to remain dispersed in C/ SiOC, the main change in the microstructure being the progressive formation of nanocrystalline graphite.

The assignment to nanocrystalline graphite is confirmed by the observed range of I_D / I_G values ($1.2 < I_D / I_G < 2.5$) for all samples prepared at 1000–1500 °C. According to Ferrari and Robertson (2000) the lateral domain size can be derived from I_D / I_G by the following equation:

$$\lambda \frac{I_D}{I_G} = C'(\lambda) \cdot L_a^2, \tag{2}$$

yielding domain sizes of 1.5–2.0 nm for C/ SiOC (13.5 vol% C). The Raman data of the graphitization process of carbon materials (amorphous carbon → graphite) has been summarized in literature (Schwan et al., 1996). Figure 3 presents the comparison of our data with literature. Evidently, the Raman features of carbon in C/ SiOC correspond to those of the nanocrystalline carbon.

3.2 Piezoresistivity

The Raman data presented above have shown that both crystalline and non-crystalline carbon exists in C/ SiOC and that their fractions depend on the preparation conditions. Considering carbon being primarily responsible for the piezoresistive behavior of C/ SiOC, a close correlation between the Raman data and the gauge factor should be observable. We therefore perform piezoresistive measurements on samples with the same I_D / I_G, i.e., samples with the same carbon microstructure.

Figure 4 summarizes the results of the Raman and piezoresistivity measurements. As it can been seen in Fig. 4, the k factor and the linewidths of D and G follow the same trend, i.e., both increase with increasing I_D / I_G. For the sake of comparison, the Raman linewidths and k factors of HOPG and graphene available in the literature have been included (shaded area). HOPG and graphene are extended, highly ordered sp^2 materials with narrow linewidth and low k values due to small amounts of defects (grain boundaries dislocations, dangling bonds, etc.). C/ SiOC is far less ordered. The non-crystalline carbon content appears to determine the k factor enhancement. Values of $k > 100$ mark C/ SiOC as a promising piezoresistive sensor material. Its most important advantage relies on its very high temperature stability in air because of the frozen-in microstructure. To our knowledge, only very few piezoresistive materials are available for elevated temperature ($250 < T < 1000\,^\circ$C) (Fraga et al., 2012; Gregory et al., 2002). Piezoresistive sensors for operations at temperatures even above $1000\,^\circ$C may be realized with composites C/SiOX (X = O, N). However, their feasibility as piezoresistive sensors for the daily use has to be proven by further work.

4 Conclusions

C/SiOC nanocomposite materials have been investigated by Raman spectroscopy of quenched samples ($1000 < T < 1500\,^\circ$C) and stress-dependent resistivity measurements (1–10 MPa) in the temperature range of $1000 < T < 1200\,^\circ$C. The observed values of the k factor (of the order of 10^2) are significantly higher than those of well-ordered carbon and decrease with increasing temperature, indicating a direct correlation with activated electronic transport ($E_A \geq 0.3$ eV for k). The comparison of the Raman and piezoresistivity data reveals that the piezoresistive effect in C/SiOC (13.5 vol% C) is mainly determined by the non-crystalline/defective carbon content.

Acknowledgements. The authors acknowledge financial support from the Deutsche Forschungsgemeinschaft (DFG, Ri510/52-1).

References

Dresselhaus, M. S., Dresselhaus, G., and Hofmann, H.: Raman spectroscopy as a probe of graphene and carbon nanotubes, Phil. Trans. R. Soc. A, 366, 231–236, 2008.

Ferrari, A. C. and Robertson J.: Interpretation of Raman Spectra of disordered and amorphous carbon, Phys. Rev. B, 61, 14095–14107, 2000.

Ferrari, A. C. and Robertson, J.: Raman spectroscopy of amorphous, nanostructured, diamond-like carbon, and nanodiamond, Phil. Trans. R. Soc. Lond. A, 362, 2477–2512, 2004.

Fraga, M. A., Furlan, H., Pessoa, R. S., Rasia, L. A., and Mateus, C. F. R.: Studies on SiC, DLC and TiO$_2$ thin films as piezoresistive sensor materials for high temperature application, Microsyst. Technol., 18, 1027–1033, 2012.

Gregory, O. J., Luo, Q., Bienkiewicz, J. M., Erwin, B. M., and Crisman, E. E.: An apparent n to p transition in reactively sputtered indium-tin-oxide high temperature strain gages, Thin Solid Films, 405, 263–269, 2002.

Kanda, Y. and Suzuki, K.: Origin of the shear piezoresistance coefficient π_{44} of n-type silicon, Phys. Rev. B, 43, 6754–6756, 1991.

Papendorf, B., Ionescu, E., Kleebe, H.-J., Linck, C., Guillon, O., Nonnenmacher, K., and Riedel, R.: High Temperature Creep Behavior of Dense SiOC-Based Ceramic Nanocomposites: Microstructural and Phase Composition Effects, J. Am. Ceram. Soc., 96, 272–280, 2013.

Pimenta, M. A., Dresselhaus, G., Dresselhaus M. S., Cançado L. G., Jorio A., and Saito R.: Studying disorder in graphite-based systems by Raman spectroscopy, Phys. Chem. Chem. Phys., 9, 1276–1291, 2007.

Riedel, R., Kleebe, H.-J., Schönfelder, H., and Aldinger, F.: A covalent micro/nano-composite resistant to high-temperature oxidation, Nature, 374, 526–528, 1995.

Riedel, R., Kienzle, A., Dressler, W., Ruwisch, L., Bill, J., and Aldinger, F.: A silicoboron carbonitride ceramic stable to $2000\,^\circ$C, Nature, 382, 796–798, 1996.

Riedel, R., Toma, L., Janssen, E., Nuffer, J., Melz, T., and Hanselka, H.: Piezoresistive Effect in SiOC Ceramics for Integrated Pressure Sensors, J. Am. Ceram. Soc., 93, 920–924, 2010.

Saito, R., Jorio, A., Souza Filho, A. G., Dresselhaus, G., Dresselhaus, M. S., and Pimenta, M.: Probing Phonon Dispersion Relations of Graphite by Double Resonance Raman Scattering, Phys. Rev. Lett., 88, 027401, doi:10.1103/PhysRevLett.88.027401, 2001.

Schwan, J., Ulrich, S., Batori, V., Ehrhardt, H., and Silva, S. R. P.: Raman spectroscopy on amorphous carbon films, J. Appl. Phys., 80, 440–457, 1996.

Terauds, K., Sanchez-Jimenez, P. E., Raj, R., Vakifahmetoglu, C., and Colombo, P.: Giant piezoresistivity of polymer-derived ceramics at high temperatures, J. Eur. Ceram. Soc., 30, 2203–2207, 2010.

Thomsen, C. and Reich, S.: Double Resonant Raman Scattering in Graphite, Phys. Rev. Lett., 85, 5214–5217, 2000.

Zhang, L., Wang, Y., Wei, Y., Xu, W., Fang, D., Zhai, L., Lin, K.-C., and An, L.: A Silicon Carbonitride Ceramic with Anomalously High Piezoresistivity, J. Am. Ceram. Soc. 91, 1346–1349, 2008.

8

A systematic MEMS sensor calibration framework

A. Dickow and G. Feiertag

Munich University of Applied Sciences, Munich, Germany

Correspondence to: A. Dickow (dickow@hm.edu)

Abstract. In this paper we present a systematic method to determine sets of close to optimal sensor calibration points for a polynomial approximation.

For each set of calibration points a polynomial is used to fit the nonlinear sensor response to the calibration reference. The polynomial parameters are calculated using ordinary least square fit. To determine the quality of each calibration, reference sensor data is measured at discrete test conditions. As an error indicator for the quality of a calibration the root mean square deviation between the calibration polynomial and the reference measurement is calculated. The calibration polynomials and the error indicators are calculated for all possible calibration point sets. To find close to optimal calibration point sets, the worst 99 % of the calibration options are dismissed. This results in a multi-dimensional probability distribution of the probably best calibration point sets.

In an experiment, barometric MEMS (micro-electromechanical systems) pressure sensors are calibrated using the proposed calibration method at several temperatures and pressures. The framework is applied to a batch of six of each of the following sensor types: Bosch BMP085, Bosch BMP180, and EPCOS T5400. Results indicate which set of calibration points should be chosen to achieve good calibration results.

1 Introduction

MEMS (micro-electromechanical systems) sensors are calibrated at one or several points at the end of the production process in order to fulfill the product specifications. Various techniques can be applied to compensate sensor nonlinearities (see Brignell, 1987). In many cases (e.g., humidity sensors, gyroscopes, pressure sensors, barometric pressure sensors) calibration relies on a polynomial for fitting the raw sensor data readout to reference signals (van der Horn and Hujising, 1997; Lyahou et al., 1996; Cerry et al., 1990; Bolk, 1985). Usually production issues limit the number of calibration points available for calibrating sensors against a reference. An adequate choice of calibration points can reduce the number of calibration points and thus sensor calibration time and cost (see Dickow and Feiertag, 2014). As far as we know, no systematic technique for MEMS sensor calibration has been published, including both optimal calibration point selection and calibration using as few points as possible. In this paper a polynomial approach for MEMS sensor calibration with only a few calibration points is proposed. An algebraic framework is used to determine all possible calibration

point combinations. In an experiment, the proposed method is applied to barometric MEMS pressure sensors. Recommendations are made for an optimal choice of calibration points. The paper is structured as follows: Sect. 2 introduces the concept of polynomial calibration with parameter extrapolation; in Sect. 3 a combinatorial framework is proposed for the selection of calibration point sets; in Sect. 4 the proposed framework is applied to barometric MEMS pressure sensors; and in Sect. 5 the results are discussed in a conclusion.

2 Calibration using polynomial regression

As sensors often show nonlinear behavior in their measurement range, calibration functions are used to linearize the raw digital sensor readout. To calibrate these sensors reference values are needed. Reference values and the sensors raw values are recorded in a calibration device at defined test conditions or calibration points. State of the art calibration uses multiple ordinary least square fits (multiple OLS) (see Seber and Lee, 2003; Martens and Naes, 2002) to calculate the sensor calibration parameters. For further reading about calibration methods refer to Martens and Naes, 2002, and Varmuza and Filzmoser, 2009. As multiple OLS is a linear regression method, calibration functions have to be linear in their parameters. In this paper a multivariable polynomial is chosen as the calibration function to linearize the sensor behavior over specification. The polynomial function is capable of linearizing continuous, differentiable sensor readouts to a given specification. In some cases, polynomial functions of very high order are needed to reach the desired calibration accuracy. Then, it is useful to switch to other functions, reaching accuracy goals with fewer parameters, or even switch to other regression methods, e.g., support vector regression (see Smola and Scholkopf, 2004).

2.1 Linear regression model for polynomial calibration

A calibration function gives the sensor output signal as a function of the sensor's raw values. Multiple regression calibration is used to generate an appropriate calibration function for given reference values. The OLS calibration method is an integral part of research and production in many fields of applications (shown in Eriksson et al., 2006). A linear regression model for N observations can be expressed as

$$y = \mathbf{X}\boldsymbol{\beta} + \boldsymbol{\varepsilon}, \tag{1}$$

where y is denoted as dependent variable (set of reference values measured at the calibration points), $\mathbf{X}$ is a matrix of independent variables (functions of sensor raw values at the calibration points), $\boldsymbol{\beta}$ is the calibration parameter vector and $\boldsymbol{\varepsilon}$ is the error term vector, describing the calibration errors at the distinct calibration points.

In matrix notation, the calibration problem transforms to

$$\begin{bmatrix} y_1 \\ \vdots \\ y_N \end{bmatrix} = \begin{bmatrix} 1 & x_{11} & \cdots & x_{1M} \\ \vdots & \vdots & \ddots & \vdots \\ 1 & x_{1N} & \cdots & x_{NM} \end{bmatrix} \begin{bmatrix} \beta_0 \\ \vdots \\ \beta_M \end{bmatrix} + \begin{bmatrix} \varepsilon_1 \\ \vdots \\ \varepsilon_N \end{bmatrix}, \tag{2}$$

where $M = N-1$ and rank $(\mathbf{X}) = N$. N calibration points are required to calculate a unique solution for the calibration parameters $\boldsymbol{\beta}$. Each single row of X describes an independent variable vector of the calibration function used. An independent variable vector $[1, x_{ij}, \ldots, x_{iM}]$ describes transformed raw sensor readouts at specific reference sensor readout y_i. When calibration uses a polynomial function with a single raw sensor readout signal, $M = N-1$ is also the polynomial order K; so that

$$y_{k,\text{poly}} = \beta_0 + \beta_1 x_{\text{raw},k} + \beta_2 x_{\text{raw},k}^2 + \ldots + \beta_K x_{\text{raw},k}^K$$
$$= \sum_{i=0}^{K} \beta_i x_{\text{raw},k}^i = \beta_0 + \sum_{i=1}^{M} \beta_i x_{k,i} \tag{3}$$

describes the polynomial approximation of y_k, where $k \in (1, \ldots, N)$. If $M < N-1$, the OLS fit produces an error ε_k. When having p raw sensor readout signals, the polynomial

approximation expands to

$$y_{k,\text{poly}} = \prod_{j=1}^{p} \left(\sum_{i=0}^{K_j} \beta_{j,i} x_{\text{raw},j,k}^i \right) = \beta_0 + \sum_{i=1}^{M} \beta_i x_{k,i}, \tag{4}$$

needing

$$N = M + 1 = \prod_{j=1}^{p} (K_j + 1) \tag{5}$$

calibration points to calculate a unique solution to the multiple sensor signal calibration problem. The parameter K_j, $j \in [1, \ldots, p]$ describes the polynomial order for each raw sensor signal used for calibration.

For example, MEMS pressure sensors with raw temperature $x_{\text{raw},T}$ and raw pressure readout $x_{\text{raw},P}$ are usually calibrated with a polynomial of order 2 in temperature and order 1 in pressure:

$$y_{\text{poly}} = \left(\beta_{1,0} + \beta_{1,1} x_{\text{raw},T} + \beta_{1,2} x_{\text{raw},T}^2 \right) \left(\beta_{2,0} + \beta_{2,1} x_{\text{raw},P} \right)$$
$$= \beta_0 + \beta_1 x_{\text{raw},T} + \beta_2 x_{\text{raw},T}^2 + \beta_3 x_{\text{raw},P} + \beta_4 x_{\text{raw},P} x_{\text{raw},T}$$
$$+ \beta_5 x_{\text{raw},P} x_{\text{raw},T}^2. \tag{6}$$

In this case, $N = (2+1)(1+1)$, six calibration points are required to determine all calibration parameters β_i, $i \in (0, \ldots, 5)$.

2.2 Calibration criteria and option selection

In the following k_i denotes a set of calibration points. For the example from Eq. (6), each k_i contains three temperature points ($T1$, $T2$, $T3$) and two pressure points ($P1$, $P2$) coupled in a row, leading to

$$k_i = \begin{bmatrix} (P1, T1) \\ (P1, T2) \\ (P1, T3) \\ (P2, T1) \\ (P2, T2) \\ (P2, T3) \end{bmatrix}. \tag{7}$$

To decide which sets of calibration points give the best sensor accuracy a reference data set is necessary. This reference data is measured at discrete values within the measurement range of the sensor. In the following we assume that all calibration points match the discrete values of the reference data. This leads to a finite number of possible sets of calibration point combinations $[k_1, \ldots, k_L]$. In Fig. 1 this is shown for the example given above. The reference data was measured for all points denoted with an "X", the group of six grey circles is one set of calibration points.

For each k_i, the OLS regression is performed using the polynomial from Eq. (4). This leads to $B = [\boldsymbol{\beta}_1, \ldots, \boldsymbol{\beta}_L]$ sets of identified polynomial parameters for a given sensor. To

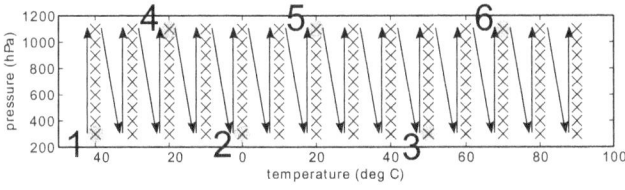

Figure 1. Six selected calibration points for digital barometric pressure sensors, chosen from a reference data set (marked with ×); recorded data is in ascending order to avoid temperature hysteresis effects.

evaluate the calibration quality, the identified parameters are inserted into y_{poly} and used to calculate the calibration error over the sensor working range. As calibration error indicator, the calibration root mean squared error (rms error),

$$e = \sqrt{\frac{1}{R-1} \sum_{j=0}^{R} \left(y_{j,poly}(\boldsymbol{\beta}_j, \boldsymbol{x}_{j,raw,1}, \ldots, \boldsymbol{x}_{j,raw,p}) - y_{j,ref} \right)^2}, \quad (8)$$

is calculated from the reference data $\boldsymbol{y}_{ref}$ and the polynomial approximation $\boldsymbol{y}_{poly}$, where R is the number of reference data points. $N < R$ describes the case, when more than one calibration option is available. In this case the calibration task has

$$O_{cal} = \binom{R_j}{K_j + 1}^K \quad (9)$$

possible calibration options. For a multidimensional calibration problem

$$O_{cal} = \prod_{j=1}^{p} \binom{R_j}{K_j + 1}^{K_j} \quad (10)$$

calibration options are possible, when the sensor is influenced by p signals/factors. From all calibration options,

$$\mathbf{C} = [k_1, k_2, \ldots, k_{O_{cal}}], \quad \mathbf{C} \in \mathbb{R}^{N \times O_{cal}}, \quad (11)$$

those are seen as optimal calibration options Γ, where

$$\Gamma = \left[e_{k_1}, \ldots, e_{k_{O_{cal}}} \right]. \quad (12)$$

2.3 Statistical evaluation

For a batch of I sensors, with evaluated rms errors at $\mathbf{C}$, there are

$$O_{\Gamma} = \{\Gamma_1, \ldots, \Gamma_I\} \quad (13)$$

optimal calibration options. It can happen that some optimal calibration options are identical. If the batch of sensors is representative for a specific sensor type, the most common optimal calibration option should be preferred. However as sensors are influenced by multiple parameters, it is not very

likely to find sensors having exactly the same calibration recommendation.

To retrieve more information about fields of attraction for best calibration options in a multidimensional optimization problem, a selection criteria weaker than Eq. (12) is proposed. The criteria

$$\Gamma_{99} \subset \left[e_{k_1}, \ldots, e_{k_{O_{cal}}} \right] \quad (14)$$

dismisses those 99 % calibration options, which have the highest rms errors according to Eq. (8). The Γ_{99} criteria is applied to each sensor of a batch. This results in a calibration recommendation for an investigated sensor type

$$\Gamma_{r99} = \left[\Gamma_{99,1}, \ldots, \Gamma_{99,I} \right], \quad (15)$$

which will be used in the following as a multidimensional indicator for close to optimal calibration points.

3 Software implementation

The proposed sensor calibration approach was implemented in a framework, written in Python language, to investigate commercial MEMS sensors with digital data readout. It uses the Fortran package LAPACK (Linear Algebra PACKage) to solve linear equations using LU (lower, upper) decomposition (Strang, 1980) with partial pivoting and row interchange. A typical calibration workflow is depicted in Table 1. Four steps are needed to retrieve meaningful statistical data out of given digital raw sensor readout and reference values: data processing, calibration, comparison and statistical evaluation. If the number of required calibration points N is much smaller than the amount of available reference points for calibration R or more than two independent sensor readouts p are used for calibration, the amount of OLS calls can get too high for desktop computers to solve within a reasonable time. In consequence, the framework should only be used for low- to medium-order polynomials, having only a few independent readout signals used for calibration.

4 Application example – barometric MEMS pressure sensors

In this section, the new framework is applied to three types of commercial barometric MEMS pressure sensors, which are calibrated against temperature and barometric pressure using polynomials (see Kim et al., 2012; Köster et al., 2003; Bosch Sensortec, 2008, 2013; EPCOS, 2013). Three batches of six barometric MEMS pressure sensors each, of the type EPCOS T5400, Bosch BMP 180 and Bosch BMP 085, are used to calculate calibration point recommendations for a multipara-

Table 1. A typical calibration workflow using the proposed framework; data processing, calibration and evaluation are implemented in independent modules.

Step 1: data processing	Step 2: calibration	Step 3: comparison	Step 4: statistical evaluation
Build containers for measurement data recorded from the sensor batch and sort recorded data into segments.	Select calibration polynomial order in each given sensor raw data readout signal, calculate all possible calibration options with Eq. (11) and run sensor multiple OLS polynomial calibrations for each sensor of the given batch.	Test the calibration parameters against the reference data set using the rms error approach from Eq. (8) with respect to Eq. (9), calculate Eq. (15) in respect to Eq. (16). Repeat this procedure for each sensor of the batch.	Use the framework for statistical evaluation (histograms, multidimensional probability plots, calibration point recommendations, nonlinearity plots, parameter and calibration function influences).

metrical second-order calibration polynomial,

$$y_{\mathrm{Baro}} = \beta_0 + \beta_1 x_{\mathrm{raw},t} + \beta_2 x_{\mathrm{raw},t}^2 + \beta_3 x_{\mathrm{raw},p}$$
$$+ \beta_4 x_{\mathrm{raw},t} x_{\mathrm{raw},p} + \beta_5 x_{\mathrm{raw},t}^2 x_{\mathrm{raw},p}, \qquad (16)$$

using the sensor's raw temperature readout $x_{\mathrm{raw},t}$ and a raw pressure readout $x_{\mathrm{raw},p}$.

4.1 Calibration setup

The test equipment uses a pressurized climate chamber, a General Electric PACE 5000 pressure controller as pressure calibration reference and a combination of Peltier elements and a type K thermocouple for reference temperature control, attached close to the sensor site. Data is recorded using a National Instruments USB-8451 I2C device, connected to the digital barometric MEMS pressure sensors, soldered on to a printed circuit board.

4.2 Calibration task

For all further investigations, it is assumed that the sensors deliver reproducible results. As barometric MEMS pressure sensors suffer from temperature hysteresis (see Waber et al., 2013), data was recorded in ascending temperature order to minimize the hysteresis effect.

Sensors used in the experiment have a measurement range from -40 to $90\,^\circ\mathrm{C}$ and from 300 to 1100 hPa. To test the calculated sensor calibration, discrete points at $[-40, -30, \ldots, 90\,^\circ\mathrm{C}]$ are chosen as test temperature points and $[300, 400, \ldots, 1100\,\mathrm{hPa}]$ are chosen as test pressure points. Within the measurement range, the sensor's raw values, the reference pressure and the reference temperature are recorded in a sequence to avoid disturbances from temperature hysteresis, as described in Fig. 1. At each point, the measurement device waits until stable pressure and temperature conditions are reached. Data recorded consists of time stamp, raw pressure, raw temperature, reference pressure and reference temperature. Six points are numerated and marked in grey. They exemplarily stand for a calibration point set, as presented in Eq. (7). In the following, this and all other possible calibration options will be evaluated to determine the best combinations possible.

4.3 Calibration point recommendation

Data is recorded according to the description in Sect. 4.2. For the calibration polynomial from Eq. (16), the most likely best calibration point combinations were calculated using the procedure taken from Table 1. The experimental results are shown in a distribution landscape plot in Fig. 2. The plot exemplarily separates the first 3 calibration points (upper row in Fig. 2) from the calibration points 4–6 (second row in Fig. 2). This clustering into two distribution plots provides information about how to combine temperature points, when calibration is restricted to only two possible pressure conditions, as it can happen in industrial sensor production to save calibration time. Figure 2 shows, that the investigated T5400 sensor should be calibrated at points at the borders of pressure range (300 and 1100 hPa), while the other sensor types investigated show a more heterogeneous calibration point recommendation (300, 900 and 1100 hPa).

5 Other applications

The proposed method can be applied to all sorts of sensors, having one output signal and a at minimum one input signal. For computational complexity reasons, the number of measurement points for searching the optimal calibration option should be low. In the pressure sensor example there are 9 pressure points and 14 temperature points, resulting in a grid of 126 measurement points for calibration. This is already a demanding task, for a polynomial order 3 in both parameters (16 parameter polynomial), according to Eq. (10). The proposed method can be applied to sensors like gyroscopes (see Aggarwal et al., 2006) or any other kind of sensor, as long as the number of parameters for calibration and the number of measurement points investigated is comparable to the numbers from the barometric pressure sensor example. For the gyroscope example in Aggarwal et al. (2006), the parameter y is a combined error including scale factor and bias in one direction, and the parameters of x include six calibration points. As directions x, y and z are assumed to be independent, three independent optimal calibration models with six calibration points can be calculated with the proposed method.

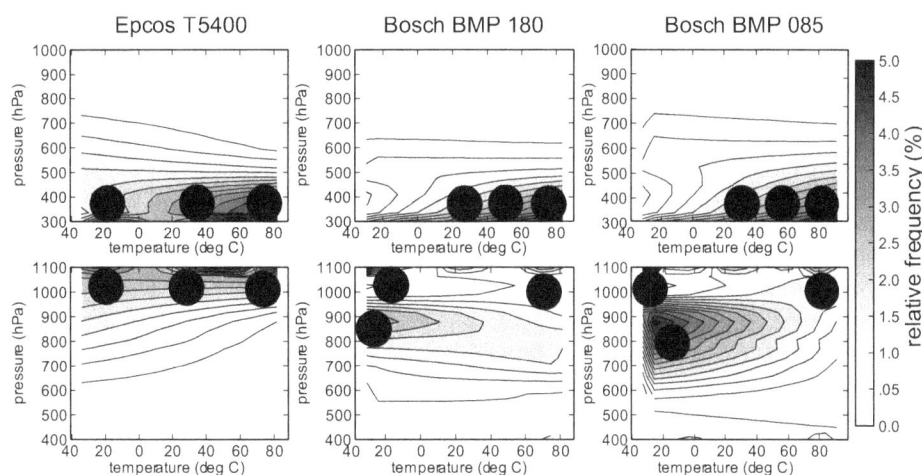

Figure 2. Recommended six calibration points for second order polynomial barometric pressure sensor calibration; frequency of the first three calibration points shown in the upper part; frequency of the last three calibration points shown in the lower part of the figure.

6 Conclusions

The proposed framework was used to calibrate barometric MEMS pressure sensors with six calibration points and a second-order calibration polynomial in temperature and first-order in pressure. For all sensors investigated, points selected at the upper and lower borders of temperature and pressure range increase the likelihood of appearing within the best 1 % of the calibration options. The proposed framework determines all possible calibration options for a given set of sensor measurement data using a linear polynomial regression approach and then applies the rms error over measured test conditions as calibration benchmark. The worst 99 % of the calibrations are dismissed to show areas of attraction for good calibration options. For further research, calibration point extrapolation will be implemented to reduce the amount of calibration points measured. This is used to achieve the sensor specification required by statistically estimating offset values for higher-order calibration polynomial parameters. Further calibration uncertainty considerations will be evaluated according to Waber et al. (2013), Heydorn and Anglov (2002) and Brüggemann and Wennrich (2002).

Acknowledgements. This work was supported by the Bavarian ministry of economic affairs within the research project MEMS-Baro.

References

Aggarwal, P., Syed, Z., Niu, X., and El-Sheimy, N.: Cost-effective Testing and Calibration of Low Cost MEMS Sensors for Integrated Positioning, Navigation and Mapping Systems, XXIII FIG Congress, Munich, Germany, 8–13 October 2006.

Bolk, W. T.: A general digital linearising method for transducers, J. Phys. E: Sci. Instrum., 18, 61–64, 1985.

Bosch Sensortec: "BST-BMP085-DS000-03", BMP 085 digital pressure sensor datasheet, Rev. 1.0, July 2008.

Bosch Sensortec: "BST-BMP180-DS000-09", BMP 180 digital pressure sensor datasheet, Rev. 2.5, April 2013.

Brignell, J.: Digital compensation of sensors, J. Phys. E: Sci. Instrum., 20, 1097–1102, 1987.

Brüggemann, L. and Wennrich, R.: Evaluation of measurement uncertainty for analytical procedures using a linear calibration function, Accred. Qual. Assur., 7, 269–273, 2002.

Cerry, S., Baer, W., Cowles, J., and Wise, K.: Digital compensation of high-performance silicon pressure transducers, Sensor. Actuators, A2I–A23, 70–72, 1990.

Dickow, A. and Feiertag, G.: A framework for calibration of barometric MEMS pressure sensors, Eurosensors conference 2014, Brescia, Italy, 7–10 September 2014.

EPCOS: "Saw Components Application Note T5400", T5400 digital pressure sensor application note, Rev 1.3.A1, February 2013.

Eriksson, L., Johansson, E., Kettaneh-Wold. N., Trygg, J., Wikström, C., and Wold, S.: Multi- and Megavariate Data Analysis – Part 1 Basic Principles and Applications, 2nd Edn., Umetrics AB, Umea, Sweden, p. 127, 2006.

Heydorn, K. and Anglov, T.: Calibration uncertainty, Accred. Qual. Assur., 7, 153–158, 2002.

Kim, Y.-K., Choi, S.-H., Kim, H.-W., and Lee, J.-M.: Performance improvement and height estimation of pedestrian dead-reckoning system using a low cost MEMS sensor, 12th International Conference on Control, 1655–1660, 2012.

Köster, O., Slotkowski, J., and Brögger, D.: Kalibrierung von integrierten Drucksensoren im Waferverbund, Technisches Messen, 70, 265–269, 2003 (in German).

Lyahou, K., Van der Horn, G., and Huijsing, J.: A non-iterative, polynomial, 2-dimensional calibration method implemented in a microcontroller, J. Proceedings of the IEEE Instrumentation and Measurement Technology Conference, Brussels, Belgium, 62–67, 4–6 June 1996.

Martens, H. and Naes, T.: Multivariate Calibration, Wiley & Sons, Hoboken, New Jersey USA, 49–59, 2002.

Seber, G. and Lee, A.: Linear Regression Analysis, Wiley series in probability and statistics, 2nd Edn., Wiley & Sons, Hoboken, New Jersey USA, 35–93, 2003.

Smola, A. and Scholkopf, B.: A tutorial on support vector regression, Stat. Comput., 14, 199–222, 2004.

Strang, G.: Linear Algebra and Its Applications, 2nd Edn., Orlando, Florida, Academic Press, p. 22, 1980.

Van der Horn, G. and Huijsing, J.: Integrated smart sensor calibration, Analog Integr. Circ. S., 14, 207–222, 1997.

Varmuza, K. and Filzmoser, P.: Introduction to Multivariate Statistical Analysis in Chemometrics, 1st Edn., Taylor & Francis Group, Boca Raton, Florida USA, p. 105, 2009.

Waber, T., Pahl, W., Schmidt, M., Feiertag, G., Stufler, S., Dudek, R., and Leidl, A.: Flip-chip packaging of piezoresistive barometric pressure sensors, Proc. of SPIE 8763, Smart Sensors, Actuators, and MEMS VI, 87632D, 17 May, 2013.

9

Is it possible to detect in situ the sulfur loading of a fixed bed catalysts with a sensor?

P. Fremerey[1,2], A. Jess[2], and R. Moos[1]

[1]Department of Functional Materials, University of Bayreuth, Bayreuth, Germany
[2]Department of Chemical Engineering, University of Bayreuth, Bayreuth, Germany

Correspondence to: R. Moos (funktionsmaterialien@uni-bayreuth.de)

Abstract. This study reports on a sensor concept to measure in situ sulfur poisoning (sulfidation) of refinery catalysts, in this case, of commercial silica pellets loaded with highly dispersed nickel. Catalyst pellets were poisoned in diluted H_2S between 100 and 400 °C and the sulfidation of the catalyst was observed. During this process, nickel sulfides are formed on the catalyst according to X-ray diffraction spectra and energy dispersive X-ray spectroscopy data. The sulfidation kinetics was quantitatively described by a shrinking core model. Representative catalyst pellets were electrically contacted, and their impedance was recorded in situ during sulfidation. At the beginning, the particles are highly insulating and behave capacitively. Their conductivity increases by decades during sulfidation. At high temperatures, an almost constant slope in the double-logarithmic representation vs. time can be found. At low temperatures, the conductivity remains constantly low for a long time but changes then rapidly by decades, which is also indicated by the phase that drops from capacitive to ohmic behavior. Since nickel sulfides exhibit a lower conductivity than nickel, the conductivity increase by decades during sulfidation can only be explained by electrically conducting percolation paths that form during sulfidation. They originate from the increased volume of sulfides compared to the pure nickel metal.

1 Introduction

In industry, heterogeneous catalysts are used to accelerate chemical reactions and to enhance product selectivity. During reaction, the catalysts are exposed to a harsh environment that may deactivate the catalyst. Sulfur compounds in the feed stream, such as hydrogen sulfide (H_2S), are a typical example. Thereby, H_2S chemisorbs either on the surface of the catalyst or it forms sulfides with the metal components. In both cases, this frequently leads to a deactivation of the active centers of the catalyst. For example, sulfur poisoning is typically observed during reforming of heavy gasoline, if the feedstock (low-octane gasoline) has not been deeply desulfurized previously (Oudar, 1980; Bartholomew et al., 1979). Another example is the nickel-catalyzed-steam reforming of hydrocarbons to syngas (Jess and Depner, 1999).

Today's standards in industry are ex situ methods to monitor the sulfur content of a fixed bed or even models that estimate it. They do not allow for an immediate response. Therefore, in situ methods are preferred to monitor the state of the catalyst directly. However, typical indirect methods based on spectroscopy are expensive and they do reflect the catalyst state only indirectly.

Therefore, a sensor that may detect the sulfur loading of a catalyst in situ was investigated in the present study. A nickel catalyst was poisoned with H_2S and the sulfidation of the catalyst was determined firstly locally resolved by EDX (energy dispersive X-ray spectroscopy). A chemical reaction model was applied to describe the sulfidation process quantitatively. After that, with a single-particle sensor, i.e., with a contacted representative catalyst particle, the sulfidation was measured in situ. In the subsequent section, it was investigated in detail how sulfur affects the particle conductivity. Finally, conductivity data, sulfidation results, and EDX data are compared.

2   Fundamentals and experimental methods

2.1   Experimental setup and parameters

Figure 1 shows the laboratory setup that was used for all measurements. A thermally isolated quartz glass reactor was heated electrically. Inside of this reactor, a single-particle sensor was fixed in a glass holder at the lower end of the tube, where also the gas inlet was integrated. With this setup, several individual sensors were measured at constant experimental parameters. A gas flow of $50 \, L \, h^{-1}$ (NTP) was applied to the specimens. A bypass was installed in order to measure and adjust the inlet gas composition. Nitrogen was always used as a carrier gas. For the subsequent sulfidation, up to 1000 ppm H_2S were admixed to N_2. Sulfidation was carried out from 100 to 350 °C. For the reduction of the catalyst, 10 % H_2 in N_2 flowed through the system. The commercial catalyst (NiSAT 200, Clariant) consisted of porous cylindrical silica particles (length, $l = 6$ mm; diameter, $d = 6$ mm) loaded with approximately 37 wt % nickel (Kernchen et al., 2007). Since the sensor catalyst particle and the particles in the fixed bed were nominally identical, i.e., they were from the same batch, the changing electrical properties of the sensor particle reflect the behavior of the particles in the fixed bed. In other words, we can consider the sensor particle as a representative for the catalyst particles.

For each sulfidation run, a fresh sensor particle was used. Prior to sulfidation, the particles were brought into the reduced state by a preconditioning step, which was identical for each run. The sensor particle was heated up to 400 °C under N_2 so that no further nickel is converted to NiO until the measured electrical signal (see below) remained constant. After heating under N_2, the catalyst was reduced with 10 % hydrogen in N_2 for 1 h to activate the entire catalyst material by reducing the existing NiO to Ni. Then, the reactor was cooled to the desired temperature and the experiment started. Such conditioning procedures are carried out also in industrial applications. Therefore, it was applied for all particles to bring them into a defined state.

2.2   Single-particle sensor and basic measurements

For the sensor, the porous cylindrical catalyst particles were contacted electrically at the flat end planes. Two additional alumina pads that were coated on one side with gold were contacted with 0.1 mm gold wires. These two pads were pressed on the flat planes of the sensor particle; compare with Fig. 2. The other side of the alumina pads insulated the sensor particle against the steel casing, in which the sensor particle was fixed. The two gold wires were fed through the reactor by a four-bore alumina tube that was also fixed at the steel casing. The gold wires connected the sensor particle with the impedance analyzer (HP 4284A Precision LCR meter). Finally, the bores of the alumina tube were locked gastight with ceramic glue. A similar setup has been applied to de-

tect coke formation on refinery catalysts in situ (Müller et al., 2010a, b).

From the recorded complex impedance data at 1 kHz (absolute impedance value Z, phase ϕ), the resistances, R, were obtained by Eq. (1) (Macdonald, 1987), which is valid for a R∥C behavior:

$$R = Z / \cos(\varphi). \tag{1}$$

First, tests were conducted to exclude that any of the utilized materials and sensor components had a significant influence (negligible parallel resistance) on the signal during sulfidation. The following components were considered: the ceramic glue, the alumina tube, and the sensor without and with a particle. All results are summarized in Table 1. Experiments with the pure catalyst were not possible, since the catalyst could be purchased only in a nickel-loaded modification. However, in-house-prepared pressed pellets consisting of silicon oxide powder show a resistance which is in the range of the sensor with a particle at 450 °C. Despite that the sensor resistance was 2 decades lower than the empty sensor holder, we could not exclude a priori that under some circumstances the parallel resistance may become non-negligible. Therefore, the empty sample holder was measured first at all temperatures and, by default, the impedances were corrected assuming a parallel impedance of sample and empty sample holder as described in Schönauer and Moos (2010).

From the resistance and the geometrical data, the conductivity, σ, was calculated.

$$\sigma = \frac{1}{R} \cdot \frac{l}{A} \tag{2}$$

Here A is the area of the flat plane ($A = \pi d^2 / 4 \approx 28.3 \, mm^2$) and l the length of the particle ($l = 6$ mm). Occasional tests with a high-impedance analyzer (Novocontrol, type Alpha-A) confirmed the results. Time-dependent impedance data were taken at 1 kHz since this was the lowest possible frequency at which the current was high enough to ensure that crosstalk from the rough electrical environment did not predominate the signal. This was of special importance if the catalyst particle was in the reduced (high-ohmic) state.

Typical impedance plots in the complex plane in Fig. 3a (in the reduced, high-ohmic state) and in Fig. 3b (sulfidized) show that an R∥C equivalent circuit is justified and that Eq. (1) can be applied.

3   Results

3.1   Sulfidation of catalyst particle

The sulfidation was at first investigated in detail ex situ at single particles by EDX analyses. The sulfidized sensor particles were ground from the flat plane to the center of the particle. Since the flat planes were covered by the gold layer,

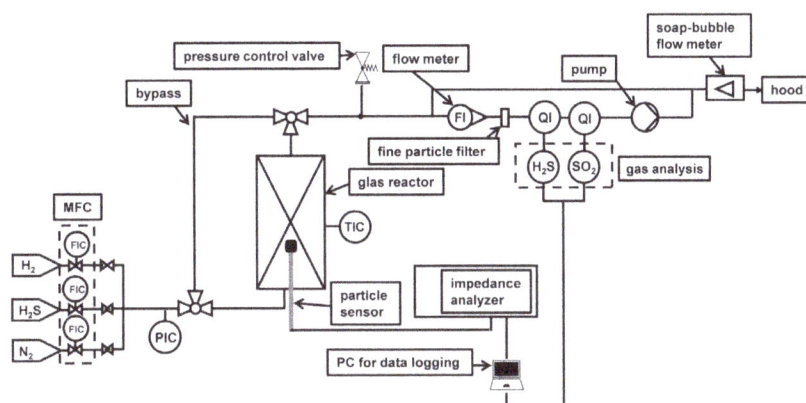

Figure 1. Scheme of the experimental setup.

Table 1. Electrical data of the sensor setup with and without a sensor particle. The resistance of the sensor particle is at least 2 decades lower than the sensor setup.

Component	Temperature	Impedance in Ω	Phase in $°$	Resistance in Ω
cables only	RT	$1.1 \cdot 10^8$	-89.99	$6.1 \cdot 10^{11}$
+ ceramic glue	RT	$4.0 \cdot 10^7$	-89.90	$2.3 \cdot 10^{10}$
+ ceramic tube	450 °C	$3.3 \cdot 10^7$	-89.90	$1.9 \cdot 10^{10}$
+ sensor without particle	450 °C	$3.0 \cdot 10^7$	-89.90	$1.7 \cdot 10^{10}$
+ sensor with particle	450 °C	$2.4 \cdot 10^7$	-82.00	$1.7 \cdot 10^8$

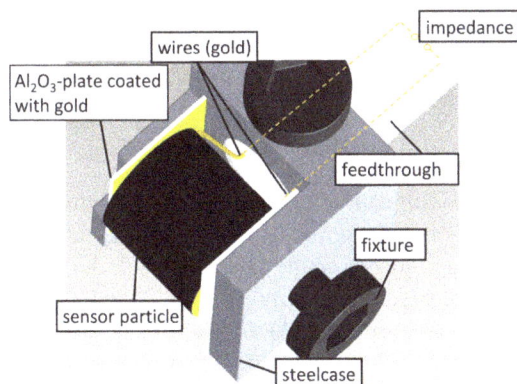

Figure 2. CAD drawing of the single-particle sensor.

the sulfidation occurred from the outer diameter of the cylindrical particles to the inner one. EDX line scans were conducted as shown in Fig. 4a. Line scans of samples sulfidized under 1000 ppm H_2S at 100 °C for 6 and 60 h are shown in Fig. 3b and c, respectively. Results for samples treated at 300 °C under 500 ppm H_2S for 2, 24, and 60 h are displayed in Fig. 5a–c. The proportions of nickel and sulfur are plotted in at. %. A percentage of 50 % nickel and 50 % sulfur is a strong indication for the occurrence of the NiS phase. In contrast, a distribution of 60 % nickel and 40 % sulfur indicates Ni_3S_2, a not fully sulfidized NiS phase. The zero points of the line scans are on the edge of the particles. As can be

obtained from Fig. 4, the sulfur content is almost uniformly distributed from the outer diameter to the center of the catalyst particle. After 6 h, a slow sulfur decrease from the outside to the inner side can be seen. After 60 h, the sulfur is homogeneously distributed in the entire particle. This Ni : S stoichiometry does not reach the 1 : 1 stoichiometry for NiS, but almost for Ni_3S_2. Overall, the sulfidation proceeds almost homogeneously on the particle cross section at this low temperature. This suggests that the chemical reaction is the limiting factor in the sulfidation process at 100 °C and not the (pore) diffusion of H_2S into the particle.

EDX data from the samples sulfidized at 300 °C (Fig. 5) look different. After 2, 24, and 60 h, an almost vertical front moves into the particle. The Ni : S ratio of nearly 1 : 1 indicates a direct formation of NiS. The front moves very slowly into the particle; even after 60 h, only an 0.8 mm thick NiS layer from the particle edge has been formed.

These observations can be explained if one assumes that formation and growth of the reaction front during sulfidation at high temperatures depends on the diffusion through the already formed nickel sulfide layer. Such a behavior can be described by a shrinking core model (Jess and Wasserscheid, 2013). Here, the loading level, X_B, can be calculated by Eq. (3):

$$\frac{t}{C_1} = X_B + (1 - X_B) \ln(1 - X_B). \tag{3}$$

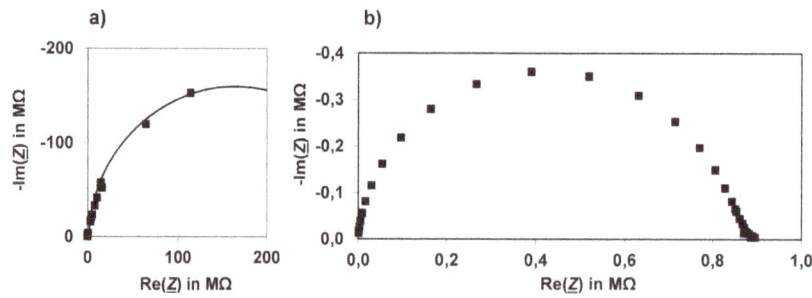

Figure 3. Typical Nyquist representation of the impedance of the single-particle sensor between 1 Hz and 1 MHz. $T = 400\,°C$. (**a**) Catalyst particle in the reduced state (10 % H_2 in N_2). (**b**) Catalyst particle is sulfidized for 1 min in 500 ppm H_2S in N_2.

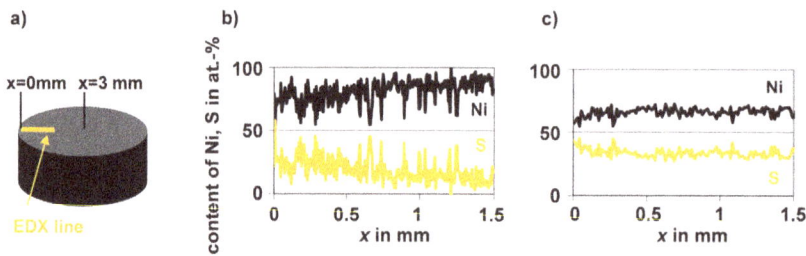

Figure 4. EDX: (**a**) sketch of the EDX line scan direction; (**b**) sample sulfidized for 6 h at 100 °C and 1000 ppm H_2S; (**c**) same conditions after sulfidation of 60 h.

In Eq. (3), t means the sulfidation time and the factor C_1 can be calculated from Eq. (4):

$$C_1 = \frac{\rho_B \cdot r_P^2}{4 \cdot M_B \cdot D_{eff} \cdot c_g}. \tag{4}$$

In Eq. (4), ρ_B is the nickel fraction of the catalyst volume, r_P is the radius of the catalyst particle, M_B is the molar mass of nickel, D_{eff} is the effective diffusion coefficient of H_2S, and c_g denotes the gas concentration, here of H_2S.

The EDX data from a sulfidation at 300 °C with 500 ppm H_2S were taken for different sulfidation periods, to verify whether the sulfur incorporation into the catalyst particle can be quantitatively described by this model. A full loading level in Fig. 6 means that all nickel is converted into NiS.

For the model, $\rho_B \approx 606\,kg\,m^{-3}$, $r_P \approx 0.003\,m$, $M_B \approx 0.059\,kg\,mol^{-1}$, and $c_g \approx 0.0105\,mol\,m^{-3}$ were used. In order to calculate D_{eff}, the equation from Froment and Bischoff (1990) was applied. It is valid if a high micro and macro porosity exists in the catalyst (Kernchen et al., 2007).

$$D_{eff} = \varepsilon_M^2 D_M + \frac{\varepsilon_\mu^2 (1 + 3 \cdot \varepsilon_M)}{1 - \varepsilon_M} \cdot D_\mu \tag{5}$$

The porosity (ε) of the catalyst was 54 % (Müller, 2011). To simplify the calculation, the porosity is split into equal parts for micro porosity (ε_μ) and macro porosity (ε_M) with the respective diffusion coefficients, D_μ (Knudsen diffusion; average pore size diameter 8 nm, measured by BET) and

D_M (gas diffusion). With these values D_{eff} is estimated to be $1.2 \cdot 10^{-6}\,m^2\,s^{-1}$. For a sulfidation with 500 ppm H_2S at 300 °C, EDX data and the shrinking core model curve coincide well (Fig. 6).

3.2   Impedance measurement results

Knowing the sulfidation kinetics, the conductivity of single-particle sensors was examined. In Fig. 7, the typical behavior of the conductivity of a single catalyst particle during sulfidation is shown, here exemplarily for 400 °C and 500 ppm H_2S. At the beginning, in the reduced state, a particle has a low conductivity, σ. When applying high H_2S concentrations up to 1000 ppm especially at a high temperature of up to 400 °C, the conductivity increases very quickly. During this process, nickel sulfides are formed on the catalyst surface according to XRD (X-ray diffraction) and EDX data. From the literature it is well-known that nickel sulfides exhibit a metallic conductivity almost as high as nickel (Dharmaprakash, 1996). Therefore, at first glance, during sulfidation one would have expected a small decrease in conductivity but not a conductivity increase by decades. A similar phenomenon for a Ni catalyst when exposed to H_2S has already been described by Bjorklund and Lundström (1983), who studied the electrical response of several supported Ni and Co catalysts during reaction.

In order to investigate this behavior further and to set up an initial theory, the sulfidation temperature was varied, always keeping the H_2S concentration at 1000 ppm. The results are

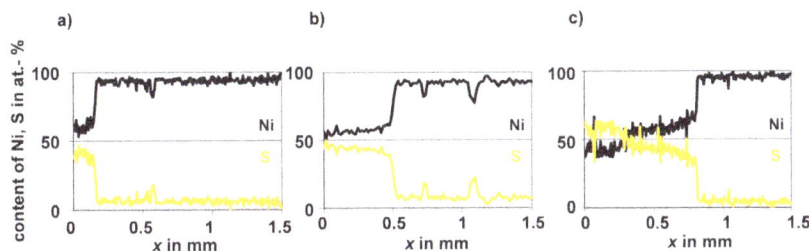

Figure 5. EDX results for samples sulfidized at 300 °C in 500 ppm H_2S in N_2; **(a)** sulfidation time 2 h; **(b)** 24 h; **(c)** 60 h.

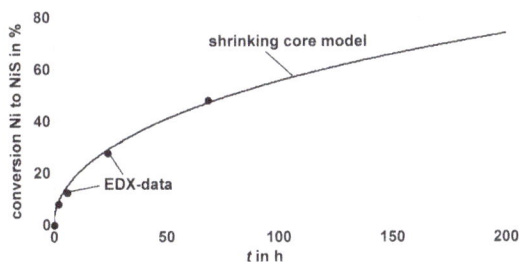

Figure 6. Amount of NiS plotted against the sulfidation time ($T = 300$ °C; $c_{H_2S} = 500$ ppm). The line is calculated with the shrinking core model (Eq. 5). Dots are determined by EDX.

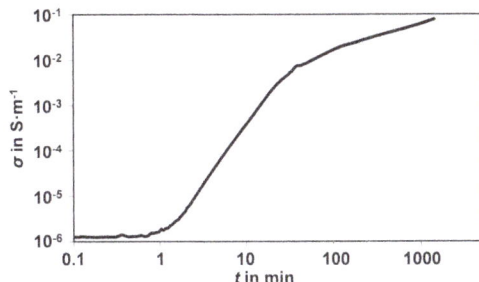

Figure 7. Example for the conductivity response of a single catalyst pellet during sulfidation (data: 500 ppm H_2S in N_2, 400 °C).

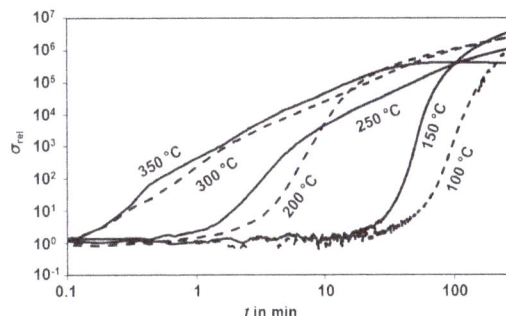

Figure 8. Relative conductivity during sulfidation under 1000 ppm H_2S in N_2 at different temperatures. For the temperature curves at 100 and 150 °C, the measured values were averaged over 10 points.

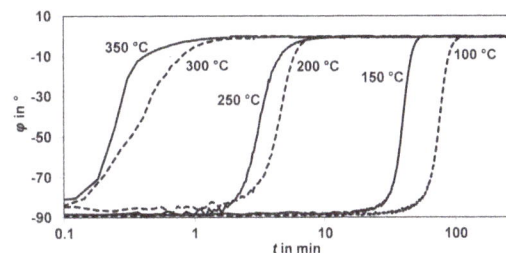

Figure 9. Phase angle during sulfidation under 1000 ppm H_2S in N_2 at different temperatures. For the temperature curves at 100 and 150 °C, the measured values were averaged over 10 points.

shown in Fig. 8. To better compare the results of different temperatures, a relative conductivity, σ_{rel}, was calculated by Eq. (6):

$$\sigma_{rel} = \frac{\sigma}{\sigma_0}, \qquad (6)$$

where σ_0 is the conductivity of each sensor particle after preconditioning, right before sulfidation. It is in the range of about 10^{-6} S·m^{-1}. The conductivity increases almost directly with the beginning of sulfur poisoning at 300 °C and above. During further sulfidation, the slopes of the curves decrease only slightly and an overall slope of almost 2 can be found in the double-logarithmic representation. In contrast, at lower temperatures, the curve shape changes to an S-like behavior. The lower the temperature, the longer the conductivity remains constant before it starts to increase. The increase in conductivity, i.e., the decreasing electrical resis-

tance, goes along with a transition from capacitive to ohmic behavior. This is reflected by the changing phases ϕ from almost -90 to almost $0°$ (see Fig. 9).

To illustrate the effect of temperature, a characteristic time, t_c, is introduced. It denotes the sulfidation start. From Fig. 8, t_c can be derived by the intersection of the baseline with the initial slope of the sulfidation. From the behavior of the phase angle (Fig. 9), the characteristic times are defined as the points in time, when the phase angles reached $-45°$. These two characteristic times are plotted for different temperatures against the reciprocal sulfidation temperature in Fig. 10 (Arrhenius-type representation).

As mentioned above, one would not have expected a conductivity increase during sulfidation, since Ni is a better electrical conductor than NiS or Ni_3S_2, although both sulfides are

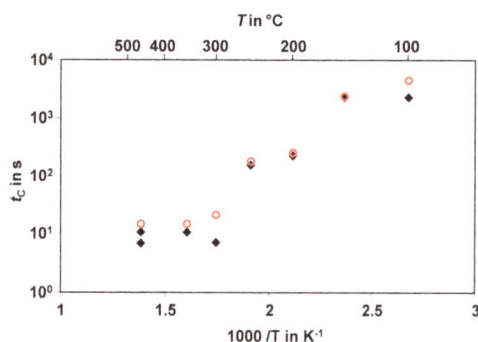

Figure 10. Characteristic time (t_C) at different temperatures. The black squares symbolize the characteristic time from the conductivity curves at the beginning of sulfur poisoning. The open circles denote the characteristic time from derived from the phase curves ($\phi = -45°$).

metal-like conductors. Hence, one needs an explanation for the conductivity increase that occurs by decades during sulfidation.

To explain the low conductivity in the reduced state, we assume that Ni is so highly dispersed on the porous catalyst particle that no conductive paths exist between the electrodes. We further assume that during sulfidation, the volume expansion from nickel to nickel sulfide forms conductive percolation paths (molar volume of nickel $\approx 6.6\,\mathrm{cm^3 \cdot mol^{-1}}$ and nickel sulfide $\approx 16.5\,\mathrm{cm^3 \cdot mol^{-1}}$; Blachnik, 1998). Similar effects are reported in literature for CuO nanofibers when exposed to H_2S (Hennemann et al., 2012). Here, CuO reacts to copper sulfide. After a certain time, a fast increase of the conductivity occurs. The authors attribute this to the formation of percolation paths between the single nanofibers. A similar effect, namely a strong conductivity increase, is observed for Pd exposed with H_2. This effect is also attributed to a volume expansion that leads to percolation effects (Dankert and Pundt, 2002). The study of Bjorklund et al. (1985) showed that metal crystallites of supported catalysts increase in O_2 and become smaller in H_2. Due to this volume change, conductive paths through the catalyst form.

The percolation is a statistical probability for building clusters in a system. At a specific limit of one substance, a cluster is building between the boundaries of the system. This limit is called the percolation threshold. During sulfidation, it can be noticed through the large conductivity increase of the sensor, which can be explained very well with the percolation theory (Cruz-Estrada and Folkes, 2002). In Fig. 7, the conductivity increases very rapidly at high temperatures. During this time, only a small nickel sulfide layer is built up; compare with Fig. 6. This small layer suffices to build enough conductive paths that the percolation threshold is reached. Also in Fig. 10 a quick sulfidation is seen by the small characteristic times at temperatures above 300 °C. If the temperature is now decreased, the loading mechanism changes. At 100 °C, sulfidation is slower and almost homogeneous over

the catalyst particle (Fig. 4). This suggests that the first conductive paths are built slower. Therefore, the characteristic time in Fig. 10 increases at lower temperatures. Based on the different loading mechanism at different temperatures it is also possible that the same sulfur amount does not result in the same conductivity value due to different numbers of conductive paths. This could be a reason for the different curve progression in Fig. 8.

4 Conclusion and outlook

The applied single-particle sensor can be used to measure directly the sulfur loading of a catalyst (e.g., of a fixed bed). This method is therefore an ideal supplement and extension of the determination of the coking degree of catalyst pellets by a similar setup (Müller et al., 2010a, c). There, coking and resistance change occur in the same timely dimensions but instead of sulfidation coke is deposited on the (internal) surface of the catalyst. This can be measured in almost the same manner with the sulfur sensor at low temperatures (< 150 °C). In contrast to that, we very precisely see the onset of the sulfidation by a pronounced conductivity increase and a phase angle change from capacitive insulating to ohmic behavior at high temperatures (> 200 °C). This is explained by a direct formation of conductive percolation paths in the nickel sulfide layer of the catalyst particles. They have their origin in the volume increase of nickel when reacting with sulfur to nickel sulfides. The sensor clearly indicates how the sulfidation process increases with temperature.

Further work will be dedicated to setup a mathematical model that describes quantitatively the correlation between conductivity and sulfidation degree. In additional experiments, samples with much lower Ni contents will be investigated. Since the volume fraction of formed NiS will then be lower, we expect not to find a strong conductivity increase, because no percolation paths should be formed. This should support the percolation theory.

In a second research direction, it shall be investigated whether the promising results with a microwave cavity perturbation method, which allows detecting the degree of coke on catalyst pellets contactless (Müller et al., 2011), can be transferred to the sulfidation process.

Acknowledgements. The authors A. Jess and R. Moos are indebted to the German Research Foundation (DFG) for financial support (Je 257/12-2, Mo 1060/5-2, respectively).

References

Bartholomew, C. H., Weatherbee, G. D., and Jarvi, G. A.: Sulfur poisoning of nickel methanation catalysts, J. Catal., 60, 257–269, 1979.

Blachnik, R.: Taschenbuch für Chemiker und Physiker: Elemente, anorganische Verbindungen und Materialien, Minerale, Band 3, Springer, Heidelberg, 644 pp., 1998.

Bjorklund, R. B. and Lundström, I.: Electrical Conductance of Catalysts in Contact with Gaseous Reactants, J. Catal., 79, 314–326, 1983.

Bjorklund, R. B., Söderberg, D., and Lundström, I.: Electrical-conductance Responses of Catalysts Exposed to Pulses of H_2 and O_2, Journal of Chemical Society, Faraday Transactions, 81, 1715–1724, 1985.

Cruz-Estrada, R. H. and Folkes, M. J.: Structure formation and modelling of the electrical conductivity in SBS-polyaniline blends: Part I: Percolation theory approach, J. Mater. Sci. Lett., 21, 1427–1429, 2002.

Dankert, O. and Pundt, A.: Hydrogen-induced percolation in discontinuous films, Appl. Phys. Lett., 81, 1618–1620, 2002.

Dharmaprakash, S. M.: Synthesis and electrical conductivity of nickel sulphide, Cryst. Res. Technol., 31, 49–53, 1996.

Froment, G. F. and Bischoff, K. B.: Chemical reactor analysis and design, John Wiley & Sons, New York, 148 pp., 1990.

Hennemann, J., Sauerwald, T., Kohl, C.-D., Wagner, T., Bognitzki, M., and Greiner, A.: Electrospun copper oxide nanofibers for H_2S dosimetry, Phys. Status Solidi A, 209, 911–916, 2012.

Jess, A. and Depner, H.: Kinetics of nickel-catalyzed purification of tarry fuel gases from gasification and pyrolysis of solid fuels, Fuel, 78, 1369–1377, 1999.

Jess, A. and Wasserscheid, P.: Chemical Technology: An Integrated Textbook, Wiley-VCH, Weinheim, 280 pp., 2013.

Kernchen, U., Etzold, B., Korth, W., and Jess, A.: Solid catalyst with ionic liquid layer (SCILL) – A new concept to improve selectivity illustrated by hydrogenation of cyclooctadiene, Chem. Eng. Technol., 30, 985–994, 2007.

Macdonald, J.: Impedance Spectroscopy, John Wiley & Sons, New York, p. 15, 1987.

Müller, N.: Direkte Bestimmung von Koksdepositen auf Festbettkatalysatoren durch elektrische Sensoren, Bayreuther Beiträge zur Sensorik und Messtechnik, Shaker Verlag, Aachen, 143 pp., 2011.

Müller, N., Kern, C., Moos, R., and Jess, A.: Direct detection of coking and regeneration of single particles and fixed bed reactors by electrical sensors, Appl. Catal. A-Gen., 382, 254–262, 2010a.

Müller, N., Jess, A., and Moos, R.: Direct detection of coke deposits on fixed bed catalysts by electrical sensors, Sensors Actuat. B-Chem., 144, 437–442, 2010b.

Müller, N., Moos, R., and Jess, A.: In-situ monitoring of coke deposits during coking and regeneration of solid catalysts by electrical impedance-based sensors, Chem. Eng. Technol., 33, 103–112, 2010c.

Müller, N., Reiß, S., Fremerey, P., Jess, A., and Moos, R.: Initial tests to detect quantitatively the coke loading of reforming catalysts by a contactless microwave method, Chem. Eng. Process., 50, 729–731, 2011.

Oudar, J.: Sulfur Adsorption and Poisoning of Metallic Catalysts, Cataly. Rev., 22, 171–195, 1980.

Schönauer, D. and Moos, R.: Detection of water droplets on exhaust gas sensors, Sensors Actuat. B-Chem., 148, 624–629, 2010.

10

Monitoring human serum albumin cell cultures using surface plasmon resonance (SPR) spectroscopy

A. Henseleit, C. Pohl, Th. Bley, and E. Boschke

Institute of Food Technology and Bioprocess Engineering, Technische Universität Dresden, 01062 Dresden, Germany

Correspondence to: A. Henseleit (anja.henseleit@tu-dresden.de)

Abstract. Continuously monitoring cell cultures is essential for both controlling critical parameters and improving understanding of key processes. An ideal technique in this context is surface plasmon resonance (SPR) spectroscopy, which essentially exploits changes in the angle of incident light that occur when molecules bind to a surface. It provides the ability to monitor real-time changes in small concentrations of various molecules, with no need for additional labels or sample preparation. Here we present an SPR-based immunoassay for monitoring concentrations of human serum albumin (HSA), and compare its sensitivity when used in conjunction with a Biacore platform and the cheaper, smaller [li]SPR system. In conjunction with either system, the immunoassay can detect HSA (a hepatocyte viability marker) at concentrations typically present in three-dimensional hepatocyte cultures mimicking the liver used to evaluate effects of drug candidates before exposure to humans or animals. Furthermore, in conjunction with the [li]SPR system, it is sufficiently sensitive to measure the much lower HSA levels present in skin–hepatocyte co-cultures.

1 Introduction

Human serum albumin (HSA) is continuously synthesized by hepatocytes. The heart-shaped, 66.5 kDa large protein is the most abundant plasma protein and is responsible, inter alia, for transporting toxins from the liver (Curry et al., 1999). The serum albumin concentration in blood (typically ca. 0.6 mM) can be used to evaluate the metabolic activity of mammals and diagnosing various diseases, e.g., liver or kidney diseases, infectious diseases and cancer (Yang et al., 2011). In clinical studies, HSA is also used to characterize specific activities of in vitro cultivated hepatocytes, and thus predict effects of pharmaceuticals. Notably, Wagner et al. monitored HSA levels, among other factors, to assess the viability of hepatocytes in liver and skin tissues co-cultivated on a multi-organ chip used to simulate processes in the human body (Wagner et al., 2013). Thus, relevant processes of a substance in the human body can be simulated. To measure the viability of the hepatocytes, they detected, among others, the HSA level during the cultivation period of 14 days. They showed that HSA was consumed by the skin tissue, and measured HSA concentrations ranged from 3.3 to 3.8 nM, far lower than concentrations in purely hepatocyte cultivations (88 to 170 nM). Thus, to monitor such cell-chip platforms, highly sensitive methods capable of detecting target proteins rapidly (ideally in real time) in minimal samples are required.

For such purposes, surface plasmon resonance (SPR) spectroscopy has highly attractive features, including capacities to continuously monitor multiple samples with minimal volumes highly sensitively (Henseleit et al., 2014). Using SPR, interactions between label-free target molecules and receptors can be detected in real time based on refractive index changes in an evanescent field arising at a metal–dielectric interface. Appropriate receptors for a target molecule are immobilized on the surface of a sensor providing such an interface and samples are injected in a microfluidic flow. Surface plasmons are then excited, usually by the Kretschmann configuration, in which a light beam of constant intensity and frequency is directed through a prism onto a gold surface and totally reflected at the metal–prism interface (Homola, 2008; Kretschmann and Raether, 1968). At a certain angle of incidence, the light transfers some of its energy to the electrons at the gold surface, thereby reducing the intensity of the reflected light (Henseleit et al., 2011). When molecules bind to (or dissociate from) the surface, the refractive index and thus the angle at which the energy transfer occurs changes immediately. The change in intensity of the reflected light is detected by an optical unit and is proportional to the mass of the binding (or dissociating) molecules (Cass et al., 1998). The resulting shifts in the intensity minima are presented as a function of time in a sensorgram.

In the study presented here, we used a low-cost ^{li}SPR system (capitalis technology GmbH) and the well-established Biacore instrument (T100, GE Healthcare Europe GmbH) for the specific and sensitive detection of HSA. Biacore instruments were the first commercially available optical SPR-based biosensors and are still widely used. However, they are costly, large and thus not ideal for point-of-care applications. In marked contrast, the ^{li}SPR instrument is a bench-top spectrometer, which affords high sensitivity and up to 180 sensing spots (Henseleit et al., 2011). We assessed the utility of SPR spectroscopy generally, and both the well-established Biacore platform and the ^{li}SPR system specifically, for immunologically measuring HSA concentrations of in vitro hepatocyte cultures. For this purpose, biotinylated HSA-specific antibodies were bound to streptavidin immobilized on the two systems' sensor surfaces, to exploit the extreme strength of non-covalent streptavidin/biotin binding (dissociation constant, ca. 4×10^{-14} M) (Cao et al., 2006; Holmberg et al., 2005; Peluso et al., 2003; Schneider et al., 2000).

2 Materials and methods

2.1 Materials

HSA-specific antibodies and streptavidin were purchased from Biomol GmbH (Germany). Biotinylated reference antibodies (BrdU-specific antibodies) were obtained from BioLegend, Inc. (USA). HSA, 11-mercaptoundecanoic acid and running buffer TBST (Tris buffered saline, with Tween® 20, pH 8.0) were purchased from Sigma-Aldrich Chemie GmbH (Germany). The amine coupling kit used to immobilize streptavidin (see below) was purchased from GE Healthcare Europe GmbH (Germany). All other chemicals were of analytical grade and obtained from VWR International GmbH (Germany).

2.2 ^{li}SPR experiments

The experiments carried out using the ^{li}SPR system (capitalis technology GmbH, Germany) were performed at 30 °C with a flow rate of $5 \mu L s^{-1}$. The sensor surface of the system, a 50 nm × 12 mm × 3 mm gold layer, is illuminated by three selectable LEDs, each covering an area of $(9 \times 0.8) mm^2$ (Henseleit et al., 2011; Mertig et al., 2009). A CCD camera with a spatial resolution of 1280 rows (spots) and an angle-dependent intensity distribution of 960 columns records the reflected light (Mertig et al., 2009). In this study, only the middle LED was used.

Levels of protein bound to the surface were measured in pixels, each corresponding to 41 resonance units (RU, $1 RU \approx 1 pg\,mm^{-2}$) (Henseleit et al., 2014). The raw data were evaluated using Microsoft Excel.

2.2.1 Preparation of the gold surface and SAM

To prepare carboxylated self-assembled monolayers (SAM), the bare gold surfaces of ^{li}SPR sensor chips were first treated with UV/Ozone (UV/Ozone ProCleaner, NanoAndMore GmbH, Germany) for 30 min and then rinsed with pure ethanol. The clean gold surfaces were immersed in 10 mM 11-mercaptoundecanoic acid overnight at 30 °C, thoroughly rinsed sequentially with ddH$_2$O, ethanol, ddH$_2$O, 100 mM HCl, 50 mM NaOH, 0.5 % (v/v) SDS, and ddH$_2$O, and then dried under a stream of nitrogen.

2.2.2 Streptavidin pH scouting

Prior to the covalent immobilization of the streptavidin, the optimal pH was identified by "scouting", as follows. Solutions of $3.6 \mu M$ streptavidin in 10 mM sodium acetate with pH values ranging from 3.5 to 5.5 were prepared, injected onto the carboxylated surface (prepared as described above) for 2.5 min, and then the bound molecules were removed by sequential injections of $500 \mu L$ of 100 mM HCl, 50 mM NaOH, 0.5 % (v/v) SDS, and ddH$_2$O.

2.2.3 Immobilization of the antibodies

For the immobilization purpose, 10 mM of sodium acetate (pH 4.5) were used as the running dielectric as well as for dilution of the molecules.

Before immobilizing the biotinylated antibodies, the streptavidin molecules were immobilized using the amine coupling kit mentioned above according to the manufacturer's instructions, as follows. The carboxylated surface was activated for 10 min by injecting EDC/NHS (400 mM 1-ethyl-3-(3-(dimethyl amino)-propyl)carbodiimide/100 mM N-hydroxysuccinimide). The activated surface was then incubated for 1 h with a $9.3 \mu M$ solution of streptavidin, followed by 1 M ethanolamine-HCl for 30 min to block the remaining active groups. This resulted in immobilization of streptavidin at surface densities equivalent to 3437 RU. A 10 mM sodium acetate (pH 4.5) was used as the running dielectric and for diluting the molecules during these immobilization procedures. Finally, $1.4 \mu M$ of biotinylated antibody was immobilized on the streptavidin layer by incubation for 1 h in TBST. All signals presented here are raw signal measurements minus signals recorded from injections with a reference surface prepared by immobilizing biotinylated BrdU-specific antibodies on the streptavidin layer using the same strategy.

2.2.4 HSA measurement

To assess the sensitivity of the [li]SPR system, a series of solutions of HSA in TBST with concentrations ranging from 0.4 to 192.5 nM were injected onto the antibody-modified surface for 10 min. The level of binding was determined by measuring SPR signals directly after this association phase, which was followed by injecting a TBST buffer for about 3 min. The resulting dissociation was monitored and incomplete. Thus, HSA was finally removed from the antibodies by injecting 100 mM glycine-HCl (pH 2.2) for 72 s.

2.3 Biacore experiments

The experiments with the Biacore T100 (GE Healthcare Europe GmbH, Germany) were performed using a CM5 sensor chip at 30 °C. Levels of protein bound to the surface were measured in resonance units (RU), where 1 RU roughly corresponded to a surface concentration of $1 \, pg \, mm^{-2}$ (Holford et al., 2012). The raw data were evaluated using BIAevaluation software 2.0.2 and Microsoft Excel.

2.3.1 Immobilization of the antibodies

The streptavidin molecules were immobilized on the CM5 chip using the amine coupling kit according to the manufacturer's instructions, and HBS-EP buffer (GE Healthcare Europe GmbH, Germany) as the running dielectric. The streptavidin molecules were dissolved in 10 mM sodium acetate (pH 4.5). Initially, the carboxylated surface was activated for 7 min by injecting EDC/NHS. Pulses of a $0.93 \, \mu M$ solution of streptavidin were then injected until a maximum immobilization level of 2747 RU was reached. The remaining active groups were blocked with the injection of ethanolamine-HCl for 7 min. The HSA-specific antibody was injected until saturation (1910 RU). All signals presented here are raw signal measurements minus signals recorded from blank injections and injections with a reference surface prepared by immobilizing biotinylated BrdU-specific antibodies on the streptavidin layer to a surface concentration equivalent to 700 RU.

2.3.2 HSA measurement

HSA samples were diluted in TBST to the desired concentrations and then injected onto the antibody-modified surface for 10 min. The degree of binding was determined by measuring the SPR signal at the end of this association phase, which was followed by injecting TBST. The resulting dissociation was monitored and incomplete. Thus, HSA was finally removed from the antibodies by injecting 10 mM glycine-HCl (pH 1.5) for 72 s.

3 Results and discussion

3.1 Assay development using the [li]SPR system

Amine coupling is one of the most frequently used methods for immobilizing receptors in biosensors (Homola, 2008; Jung et al., 2008), especially proteins, since most proteins contain available primary amine groups. The procedure starts with activation of the sensor surface using a mixture of EDC and NHS. The product is a reactive succinimide ester, which reacts spontaneously with the amine group of the ligand. When the desired immobilization level is reached, a solution of ethanolamine is injected to wash away non-covalently bound molecules and deactivate the unreacted NHS-ester (Löfås and Johnsson, 1990).

To immobilize streptavidin at high densities, a high concentration near the active surface is required. This can be achieved by optimizing electrostatic interactions between the negatively charged carboxylated surface and the ligand. Ligands are commonly dissolved in 10 mM sodium acetate to maintain low ionic strength (Drescher et al., 2009; Johnsson et al., 1991). The pH of this solution should be at least one pH unit below the ligand's pI (isoelectric point) to ensure that the ligand is positively charged and thus present at high concentrations at the surface. However, the amine coupling procedure exploits uncharged amine groups, which prevail at high pH (Johnsson et al., 1991). Thus, the optimal pH for immobilization is always a compromise, and should be identified by pH scouting.

Therefore, solutions of streptavidin (pI ~ 5–6; Diamandis and Christopoulos, 1991) in 10 mM sodium acetate at several pH values were prepared and incubated with the non-activated carboxylated surface (Fig. 1). The responses increased with increasing pH up to 4.5, then decreased. Thus, pH 4.5 was identified as optimal for the streptavidin immobilization. Under these conditions, streptavidin could be immobilized at densities equivalent to 3437 RU, somewhat lower than the densities detected during pH scouting (see Fig. 1). As mentioned above, non-covalently bound molecules were washed away by the ethanolamine injection during the immobilization procedure.

The surface concentration of the biotinylated antibody (2665 RU) was kept as high as possible to work under mass transport limitations to allow concentration-dependent binding (Fägerstam et al., 1992). The specificity of the antibody has been previously demonstrated in experiments, including controls with BSA (bovine serum albumin), which has 76 % sequence homology with HSA (Henseleit et al., 2014). Furthermore, we performed regeneration scouting to identify optimal regeneration conditions, in which the antibodies' binding efficiency remained constant during several binding cycles (data not shown).

HSA concentrations ranging from barely detectable up to near saturation (0.4 to 192.5 nM) were injected sequentially onto the sensor surface (Fig. 2).

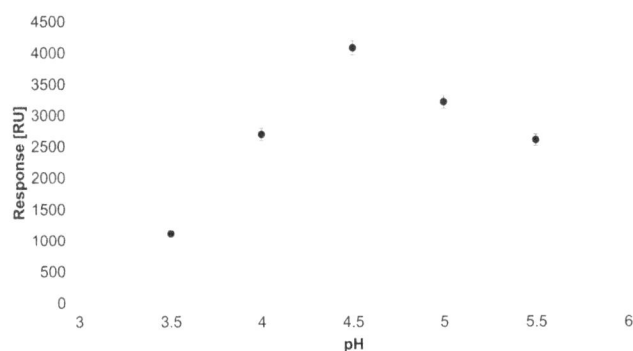

Figure 1. ^{li}SPR system responses of the electrostatically adsorbed streptavidin molecules vs. pH. The data represent mean values of 1280 monitored spots across a $(9 \times 0.8)\,mm^2$ measuring area; error bars indicate the corresponding standard deviations.

Deviation of the signals from the mean value resulting from triplicate analysis of 0.4 and 0.8 nM HSA were substantial: 21 and 29 %, respectively. This is not surprising, as the limit of detection (LOD), calculated following recommendations by the International Union of Pure and Applied Chemistry (IUPAC) as 3 times the machine noise (determined as the deviation of signals resulting from three blank injections), was 30 RU, corresponding to 0.4 nM HSA. The deviation from the mean value decreased with increases in HSA concentration (192.5 nM up to 7 %).

A typical response curve was recorded (Fig. 2, left panel), including a log-linear phase for HSA concentrations ranging from 0.8 to 48 nM, fitting the regression equation $\Delta RU = 60.517 \times \ln[HSA] + 66.97$ with a correlation coefficient (R^2) of 0.9929.

3.2 ^{li}SPR vs. Biacore sensitivity

To assess the ^{li}SPR system's utility for measuring HSA concentrations of in vitro hepatocyte cultures, we obtained comparative measurements using the Biacore platform. The results show that the immobilization level was lower on the Biacore CM5 sensor chip than on the ^{li}SPR sensor. This is because the gold surface of the CM5 chip is functionalized with a hydrophilic hydroxyalkylthiol SAM, modified with a matrix of carboxymethylated dextran (Löfås and Johnsson, 1990). The carboxymethylated dextran matrix forms a protective layer (ca. 100 nm, depending on pH and ionic strength) between the biomolecules and gold surface that is almost 50 times thicker than the SAM formed in the ^{li}SPR system. This leads to signal masking due to evanescent waves decaying as a function of the distance from the gold surface (Frederix et al., 2003). Furthermore, the three-dimensional structure increases steric hindrance during the diffusion of the proteins inside the matrix.

HSA solutions with concentrations ranging from 0.2 nM to 27.1 μM were sequentially injected into the Biacore instru-

ment, in duplicates, and a log-linear calibration curve ranging from 15 nM to 9 μM was obtained, fitting the regression equation $\Delta RU = 33.368 \times \ln[HSA] - 55.394$ with a correlation coefficient (R^2) of 0.9992.

Like the ^{li}SPR measurements, the deviation of the signals from the mean value decreased, from 24 to 2 %, with increasing HSA concentrations.

As a further control, risks of false-positive results were minimized by subtracting signals arising from the reference surface (constructed to be as similar as possible to the sensing surface in the immunoassay by immobilizing unrelated antibodies via streptavidin) from the binding signals. Signals obtained from injecting 1.5 μM HSA onto both the reference and measurement surfaces (showing that non-specific binding of HSA and the sensor surface was close to zero) are presented in Fig. 3. The fluctuations of the signals resulting from the ^{li}SPR system are due to effects of the samples being pumped backwards and forwards over the sensor surface to enhance contact time without wasting sample solution.

Figure 4 depicts the calibration curves obtained from the two systems in relation to HSA concentrations relevant to in vitro cultivated hepatocytes. In the multi-organ-chip system presented by Wagner et al., HSA concentrations reportedly range from 88 to 170 nM HSA for purely hepatocyte cultures and from 3.3 to 3.8 nM HSA for co-cultivated skin and liver tissue cultures (Wagner et al., 2013). Only 40 % of the total on-chip volume is renewed every 12 h during the first 7 days and every 24 h during the next 7 days. Thus, HSA accumulates during the cultivation period.

Using the Biacore T100 in conjunction with our SPR-based immunoassay, the HSA concentrations in pure hepatocyte cultures could be detected easily by the presented immunoassay, but additional steps would be required to enhance signals sufficiently to monitor them in co-cultures, with accompanying increases in analysis time and costs.

In contrast, using the ^{li}SPR system in conjunction with our SPR-based immunoassay, HSA concentrations ranging from 0.8 to 48 nM can be measured. Thus, samples from both types of cultures can be diluted, if necessary, for HSA determination. The system requires ca. 60 μL samples. Hence, the presented results indicate that $\leq 14.6\,\mu$L samples from co-cultivated liver and skin cells, and $\leq 0.6\,\mu$L samples from pure hepatocyte cultures, are needed.

Furthermore, dilution offers the possibility of adding surfactants like Tween$^\circledR$ 20 to the samples to minimize the non-specific binding of interfering sample components, which may occur by replacement of water molecules on the sensor surface by proteins (Vogler, 2012). Thus, adsorption is more likely on hydrophobic surfaces (with a water contact angle greater than 65°). SAMs are applied to form hydrophilic protective layers (Jans et al., 2008; Kausaite-Minkstimiene et al., 2010). However, gaps in these layers due to contaminants, temperature effects or incomplete self-assembling cannot be excluded (Love et al., 2005). Tween$^\circledR$ 20 is a non-ionic molecule consisting of a hydrophilic head group and

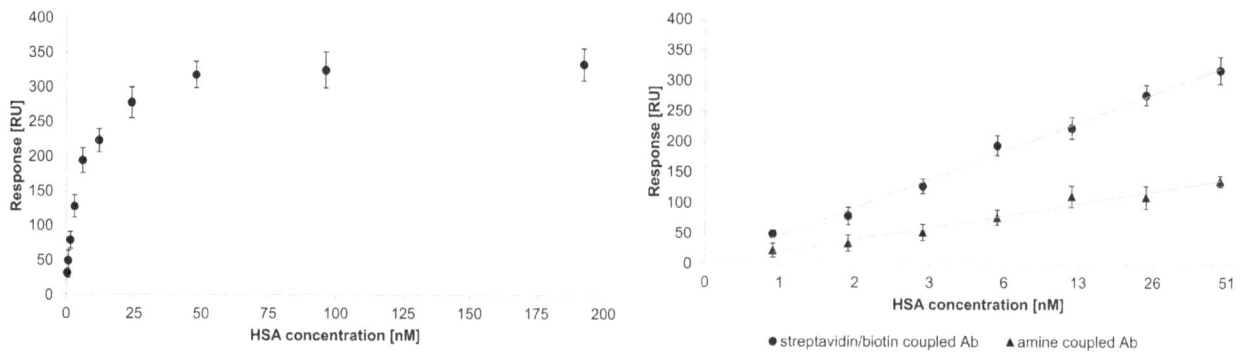

Figure 2. [li]SPR system binding responses obtained from sequentially injections of HSA (from bottom to top: 0.4, 0.8, 1.5, 3 , 6, 12, 24, 48, 96 and 192.5 nM; left panel) and the resulting calibration curve, with a correlation coefficient of 0.9929, obtained from trails with HSA concentrations ranging from 0.8 to 48 nM (right panel). Signals arising from reference antibodies were subtracted from those arising from the HSA-specific antibodies. Error bars indicate the deviations of means from four replicates.

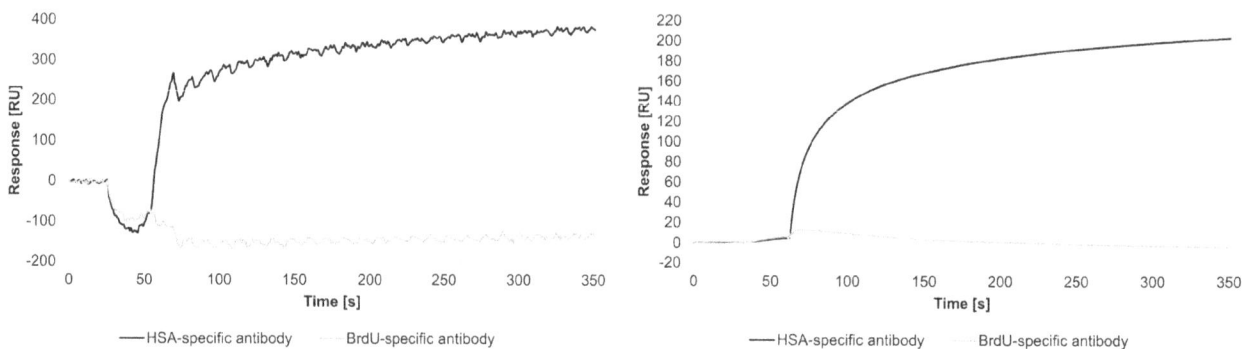

Figure 3. Sensorgrams showing the interaction of 1.5 μM HSA with immobilized HSA-specific antibodies and BrdU-specific antibodies. The left and right panels depict the signals obtained using the [li]SPR and Biacore T100 systems, respectively.

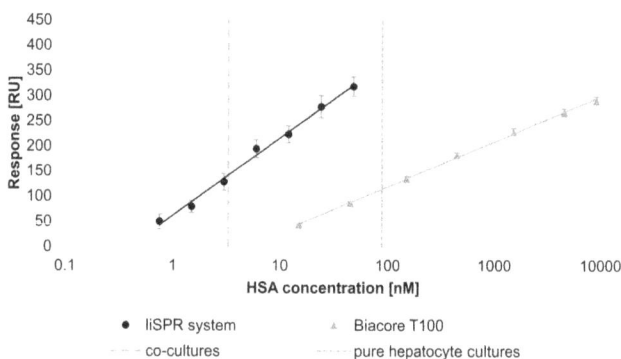

Figure 4. Comparison of the calibration curves obtained with the [li]SPR system and Biacore T100 in relation to the minimal HSA concentrations reportedly detected in co-cultivated and pure hepatocyte cultures in the multi-organ chip presented by Wagner et al. (2013). The signals arising from the reference antibodies were subtracted from those arising from the HSA-specific antibodies. Error bars indicate the deviations of means obtained from four replicates using the [li]SPR system and two replicates using the Biacore T100.

a hydrophobic tail (Kerwin, 2008) that suppresses nonspecific binding of proteins, by binding to the hydrophobic areas of the sensitive surface, thus displaying its hydrophilic head group to the samples.

4 Conclusions

Surface plasmon resonance (SPR) spectroscopy is being increasingly widely used in optical biosensors for detecting diverse clinical, food safety and environmental markers. SPR biosensors are also highly suitable for controlling biotechnological processes, due to their high sensitivity, rapidity and capacity for parallel measurements (which allows the use of control and reference surfaces to avoid false-positive results due to interfering sample components in simultaneous, real-time analyses).

Here, we present an SPR-based immunoassay for specific detection of HSA, which we tested in comparative trials with both the well-known Biacore platform and the cheaper bench-top [li]SPR system. The results show that the assay is capable of monitoring in vitro cultivated hepatocytes using both systems. However, only the [li]SPR system appears to be capable of monitoring HSA levels in co-cultivated liver and skin cells, which may be as low as 3.3 nM in samples with very small volumes (Wagner et al., 2013). Using the ^{li}SPR system and our (re-usable) sensor surface, we obtained a log-linear calibration curve from 0.8 to 48 nM HSA, theoretically allowing determinations of HSA in just 14.6 μL samples from co-cultivated liver and skin cells, and 0.6 μL samples from purely hepatocyte cultures.

In future work, we aim to measure HSA concentrations in effluents from hepatocyte cultures in the multi-organ chip presented by Wagner et al. (2013) to test the true potential of the developed immunoassay for monitoring liver cells.

Author contributions. Anja Henseleit, Elke Boschke and Thomas Bley designed the experiments and Carolin Pohl and Anja Henseleit carried them out. Anja Henseleit and Carolin Pohl analyzed the data. Anja Henseleit prepared the manuscript with contributions from all co-authors.

Acknowledgements. The authors thank D. Scharnweber and V. Hintze of the Institute of Materials Science TU Dresden for the opportunity to perform the Biacore experiments. We acknowledge the valuable advice on data interpretation provided by U. H. Yildiz of the Izmir Institute of Technology Department of Chemistry, Turkey, and both J. Homola and M. Bocková of the Institute of Photonics and Electronics AS, Czech Republic. We also thank E.-M. Materne of the Institute of Biotechnology TU Berlin and F. Sonntag of the Fraunhofer Institute for Material and Beam Technology (IWS) for useful discussions regarding the multi-organ-chip platform. This work was financially supported by the European Union and the Free State of Saxony (SAB project UNILOC).

References

Cao, C., Kim, J. P., Kim, B. W., Chae, H., Yoon, H. C., Yang, S. S., and Sim, S. J.: A strategy for sensitivity and specificity enhancements in prostate specific antigen-α1-antichymotrypsin detection based on surface plasmon resonance, 21, 2106–2113, 2006.

Cass, A. E. G., Cass, T., and Ligler, F. S.: Immobilized Biomolecules in Analysis: A Practical Approach, Oxford University Press, 1998.

Curry, S., Brick, P., and Franks, N. P.: Fatty acid binding to human serum albumin: new insights from crystallographic studies, Biochim. Biophys. Acta, 1441, 131–140, 1999.

Diamandis, E. P. and Christopoulos, T. K.: The biotin-(strept)avidin system: principles and applications in biotechnology, Clin. Chem., 37, 625–636, 1991.

Drescher, D. G., Ramakrishnan, N. A., and Drescher, M. J.: Surface plasmon resonance (SPR) analysis of binding interactions of proteins in inner-ear sensory epithelia, Methods Mol. Biol., 493, 323–343, 2009.

Fägerstam, L. G., Frostell-Karlsson, Å., Karlsson, R., Persson, B., and Rönnberg, I.: Biospecific interaction analysis using surface plasmon resonance detection applied to kinetic, binding site and concentration analysis, J. Chromatogr. A, 597, 397–410, 1992.

Frederix, F., Bonroy, K., Laureyn, W., Reekmans, G., Campitelli, A., Dehaen, W., and Maes, G.: Enhanced performance of an affinity biosensor interface based on mixed self-assembled monolayers of thiols on gold, Langmuir, 19, 4351–4357, 2003.

Henseleit, A., Schmieder, S., Bley, T., Sonntag, F., Schilling, N., Quenzel, P., Danz, N., Klotzbach, U., and Boschke, E.: A compact and rapid aptasensor platform based on surface plasmon resonance, Eng. Life Sci., 11, 573–579, 2011.

Henseleit, A., Stürmer, J., Pohl, C., Haustein, N., Sonntag, F., Bley, T., and Boschke, E.: Surface plasmon resonance based detection of human serum albumin as a marker for hepatocytes activity., Intell. Sens. Sens. Netw. Inf. Process. ISSNIP 2014 IEEE Ninth Int. Conf., 1–5, 2014.

Holford, T. R. J., Davis, F., and Higson, S. P. J.: Recent trends in antibody based sensors, Biosens. Bioelectron., 34, 12–24, 2012.

Holmberg, A., Blomstergren, A., Nord, O., Lukacs, M., Lundeberg, J., and Uhlén, M.: The biotin-streptavidin interaction can be reversibly broken using water at elevated temperatures, Electrophoresis, 26, 501–510, 2005.

Homola, J.: Surface plasmon resonance sensors for detection of chemical and biological species, Chem. Rev., 108, 462–493, 2008.

Jans, K., Bonroy, K., De Palma, R., Reekmans, G., Jans, H., Laureyn, W., Smet, M., Borghs, G., and Maes, G.: Stability of mixed PEO-thiol SAMs for biosensing applications, Langmuir ACS J. Surf. Colloids, 24, 3949–3954, 2008.

Johnsson, B., Löfås, S., and Lindquist, G.: Immobilization of proteins to a carboxymethyldextran-modified gold surface for biospecific interaction analysis in surface plasmon resonance sensors, Anal. Biochem., 198, 268–277, 1991.

Jung, Y., Jeong, J. Y., and Chung, B. H.: Recent advances in immobilization methods of antibodies on solid supports, The Analyst, 133, 697–701, 2008.

Kausaite-Minkstimiene, A., Ramanaviciene, A., Kirlyte, J., and Ramanavicius, A.: Comparative study of random and oriented antibody immobilization techniques on the binding capacity of immunosensor, Anal. Chem., 82, 6401–6408, 2010.

Kerwin, B. A.: Polysorbates 20 and 80 used in the formulation of protein biotherapeutics: structure and degradation pathways, J. Pharm. Sci., 97, 2924–2935, 2008.

Kretschmann, E. and Raether, H.: Radiative decay of nonradiative surface plasmons excited by light, Z. Naturforsch A., 23, 2135–2136, 1968.

Löfås, S. and Johnsson, B.: A novel hydrogel matrix on gold surfaces in surface plasmon resonance sensors for fast and efficient covalent immobilization of ligands, J. Chem. Soc. Chem. Commun., 21, 1526–1528, 1990.

Love, J. C., Estroff, L. A., Kriebel, J. K., Nuzzo, R. G., and Whitesides, G. M.: Self-assembled monolayers of thiolates on metals as a form of nanotechnology, Chem. Rev., 105, 1103–1169, 2005.

Mertig, M., Kick, A., Bonsch, M., Katzschner, B., Voigt, J., Sonntag, F., Schilling, N., Klotzbach, U., Danz, N., Begemann, S., Herr, A., and Jung, M.: A novel platform technology for the detection of genetic variations by surface plasmon resonance, in 2009 IEEE Sensors, 392–395, 2009.

Peluso, P., Wilson, D. S., Do, D., Tran, H., Venkatasubbaiah, M., Quincy, D., Heidecker, B., Poindexter, K., Tolani, N., Phelan, M., Witte, K., Jung, L. S., Wagner, P., and Nock, S.: Optimizing antibody immobilization strategies for the construction of protein microarrays, Anal. Biochem., 312, 113–124, 2003.

Schneider, B. H., Dickinson, E. L., Vach, M. D., Hoijer, J. V., and Howard, L. V.: Highly sensitive optical chip immunoassays in human serum, Biosens. Bioelectron., 15, 13–22, 2000.

Vogler, E. A.: Protein adsorption in three dimensions, Biomaterials, 33, 1201–1237, 2012.

Wagner, I., Materne, E.-M., Brincker, S., Süßbier, U., Frädrich, C., Busek, M., Sonntag, F., Sakharov, D. A., Trushkin, E. V., Tonevitsky, A. G., Lauster, R., and Marx, U.: A dynamic multi-organ-chip for long-term cultivation and substance testing proven by 3D human liver and skin tissue co-culture, Lab. Chip, 13, 3538–3547, 2013.

Yang, M.-H., Jong, S.-B., Chung, T.-W., Huang, Y.-F., and Ty, Y.-C.: Quartz crystal microbalance in clinical application, in: Biosensors for Health, Environment and Biosecurity, edited by: Serra, P. A., InTech., 257–272, 2011.

11

Luminescent determination of nitrite traces in water solutions using cellulose as sorbent

S. G. Nedilko[1], **S. L. Revo**[1], **V. P. Chornii**[1], **V. P. Scherbatskyi**[1], **and M. S. Nedielko**[2]

[1]Taras Shevchenko National University of Kyiv, Kyiv, Ukraine
[2]E. O. Paton Electric Welding Institute of NASU, Kyiv, Ukraine

Correspondence to: S. G. Nedilko (snedilko@univ.kiev.ua)

Abstract. Morphology and photoluminescence (PL) properties for microcrystalline cellulose (MCC), microcrystalline nitrite powders of common formulae MNO_2 ($M = Na, K$) (MCN) and two-component materials composed of both MCC and MCN have been prepared and characterized by means of optical microscopy and luminescence spectroscopy. This study aimed to clarify a possibility of low-limit determination of the nitrite traces in water solutions by luminescent method.

1 Introduction

The fast progress of techniques is accompanied by negative effects on the environment. In this context, detection of hazardous material traces (e.g., drugs, heavy metals, petroleum oils, polycyclic hydrocarbons) in water and soil is of high importance. Nitrites and nitrates are also known as toxic pollutants of environment, plant-based food and biological systems (Fanning, 2000; Zhou and Wang, 2012). In fact, the high level of nitrites in the blood promotes the cancerization of hemoglobin. So, it is undoubtedly important to use a simple, easy and convenient method for removing these compounds (e.g., by sorption), followed by a quick estimation of their concentration in solutions.

According to recommendations of the World Health Organization, the amount of nitrites in water should not exceed $3 \, mg \, L^{-1}$, while for nitrates the corresponding value is $50 \, mg \, L^{-1}$ (*Guidelines for drinking-water quality* – 4th ed. Geneva, Switzerland, 2011). Available methods satisfy the parameters of determination of nitrite content in solutions. The limits for detection are 0.005–$0.01 \, mg \, L^{-1}$ for the molecular absorption spectrometric method, $0.035 \, mg \, L^{-1}$ for ion chromatography and $0.05 \, mg \, L^{-1}$ for liquid chromatography. However, the methods listed above require pretreatment involving application of additional chemical reagents. LED (light emitting diode)-sourced optical fiber sensor, based on evanescent wave absorption, allows measur-

ing a nitrite concentration higher than $10^{-3} \, mg \, L^{-1}$ (Suresh Kumar et al., 2002). The main disadvantage of absorption-based methods of nitrite traces determination is associated with the absence of characteristic peculiarities on the absorbance spectra of water solutions. The presence of some other contaminant in a solution can lead to absorption changes and affect the accuracy of nitrite concentration estimation.

It has been shown in earlier works (e.g., Kononenko and Kushnirenko, 1980) that the series of sharp emission peaks in luminescence spectra of alkali metal nitrites can be observed. Spectral positions and distances between these peaks are practically insensitive to the type of the host where nitrites are incorporated. This feature can be used for the exact identification of nitrates present in various materials. This work extends the approach on nitrite immobilization from water solutions in cellulose to reveal their presence through the luminescent method.

It is common knowledge that cellulose is a very effective sorbent. Really, synthetic sorbents based on polyethylene, propylene and other polymers show better performance, but they are not eco-friendly enough in comparison to cellulose. Cellulose is a material widely used and it is also considered as an inexhaustible source for manufacturing biocompatible products. The range of cellulose's promise spreads from "paper electronics" ("paper transistor" on a carbon nanotube or organic LED), based on luminescent cellulose (Yun et al.,

2009; Karakawa et al., 2007; Pikulev et al., 2012) to forensic examination and eco-friendly sorbents suitable for sorption and stabilization of a wide range of various types of materials (Konstantinou et al., 2006; Chukova et al., 2005). This variety of possible applications is determined by the porous, micro- and nanostructured morphology of the cellulose host and by the unique nature of its interaction with other chemical compounds (Glikman and Somova, 1964; Zugenmaier, 2008; Kovalenko, 2010). The physical properties (in particular, optical properties) of cellulose are dependent on its degree of crystallinity. It is worth noting that cellulose is, like many other polymers, partially crystallized. Its crystallites are long-ordered but its molecular chains do not have an ideal tridimensional structure; in other words, the crystallites are defective. The degree of crystallinity and the chain's order depend on conditions of cellulose preparation (coagulation procedure, drying temperature if $T > 200\,^\circ\text{C}$, type of fixed gas, mechanical and ultrasonic effects) and is lower after cellulose regeneration from solutions (Glikman and Somova, 1964; Zugenmaier, 2008; Kovalenko, 2010; Yun et al., 2009; Karakawa et al., 2007; Pikulev et al., 2012). Taking into account that the proposed method of nitrite traces determination deals with a composed material, cellulose + nitrite, it is clear that luminescence properties of pure cellulose should be studied in the first stage of investigation.

2 Samples and experimental details

There were four types of cellulose samples used for the study. Sample #1, labeled CT, was the starting tablet of chemically pure microcrystalline cellulose (MCC) manufactured at ANCYR-B, Ukraine. The next samples, CD, were prepared from the CT samples by their dispersion using mechanical milling of the cellulose tablet followed by ultrasonic treatment ($f = 4.2\,\text{kHz}$). Then, the powder was pressed into a disk using light pressure. Samples #3, or CDW, were made from the CD samples in a two-stage treatment. The first stage of the preparation was soaking of the CD sample in distilled water, followed by applying sufficient light pressure to obtain a durable sample with disc thickness of close to 1 mm. Afterwards, the CDW samples were dried in ambient air conditions at a temperature of $60\,^\circ\text{C}$ for 16 h. The soaking of the CD sample in a water solution of nitrites, MNO_2 (M = Na, K), was followed by the described-above pressing and drying procedures used to make the disk of composite material, cellulose + nitrite, producing sample #4, named Cell-NO. The luminescence of the microcrystalline $NaNO_2$ powders (MCN) was investigated for comparison, too.

A set of the microphotos of the above samples was made with the optical microscope. The PL (photoluminescence) spectra and PL excitation were measured using the spectrometric equipment SDL-2M. The PL spectra were studied as a function of the exciting radiation wavelength (λ_{ex}) and were analyzed over a wide range of excitation and emission wavelengths (200–800 nm) and sample temperatures (77–300 K). The PL emission spectra were analyzed using a single-grating (1200 grooves mm^{-1}) monochromator MDR-23 (linear dispersion 0.5 mm nm^{-1}) equipped with a FEU-100 photomultiplier. The N_2 laser ($\lambda_{ex} = 337.1$ nm), two diode-pumped lasers ($\lambda_{ex} = 473$ and 532 nm, respectively) and a xenon lamp (DKsSh-150) were used as sources of PL excitation. The results were compared with known literature data and with results of our own investigation of the same luminescent analytes incorporated to some other solid matrixes (see Kushnirenko et al., 1984; Belii et al., 1984; Kononenko et al., 1985). During the measurements at $T = 77$ K, the samples were put into liquid nitrogen cryostat. A fast plunge from nitrogen vapors at a higher part of the cryostat to liquid nitrogen at the bottom of it allowed avoiding icing at the surface of the samples.

3 Experimental results and discussion

3.1 Morphology of the samples under study

Figure 1a shows an image of a cellulose tablet surface (sample #1, CT). It is clearly seen that the tablet is a conglomerate of amorphous-phase and evenly distributed microcrystals with a predominant size of ~ 20–40 μm. Dissolution in water after dispersion (sample #3, CDW) leads to cellulose fluff-out and following disassimilation into an ensemble of flakes with sizes from 10 to 100 μ (Fig. 1b). The flakes consist of a great number of 2–3 μ microglobules.

The MCN samples represent a set of microcrystals randomly distributed over the surface of glass substrate (Fig. 1c). The predominant size of the crystal grains is approximately 20–40 μ.

Finally, a two-component composite Cell-NO (sample #4) is strongly heterogenic. The areas similar to the mixture of the $NaNO_2$ particles and cellulose (Fig. 1d) and the areas that consist of large transparent jelly-like cellulose grains (their size is near 200 μ), with a great number of smaller 5–20 μ inclusions (Fig. 1e), are also found. We can expect that the last ones are the inclusion of the $NaNO_2$ particles into the cellulose host. It is easy to see that the morphology and sizes of MCN grains in the "free" powder and in the composite are different. In the first case, the crystals have similar sizes in all three directions, while in the second case the crystals have various shapes and sizes. These changes are obviously the result of the cellulose lattice effect on MCN crystals. This fact shows that the cellulose + $NaNO_2$ composite is not only a mechanical mixture of two components – cellulose and $NaNO_2$ crystallites.

3.2 Luminescence properties

Physical properties as well as the degree of crystallinity and chain order of cellulose-based materials depend highly on conditions of the manufacturing and treatment procedure.

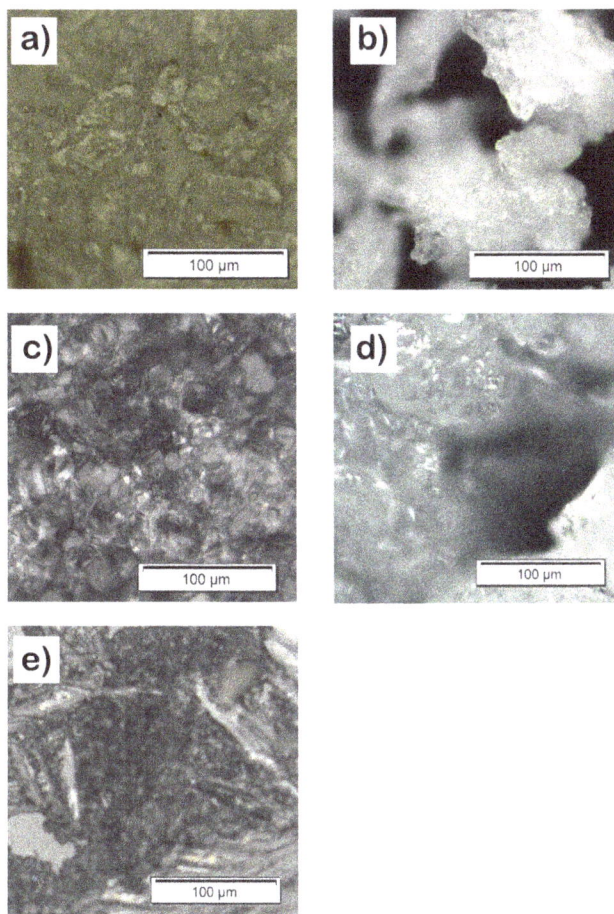

Figure 1. Optical microscopy images of the cellulose tablet, CT sample (**a**); dispersed cellulose, CDW sample (**b**); NaNO$_2$ powder (**c**); and a composite Cell + NaNO$_2$ sample taken in two different areas (**d, e**).

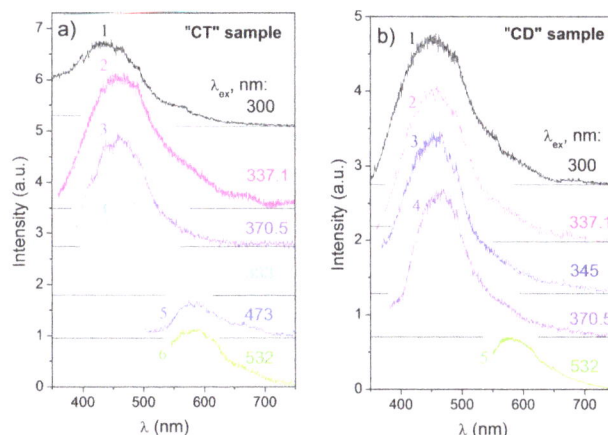

Figure 2. The PL spectra of the samples CT (**a**) and CD (**b**); λ_{ex} = 300 (**a**, 1; **b**, 1), 337.1 (**a**, 2; **b**, 2), 345 (**b**, 3), 370.5 (**a**, 3; **b**, 4), 393 (**a**, 4), 473 (**a**, 5) and 532 nm (**a**, 6; **b**, 5); T = 300 K.

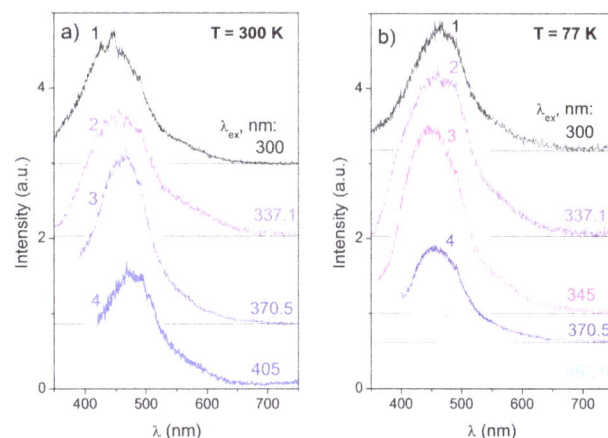

Figure 3. The PL spectra of the CDW samples at 300 (**a**) and 77 K (**b**); λ_{ex} = 300 (**a**, 1; **b**, 1), 337.1 (**a**, 2; **b**, 2), 345 (**b**, 3), 370.5 (**a**, 3; **b**, 4), 405 (**a**, 4), and 465.5 nm (**b**, 5).

Thus, we have performed experiments on luminescence spectroscopy for our cellulose samples.

All the cellulose samples are characterized with intensive PL. The widely composed emission band located in the 325–750 nm range is observed in the PL spectra of cellulose samples #1–3 under excitation in a wide range of excitation wavelengths, λ_{ex}: 300–532 nm (Figs. 2, 3). The shape, peak position and intensity of the PL band depend on the λ_{ex}. The dependences of the spectra on λ_{ex} have a similar character for the mentioned cellulose samples: as λ_{ex} increases as the peak position, λ_{max}, reveals a tendency to the long wavelength side shifting of the spectra (see Figs. 2, 3). Mainly, the component with the peak position, λ_{max}, near 430–470 nm dominates in the spectra at UV excitation. It is clearly seen for the dispersed cellulose sample CD (Fig. 2b). An additional band is also observed if visible radiation (λ_{ex} = 450–530 nm) is used for PL excitation. The band is situated at longer wavelengths than UV excitation is. This is a range of 500–700 nm and the λ_{max} of the band is near 680–590 nm (Figs. 2, 3b, spectra 5).

As luminescence of alkali metal nitrite and impure NO$_2^-$ molecular anions are more distinctive at low temperatures (T < 300 K), the PL properties of cellulose samples taken after dissolution in water, CDW samples, were monitored at both room and liquid nitrogen temperature (77 K). When temperature of the samples decreases from 300 to 77 K, the change of PL spectra is small (compare Fig. 3a and b). At observed low temperatures the PL is mainly excited in the UV–V region of light: 300–405 nm.

The PL spectra of the powder microcrystalline salts, NO$_2$ (M = Na, K), were recorded at 300 and 77 K (Fig. 4). These spectra at RT (room temperature) show fine-structure details like the series of narrow bands placed on the wide-band background. The bands are clearly distinguished for the NaNO$_2$ case (Fig. 4, curve 1). That is why posterior studies were performed using sodium nitrite, NaNO$_2$, compounds. The contrast of the mentioned bands strongly increases if temperature

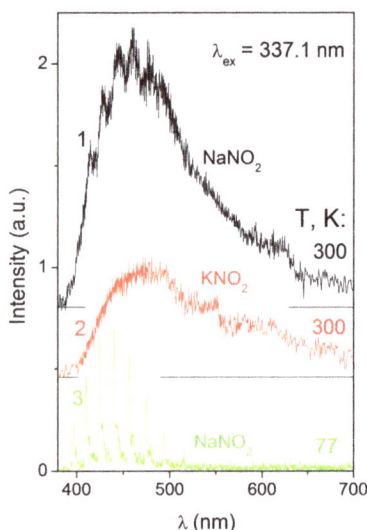

Figure 4. The PL spectra of the $NaNO_2$ (1 and 3) and KNO_2 (2) samples; $T = 300$ (1, 2) and $77\,K$ (3); $\lambda_{ex} = 337.1\,nm$.

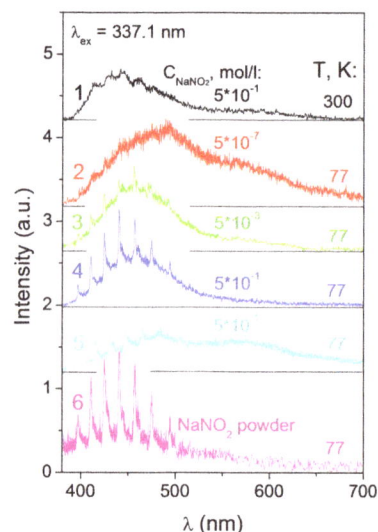

Figure 5. The PL spectra of the cellulose–$NaNO_2$ composite samples #4, Cell-NO, (1–5) and $NaNO_2$ powder (6); $\lambda_{ex} = 337.1\,nm$; $T = 300$ (1) and $77\,K$ (2–6). The concentrations of the $NaNO_2$ salts in water solutions were 5×10^{-1} (1, 4, 5), 5×10^{-3} (3), and $5 \times 10^{-7}\,mol\,L^{-1}$ (2). The spectra 4 and 5 were taken for the sample areas shown in Fig. 1d and 1e, respectively.

of the sample decreases. Then, the spectrum shows the series of narrow lines, each of them accompanied with a longer wavelength shadow (Fig. 4, curve 3).

The PL spectra of the composite samples containing $NaNO_2$, Cell-NO, show a weak fine structure at $T = 300\,K$. When temperature decreases down to $77\,K$, distinctive series of the lines appear (Fig. 5, spectra 3–5). It should be noted that concentrations of the $NaNO_2$ salts in the initial water solutions were high: $5 \times 10^{-1}\,mol\,L^{-1}$ (Fig. 1, spectra 3–5). So, the $NaNO_2$ concentration was the same for three spectra – 3, 4 and 5 – taken at the same condition. At the same time, the $NaNO_2$ concentration in initial water solutions was only $5 \times 10^{-7}\,mol\,L^{-1}$ for the case of the spectrum 2 in Fig. 5.

The previously escribed luminescence peculiarities of cellulose samples #1–3 are similar to those reported earlier in the literature (Gavrilov and Ermolenko, 1966). For example, our findings prove that there are several luminescence centers in the studied cellulose samples. This circumstance determines the fact that PL spectra contain several components with peak positions near 360, 430, 465 and 580 nm and their contributions to the total spectrum depend on the λ_{ex} (Figs. 2, 3). The mentioned multicomponent structure of the PL bands is usually related to the presence of several types of organic chromospheres in a host matrix of cellulose. There can be carbonyl groups and different kinds of low-molecular derivatives of cellulose destruction. Taking this into account, we are not surprised that the incorporation of nitrite compounds into the cellulose host matrix will change the PL properties (compare Figs. 3 and 5).

The fine-structure details obtained for MNO_2 and Cell-NO samples correspond to the phononless radiation $^1B_1 \rightarrow {}^1A_1$ transitions in NO_2^- molecular anions. The periods of these details, so-called zero-phonon lines (ZPL) accompanied with

phonon wings, in the series are determined by the value of NO_2^- intra-molecular frequency – v_2. It is worth noting that there is one similar series of lines in the cases of the 1–4 spectra in Fig. 5, and that the positions of the lines are close to the ones for the $NaNO_2$ powder sample (Fig. 5, curve 6). Moreover, in the case of the spectrum 5 in Fig. 5 additional series dominate in the spectrum. The ZPL positions for each series, $v_n^{1,2}$, are described with formula $v_n^{1,2} = v_{00}^{1,2} - n v_2^{1,2}$, where v_{00} is the energy of pure electronic transition, and $n = 1, 2, 3, \ldots$. It was found that $v_2^1 = 826\,cm^{-1}$ and $v_{00}^1 = 26\,840\,cm^{-1}$ for the first series of ZPL and $v_2^2 = 868\,cm^{-1}$ and $v_{00}^2 = 27\,570\,cm^{-1}$ for the second one. It is worth noting the last parameters are very close to the ones for the series of ZPL in the PL spectrum of the $NaNO_2$ powder sample: $v_2 = 824\,cm^{-1}$ and $v_{00} = 26\,838\,cm^{-1}$.

Thus, we can conclude that the PL spectra of the two-component samples show both cellulose and NO_2^- emission simultaneously. Contribution of these two different emissions depends on the $NaNO_2$ concentration in the initial water solution and of the sample #4 area taken for monitoring (Fig. 5). It is known that luminescence spectra of the MA (molecular anion) NO_2^- in solid water solutions reveal several components caused by NO_2^- ions located in various structurally different phases of solid solutions. There are several types of PL centers where MA NO_2^- are the core, but are located in different neighboring spaces (Kononenko et al., 1985; Belii et al., 1985, 1988). It is obvious that the formation of the several types luminescence centers based on NO_2^- anions is possible in the case of the cellulose compositesample #4. First, there can be the centers formed by MA NO_2^- located in the volume of the MNO_2 particles. These centers have to reveal spectral features similar to the cases of metal alkali nitrite salts and alkali–halide crystals doped with MA NO_2^- (Sidman, 1957; Brooker and Irish, 1971; Kushnirenko et al., 1984; Belii et al., 1984, 1988; Kononenko et al., 1985). Such type centers dominate PL emission of the area shown in Fig. 1d. Second, there can be the centers formed by NO_2^-, which are located near surface of mentioned particles. These NO_2 groups can be related to nearby atoms that belong to the cellulose matrix. Finally, the third type of luminescence centers is formed by the "isolated" NO_2^- molecular anions that are adsorbed by the cellulose host. The first and second types of centers can be responsible for the two observed series of ZPL in the PL spectra. The third type of centers can create a structureless band that manifests in the range 520–620 nm in the PL spectra of the Cell-NO sample #4 (Fig. 5). The facts mentioned above indicate interaction between the host matrix and nitrite component of the composite material.

From the analytical point of view, it is easy to conclude that spectra like #4 in Fig. 5 are the most suitable for low-limit determination of sodium nitrite in water solutions.

Furthermore, we would like to emphasize the "energetic" interaction between the mentioned components of the composite sample: the cellulose host matrix luminescence covers the spectral region where excitation of the NO_2 groups takes place (Kononenko et al., 1985; Belii et al., 1985, 1988). So, there is an opportunity for excitation energy transfer from cellulose host to luminescent NO_2^- molecular anions. This fact promotes the enhancement of the NO_2^- molecular anion luminescence intensity. Therefore, we were able to detect NO_2^- emission for the samples Cell-NO obtained from a $NaNO_2$ water solution where the concentration was 5×10^{-7} mol L^{-1} or 0.035 mg L^{-1} for $NaNO_2$ nitrite compounds (Fig. 5, curve 2). Obtained results show the proposed luminescence-based method for nitrite trace determination has sensitivity similar to methods reported in the Introduction. The proposed method is simple and can be applied in some cases when the possibility of using other techniques is restricted for some reason (e.g., absence of required chemical reagents). We suppose that specially before prepared "cellulose serviette" can be used as a selective device, which reacts by means of luminescence signal, change on the nitrite compounds incorporation. The obtained lower limit for a nitrite trace to be identified is high enough (in comparison with the usual methods); so, we think cellulose can be used not only as sorbent but also as sensing material for nitrites. Therefore, such a device can be regarded as a chemical sensor and can be of interest for the sensor systems community. It is worth noting that luminescence spectra should be obtained for various areas of a cellulose + nitrite sample in order to avoid influence of partial crystallinity effects.

At present we have no certain data of other substances such as heavy or transition metal ions showing wide luminescence spectra and influence over their own cellulose luminescence. As for RE (rare earth) ions, they manifest well-known linear spectra at the UV region of the PL excitation and we did not find any trace of the RE ions' luminescence. The possibility that the luminescence determination of other hazardous materials which can be absorbed by cellulose will be a subject of further investigations.

4 Summary

The method of determining the nitrite compound traces via their sorption by cellulose using luminescent properties of the NO_2^- molecular ion has been developed, and the low limit of the $NaNO_2$ determination in water solution was evaluated as 5×10^{-7} mol L^{-1} (0.035 mg L^{-1}). Further studies of cellulose + MNO_2 (M = Na, K, Cs) composite materials have to be performed to determine the most suitable conditions of analytical performance.

References

Belii, M. U., Bojko, V. V., Kushnirenko, I. Ya., Nedilko, S. G., and Sakun, V. P.: Structure of the luminescence and absorption spectra in KBr single crystals doped with NO_2^- anions and co-doped with Ca^{2+} cations, Dopovidi AS USSR, 2, 53–57, 1984 (in Ukrainian).

Belii, M. U., Kushnirenko, I. Y., Nedilko, S. G., and Sakun, V. P.: Spectral-luminescent properties of the luminescence centers in frozen water solution of alkali-halide salts doped with NO_2^- anions, Optics Spectr., 58, 367–372, 1985.

Belii, M. U., Kushnirenko, I. Ya., Nedilko, S. G., and Sakun, V. P.: Spectroscopic appearances of inter-ionic interactions in water solutions of alkali metal nitrites, J. Appl. Spectr., 48, 835–839, 1988 (in Russian).

Brooker, M. H. and Irish, D. E.: Raman and Infrared Spectral studies of solid alkali metal nitrites, Can. J. Chem., 49, 1289–1295, 1971.

Chukova, O., Krut, O., Nedilko, S., Sakun, V., and Scherbatskyi, V.: Luminescent determination of automobile petrol in hexane solutions, Annali di Chimica, 95, 885–895, 2005.

Fanning, J. C.: The chemical reduction of nitrate in aqueous solution, Coord. Chem. Rev., 199, 159–179, 2000.

Gavrilov, M. Z. and Ermolenko, I. N.: A study of cellulose luminescence, J. Appl. Spectr., 5, 542–544, 1966.

Glikman, S. A. and Somova, A. I.: Chemistry and Technology of Cellulose Derivatives, Verkhne-Volzhskoe Knizhnoe Izdatelstvo, Ghorkii, 169 pp., 1964 (in Russian).

Karakawa, M., Chikamatsu, M., Nakamoto, C., Maeda, Y., Kubota, S., and Yase, K.: Organic light-emitting diode application of fluorescent cellulose as a natural polymer, Macromol. Chem. Phys., 208, 2000–2006, 2007.

Kononenko, Y. T. and Kushnirenko, Y. I.: Luminescent method of NO_2^- ions determination in aqueous solutions of halide salts under low temperatures, Vestnik Kyiv University, Ukraine, Physics, 21, 49–52, 1980 (in Russian).

Kononenko, Y. T., Kushnirenko, I. Y., Nedilko, S. G., and Sakun, V. P.: Spectral trends and radiativeless transitions in solution and crystals of alkali-halide salts doped with molecular anions, J. Appl. Spectr., 42, 85–89, 1985 (in Russian).

Konstantinou, I. K., Hela, D. G., and Albanis, T. A.: The status of pesticide pollution insurface waters (rivers and lakes) of Greece – Part I: Review on occurrence and levels, Environ. Pollution, 141, 555–570, 2006.

Kovalenko, V. I.: Crystalline cellulose: structure and hydrogen bonds, Rus. Chem. Rev., 79, 231–242, 2010.

Kushnirenko, I. Y., Kononenko, Y. T., and Nedilko S. G.: Formation of the complex in water solutions of zinc and cadmium halides doped with molecular anions NO_2^-, Ukr. Phys. J., 29, 673–675, 1984 (in Ukrainian).

Pikulev, V., Loginova, S., and Gurtov, V.: Luminescence properties of silicon-cellulose nanocomposites, Nanoscale Res. Lett., 7, 426, 2012.

Sidman, W. J.: Electronic and Vibrational States of the Nitrite Ion. I. Electronic States, J. Am. Chem. Soc., 79, 2669–2674, 1957.

Suresh Kumar, P., Thomas Lee, S., Vallabhan, C. P. G., Nampoori, V. P. N., and Radhakrishnan, P.: Design and development of an LED based fiber optic evanescent wave sensor for simultaneous detection of chromium and nitrite traces in water, Optics Commun., 214, 25–30, 2002.

Yun, S., Jang, S. D., Yun, G. Y., Kim, J. H., and Kim, J.: Paper transistor made with covalently bonded multiwalled carbon nanotube and cellulose, Appl. Phys. Lett., 95, 1–3, 2009.

Zhou, Z. and Wang, Q.: Two emissive cellulose hydrogels for detection of nitrite using terbium luminescence, Sensor. Actuat. B-Chem., 173, 833–838, 2012.

Zugenmaier, P.: Crystalline Cellulose and Derivatives: Characterization and Structures, Springer, Heidelberg, Germany, 285 pp., 2008.

12

Micro-structured electron accelerator for the mobile gas ionization sensor technology

C. M. Zimmer, K. T. Kallis, and F. J. Giebel

Intelligent Microsystems Institute, Faculty of Electrical Engineering and Information, Technische Universität Dortmund, Dortmund, Germany

Correspondence to: C. M. Zimmer (cordula.zimmer@tu-dortmund.de)

Abstract. Mobile and economically priced gas monitoring and warning systems will become increasingly important for civil security, such as in fire brigade operations in undefined hazardous environments (Daum et al., 2006). Normally, photoionization detectors (PIDs) are used for the detection of gases. Hereby, the principle is based upon the ionization of the measured gas by photons, which are generated by a high-energetic gas discharge lamp with energy of 10–11 eV. Besides the detrimental unspecific gas detection because of the ionization of all gases with ionization potential (IP) below the provided photon energy, sensors also have a short lifetime combined with a high cost (http://www.intlsensor.com/pdf/photoionization.pdf).

This can be remedied by the concept of an electronic supported photoionization detector (ePID; Zimmer et al., 2012) consisting of a durable UV-LED with an above-positioned electron accelerator chip manufactured on a glass substrate by planar technology. Photoelectrons are extracted by UV illumination out of the bottom electrode and will be accelerated to an energy matching the ionization potential of the gas by a downstream acceleration grid. Thereby, the stable honeycomb-structured grid acts as a porous separator between the evacuated electron acceleration path due to nm scaling and the ionization area of the detector. To enhance the emitting area yielding a higher photoelectron current, the grid structure almost levitates, realized by the use of compatible planar technological processes such as reactive ion etching (RIE) and isotropic wet etching of sacrificial layers, which will be explained in detail in this paper. Furthermore, the tunability of the grid's acceleration voltage would enable a substance-specific determination of the gas composition, where the ionization of the analytes is clearly performed by photoelectrons instead of photons.

1 Conventional possibilities and methods for gas detection

Measuring instruments with different mechanisms are currently used for the detection of volatile gases, including a wide variety of potentially available chemical substances. Examples initially comprise the ion mobility spectrometer (IMS), the mass spectrometer (MS; Eiceman and Karpas, 2013), as well as Fourier transform infrared spectroscopy (FTIR). But, these measuring devices are partly suitable for on-site use. Alternative and portable measuring methods are presupposed by known analytes and limited by the TLV (threshold limit value) measurement done with the test tubes of the Dräger company. Another detection method is presented by the LIMS (laser ion mobility spectrometer), whereby substances are identified by their characteristic ionization potential detected by a tunable laser (Constapel et al., 2005). But, the expenditure for the realization of such a measuring system is tremendous. According to this, small and portable detecting systems such as conventional PIDs are increasingly in demand. All the above-mentioned methods are built up from discrete parts and are not suitable for realization as a single-chip system by planar technology. Lately, this is also reflected by the price of these systems.

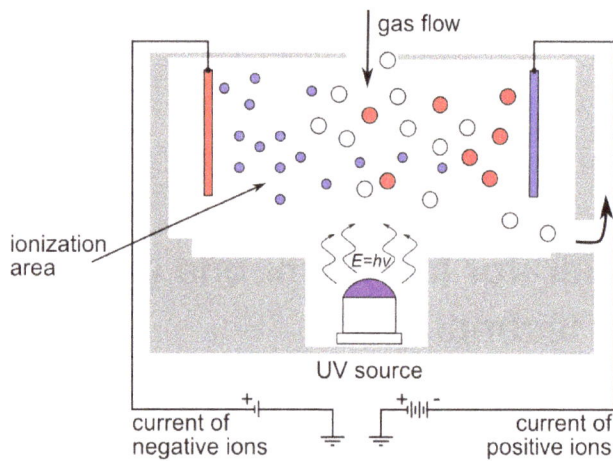

Figure 1. Basic principle of a conventional photoionization detector (PID).

2 Gas sensor technology by photoionization detectors

2.1 Conventional photoionization detectors (PIDs)

The structure of the conventional photoionization detector (PID) is shown in Fig. 1, and consists of an ionization chamber that includes a UV source and two capturing electrodes.

Hereby, the one-photon ionization based on the photoelectric effect is used by the PID for detecting the analytes. During ionization, the complete energy of the photon (hv) is transferred to the gas atom (B) as inelastic impact, whereupon the gas atom (B^+) is ionized by extracting an electron (e^-; Eiceman and Karpas, 2013).

$$B + hv \xrightarrow{\text{Ionization}} B^+ + e^-$$

In case of gas ionization by a photon, the provided photon energy has to be larger than the ionization potential (IP) of the atom or molecule ($hv > $IP). According to this, substances in the air such as aromatic hydrocarbons and organic compounds are detected by the PID in terms of their ionization potentials. Thereby, substances are introduced into the ionization area of the detector by a pump or easily by diffusion. Ionized atoms or molecules resulting from impacts by photons and free electrons are attracted by the biased electrodes, respectively. Afterwards, a discharge and current flow accordingly occurs, which is converted into a signal by processing electronics. The higher the available substance concentration, the higher the detectable current. Hence, system-specific parameters like the dimension of measurement cells as well as biased voltage at electrodes have to be considered and calibrated on the particular gas for converting into the parts-per-million range. Typically, a gas discharge lamp filled with inert krypton gas for generating photon energy of 10.6 eV is used, whose energy is high enough to ion-

ize volatile organic compounds (VOCs) without ionizing the so-called background molecules such as oxygen, nitrogen and water vapor that have an ionization potential of over 10.6 eV. Therefore, the ionization of substances is not only dependent on the introduced energy, but also on the cross section of ionization (http://www.intlsensor.com/pdf/photoionization.pdf, Eiceman and Karpas, 2013, Freedman, 1980). Only a certain part of the molecule structure is ionized and registered by the PID, which is dependent on substance specification but not on substance concentration. For that reason, so-called response factors are necessary, which are detected as a reference to a gas like 2-methylpropene. These response factors are determined in time-consuming experiments, but are indispensable to making a relevant statement about the ionization probability.

2.2 Disadvantages of photoionization detectors (PIDs)

Unfortunately, the PID cannot exactly identify lightly or heavily ionizable substances in gas compounds, which are introduced in the measurement cell. It is also difficult to make a statement about its concentration. Lately, a PID can only determine the amount of ionized gaseous substances (Constapel et al., 2005).

During the use of PIDs, some mistakes took place, resulting from different factors and gaining more and more relevance with increasing operation time. One of the key factors is the influence of dust and humidity and the degradation of components, which cannot be corrected by processing electronics.

Before using the PID, a reference routine is made, where the test gas, usually 2-methylpropene with a known concentration, is pumped into the PID. The 2-methylpropene is ionized without any influences and its concentration is displayed. Normally the test gas is a dry gas, but unfortunately air humidity is present during on-site usage under ambient air, and ionizations can be hindered.

This is a result of water attraction on formed ions or electrons decreasing the mobility and ionizing capabilities (quench effect; RAE Systems Inc., 2013). With this effect, an offset of 20–30 % to the reference value occurs. To minimize the influences of air humidity, a humidity sensor is installed in the PID, separately determining the air humidity and contributing it to the measuring process (http://www.intlsensor.com/pdf/photoionization.pdf, RAE Systems Inc., 2013).

Other influences on measuring signals are equally performed by contaminations in the air, meaning electrodes are contaminated by dust particles penetrating into the measuring cell, causing a capillary condensation. Due to existing salts in the dust layer, a current flow proceeds. This effect is called parasitical conductivity and will be substantially enhanced by the humidity in ambient air.

Preventing the penetration of dust particles, the PID is equipped with dust filters, which are cleaned up or replaced at periodic time intervals to avoid measuring errors (RAE

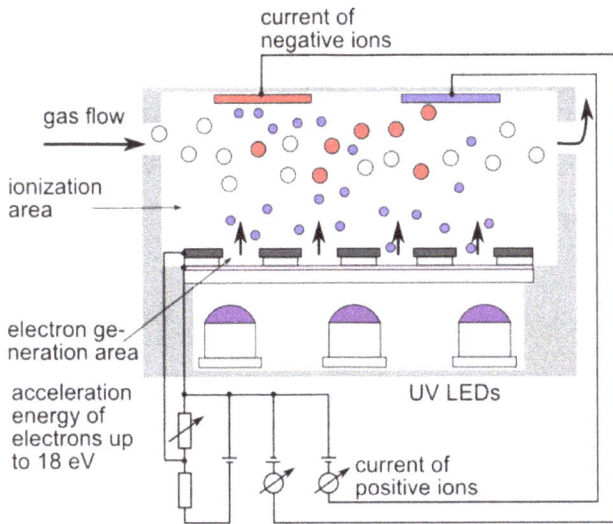

Figure 2. Basic principle of an electronic supported photoionization detector (ePID; Zimmer et al., 2012).

Systems Inc., 2013). Another negative influence on the detection by PID is represented by bulb degradation of the UV light source. Here, the glass bulb of the UV lamp consisting of magnesium oxide or magnesium fluoride is roughened by the extraction of high-energetic photons. This damage supports the reflection of photons at the glass, leading to a complete blindness of the UV light source, performing no further gas ionizations (Zimmer et al., 2012).

2.3 Electronic supported photoionization detector (ePID)

Not all the above-mentioned disadvantages of the PID can be compensated for the use of the electronic supported photoionization detector (ePID) presented here. Besides the quench effect, the main problems such as high maintenance costs due to the replacement of a degenerated discharge lamp as well as the imprecise detection of volatile organic substances can be solved by using an optimized detector via a micro-structured accelerator. The ePID works on the same principle as a PID. The substances to be detected in air are aspirated and ionized by the detector. These gases are measured as an ionization current due to biased electrodes and registered, analyzed and displayed by the processing electronics. The most significant difference between the ePID and the PID explained above is the type of gas ionization. Instead of a direct ionization by high-energetic photons, an indirect ionization by photoelectrons with the help of an electron emitter being subjected to the photoelectric effect and a downstream honeycomb-structured acceleration path fabricated by planar technology is performed. The system of such an ePID is schematically shown in Fig. 2.

Hereby, the detrimental rapid degradation of the UV source due to radiation damage is prevented by this modi-

fication of the PID. And the use of an expensive and high-energetic UV lamp becomes invalid. The electron emitter consisting of lanthanum hexaboride (LaB_6) exhibits a low crystal-dependent work function of 2.3 eV (Nishitani et al., 1980). As a primary photon source, a comparatively inexpensive UV-LED is chosen whose wavelength is matched to the work function of the emissive layer. Due to the illumination, photoelectrons are extracted from the emitter that has a kinetic energy of almost 0 eV (Zimmer et al., 2012) due to optimum adjustment of the light source and emissive layer, meaning the photoelectrons only perform disordered thermal movements.

Afterwards, photoelectrons are accelerated by a downstream acceleration path towards the ionization area to perform the gas ionization. Due to the application of different voltages at the acceleration grid, a wide range of electron energies can be realized. Besides a better selectivity, ionization energies above the typical 10.6 eV from commercial PIDs can also be implemented. But, to preserve the measurement of significant results, an acceleration energy of 25 eV should not be exceeded because of the formation of fragment ions.

Hereby, the geometric expansion of this acceleration path is another important aspect, which should be smaller than the mean free path of electrons in air. Thus, it is guaranteed that the use of expensive vacuum components can be waived. Furthermore, it is not advantageous to position the UV-LED directly into the ionization area because of the indirect gas ionization. Therefore, a backside illumination through a UV transmissible substrate such as borosilicate glass (BOROFLOAT® 33) is considered to benefit the non-interaction between photons generated by UV-LED and gas molecules in the ionization area, gaining a higher yield of photoelectrons from the emitter (Zimmer et al., 2012). This means that, besides the lanthanum hexaboride, another material has to be chosen as the counter-electrode to the acceleration grid, having high UV transmission behavior and good electrical properties. For this purpose, indium tin oxide (ITO) is a suitable candidate known from the display and solar cell industry.

3 Fabrication of ePIDs by planar technology

3.1 Process flow

The process of an ePID chip starts with the deposition of an indium tin oxide layer on borosilicate glass by the magnetron sputtering technique. Then, lanthanum hexaboride (LaB_6) is sputtered as the emitter layer on top of it. The deposition of an insulating layer of silicon dioxide in a PECVD (plasma enhanced chemical vapor deposition) reactor is followed using silane thinned by argon (SiH_4/Ar) and dinitrogen monoxide (N_2O) as processing gases. The front electrode consists of titanium nitride (TiN), a mechanically stable material. Here, the magnetron sputtering technique is also performed to gain a high layer quality of TiN. Hence, the deposition of tita-

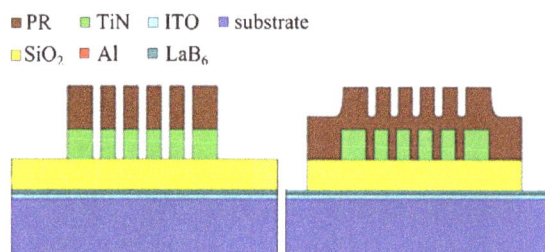

Figure 3. Schematic cross section of the detector chip after the patterning of a front electrode (left) and after etching the mesa (right) (Kallis et al., 2014).

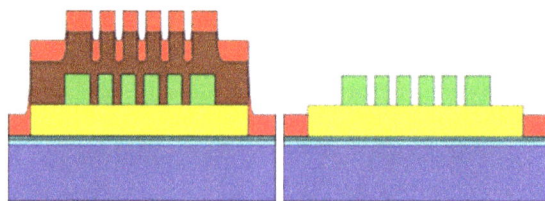

Figure 4. Cross section of the detector chip before (left) and after (right) the lift-off process of aluminum contacts (Kallis et al., 2014).

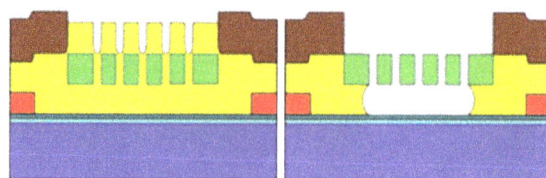

Figure 5. Cross section of the detector chip before (left) and after (right) wet chemical etching of silicon dioxide (Kallis et al., 2014).

Figure 6. Cross section of the detector chip before (left) and after (right) the exposure of contact holes (Kallis et al., 2014).

nium is reactively executed with the addition of nitrogen. Then, this front electrode is anisotropically etched by means of conventional photo lithography in a parallel plate reactor. The appropriate processing gases are nitrogen (N_2), chlorine (Cl_2), silicon tetrachloride ($SiCl_2$) and methane (CH_4) for the sidewall passivation. The cross section of the sensor chip after the patterning of the front electrode is shown in Fig. 3.

Afterwards, an anisotropic etching step forming a mesa structure took place to isolate the chips from each other and contact the bottom electrode (Fig. 3, right). This step is also performed by a parallel plate reactor of a reactive ion etching system using trifluormethan (CHF_3) and argon (Ar) as process gases. The contact with the bottom electrode including lanthanum hexaboride and indium tin oxide, respectively, is realized by lift-off structured aluminum. The cross section of the detector chip before (left) and after (right) the removal of photoresist is shown in Fig. 4.

In the following, a deposition of another PECVD silicon dioxide layer is performed as a passivation film for the aluminum metal circuit paths. As a next step, the front electrode has to be levitated with the help of buffered hydrofluoric acid (BHF) (cf. Fig. 5, right). To protect the passivating oxide and circuit paths against the etching procedure, a resist mask is used during the wet chemical etching (cf. Fig. 5, left).

The process is concluded by exposing the connections of the still fully passivated bottom electrode due to contact holes. A cross section of the complete processed detector chip is shown right in Fig. 6.

3.2 Dimensioning and boundary conditions

In addition to modeling of the front electrode, the optimization of deposited thin films' quality and their dimensioning is also of importance.

3.2.1 Bottom electrode

First, the ITO layer on the glass substrate has to be optimized in terms of its transmission and conductivity. The transmission of ITO in relation to the used wavelength as well as deposited film thickness is shown in Fig. 7. A better transmission is obviously recognizable in thin films for the deep ultraviolet wavelength range.

Furthermore, an increase in transmission is shown with increasing film thickness at 50 nm at a wavelength of approx. 365 nm, which is explainable due to the interference effects at the interface between the substrate and the ITO layer (Zimmer et al., 2013). In accordance with the adjustment of the UV source and the emissive layer, an optimum between transmission and conduction behavior of ITO can be achieved.

3.2.2 Photoelectron emitter

The following layer used as the electron emitter has to be deposited very thinly, guaranteeing the escape of photoelectrons from the front side by backside UV illumination. It is also important to consider that the used lanthanum hexaboride (LaB_6) as emitter material is not totally inert in normal environments, resulting in a slight oxidation of the surface, because adsorbates are collected on the surface during the operation of the ePID chip, inducing the increase in the electron's work function of up to 1 eV (Nishitani et al., 1980). Besides these degenerative effects, the work function is also influenced by the crystallinity of the thin film. But, with an illumination of the emitter layer by a wavelength of 365 nm,

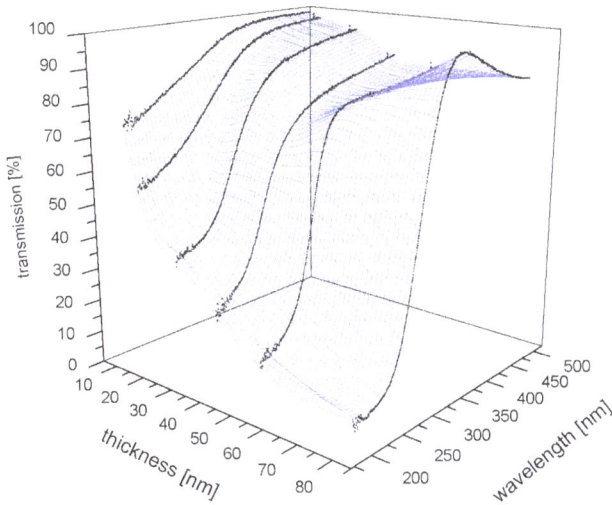

Figure 7. Transmission of the bottom electrode ITO in relation to film thickness and used wavelength (Zimmer et al., 2013).

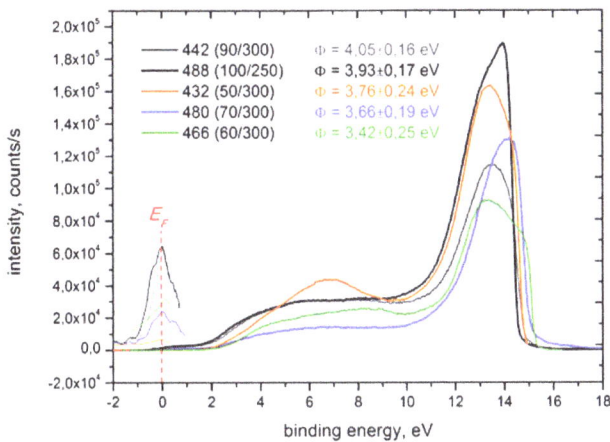

Figure 8. Ultraviolet photoelectron spectra of analyzed LaB$_6$ films (including sample numbers and deposition parameters; argon flow/DC power) with corresponding measured work functions (Zimmer et al., 2014).

the work function should be lower than 3.39 eV. Although work function values below 2.5 eV are specified in the literature (Yamauchi et al., 1976), higher values are presented by both evaporated and sputtered as well as PLD (pulsed laser deposition) deposited layers after storage in air (Zimmer et al., 2013). In this work, LaB$_6$ thin films are deposited by the magnetron sputtering technique using DC power of 300 W and a processing argon flow of 60 sccm. Thereby, as a result, a (100) crystalline texture is reached, benefiting the work function of the material (Yamauchi et al., 1976). Due to decreasing argon flow rate at constant DC power, a reduction in the work function can be seen in Fig. 8, forwarding a higher atomic ratio of boron to lanthanum (≥ 6) and decreasing the internal stress (Zimmer et al., 2014).

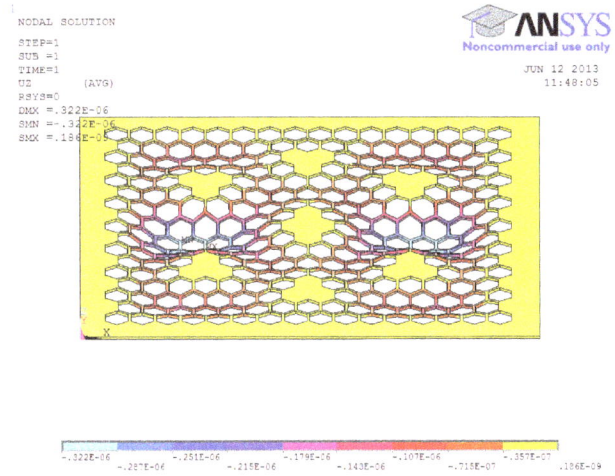

Figure 9. Deformation of a TiN grid structure with supporting elements by maximum electrostatic stress.

On the other hand, a change to other materials is possible, such as samarium oxide (Constantinescu et al., 2012). But, experiences concerning long-term stability and high-quality film depositions are currently still missing.

3.2.3 Insulator

The following deposited silicon dioxide film is used as the insulation layer for the front electrode. In particular, the optimized PECVD oxides are ideally suited for this purpose, because of their dielectric strength in the range of $100\,\mathrm{mV\,nm^{-1}}$. But, for sufficient dimensioning of this area, the mechanical deformation of the front electrode is also a decisive factor, as well as the dielectric strength. Since the acceleration path is used in air, a maximum layer thickness of 500 nm must not be exceeded, which corresponds to the free mean path of electrons in air (Seah and Dench, 1979). The insulator is also used as a buffer layer against the mechanical stress that occurred between the ITO and the TiN electrode. The deposition of SiO$_2$ is performed by a combination of high- and low-frequency plasma excitation to decrease the intrinsic layer tension at greater film thicknesses (Tarraf et al., 2004).

3.2.4 Front electrode

From the mechanical point of view, the front electrode is the most stressed part of the detector chip. To gain the highest electron yield possible, the grid has to be realized as almost self-supporting. Furthermore, it has to carry its own weight completely and has to compensate for the electrostatic forces due to the applied acceleration voltage. For this purpose, a honeycombed grid structure with supporting elements of silicon dioxide is used, which is optimized with the help of the finite element method (see Fig. 9, Kallis et al., 2014).

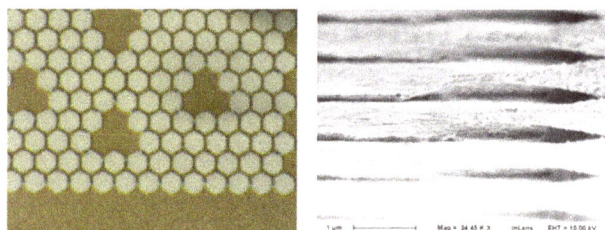

Figure 10. Image of a collapsed ePID chip taken by microscope (left) and scanning electron microscope (right).

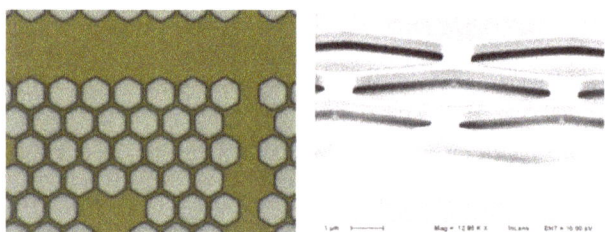

Figure 11. Image of an optimized ePID chip taken by microscope (left) and scanning electron microscope (right, Kallis et al., 2014).

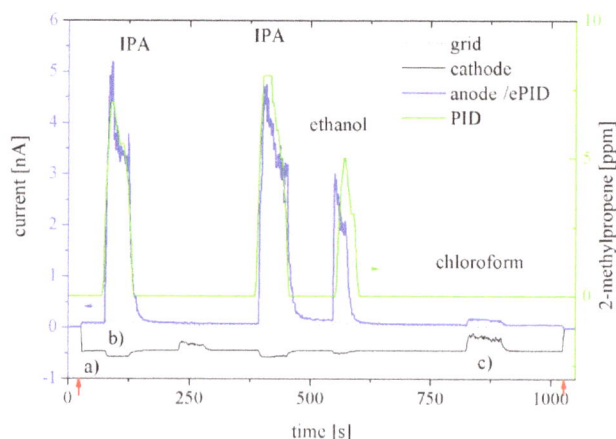

Figure 12. Measured ionization signal of PID calibrated to 2-methylpropene (green signal) and compared to an ePID (blue signal). (**a**) Emission of photoelectrons out of the cathode, (**b**) ionization by photons, and (**c**) ionization by emitted photoelectrons (Zimmer et al., 2012).

However, the biggest challenge of the front electrode is described by its fabrication considering the operation conditions. Due to an incorrect dimensioning or the use of unsuitable materials, the grid structure can be collapsed irreversibly either during fabrication or operation by applying acceleration voltage. Thereby, an optimum adjustment between the grid thickness and the applied voltage has to be made.

By applying excessive voltages, the grid structure collapses, which is shown exemplarily in Fig. 10. Here, strong deformations are already divined by the microscope (left) based on the irregular reflection behavior of the front electrode. This assumption is confirmed by the oblique view of the chip surface taken by a scanning electron microscope (SEM). Obviously, a short-circuit is performed due to adhesion of the front electrode on the lower electrode by leaving the area of reversible elastic deformation. Besides the exact geometric dimensioning, the use of complete stoichiometric titanium nitride is essential because of its optimal mechanical stiffness, preventing the collapses of the front electrode. The image of a perfect ePID chip taken by both a microscope (left) and a scanning electron microscope (right) is shown exemplarily in Fig. 11. The almost levitating grid is only stabilized by its frame structure and the supporting points, where no deformation occurred.

4 Discussion

The processes presented in the previous sections were investigated with the help of electrical and mechanical simulations at the Technology of Research and Application Development of TU Dortmund, and required that processes were optimized continuously. Hereby, the technical feasibility as well as the superiority of such a detector is already presented by given prototypes (Zimmer et al., 2012).

First, the ionization behavior of the ePID is registered and compared to the PID used as the reference signal. During ionization measurement, the same gas discharge lamp with a wavelength of 116 nm is used as the UV source to guarantee the photoelectric effect in the ePID. The detected signal of evaporated solvents by both detectors is shown in Fig. 12. Hereby, the majority of gaseous substances is detected by an ePID (see the blue signal in Fig. 12), having a downstream acceleration path instead of a classical PID.

Both 2-propanol (IPA, C_3H_7OH) and ethanol (C_2H_5OH) with ionization potentials of 10.15 and 10.5 eV, respectively, are detected by PID and ePID due to the dominating ionization process by photons. But, only the ePID is able to deliver a signal by introducing evaporated chloroform ($CHCl_3$) into the measuring cell, which has an ionization potential of 11.4 eV, exceeding the used photon energy of 10.6 eV (Zimmer et al., 2012). Here, only photoelectrons emitted by the gas discharge lamp (10.6 eV) ionize chloroform molecules shown as a signal in Fig. 12c. The ionization behavior of the ePID is also successfully performed by the use of a UV-LED ($\lambda = 260$ nm) as a radiation source (Zimmer et al., 2012). But, the achieved signal strength is still too low and has to be improved by further optimization of used masks and materials according to Sect. 3.2. Thus, additional investigations have to be done, particularly concerning the dependence of gas selectivity on the variation of acceleration voltage. Presumably, there is an offset between the applied grid voltage and the gas' ionization potential due to existing contaminations on diverse surfaces performing uncontrollable ionization effects. In particular, the possible formation of ions caused by clustering with water molecules has to be analyzed in detail by spectrometric methods.

5 Conclusion and outlook

In this paper, the basics for the dimensioning and fabrication of novel and innovative electronic supported photoionization detectors (ePID) has been presented.

Here, ePIDs point out further advantages, including the enormous cost reduction due to the application of planar technological processes known from chip manufacture compared to already existing sensors. A wide spectrum of application fields is especially feasible due to a higher selectivity by the tunability of the sensor chip as well as the expanded ionization range by a factor of 2 compared with UV lamps. Despite the proven basic functionality of an ePID on a laboratory scale, a long way remains for optimization and further development until its realization in a finished product, in detail, mainly the optimization of the electronic work function of the emissive layer regarding the long durability for the possible use of low-cost UV-LEDs as light sources. Another important aspect in further investigations is the additional integration of capturing electrodes into the chip, which still have to be built up from discrete parts.

Acknowledgements. The authors would like to thank Karola Kolander, Marko Kremer and Achim Wiggershaus for the technical support with the manufacture of prototypes.

In addition, the persons concerned would like to thank the Federal Ministry of Education and Research for the financial support of the study.

References

Constantinescu, C., Ion, V., Galca, A. C., and Dinescu, M.: Morphological, optical and electrical properties of samarium oxide thin films, Thin Solid Films, 520, 6393–6397, 2012.

Constapel, M., Schellenträger, M., Schmitz, O. J., Gäb, S., Brockmann, K. J., Giese, R., and Benter, Th.: Atmospheric-pressure laser ionization: A novel ionization method for liquid chromatography/mass spectrometry, Rapid Commun. Mass Sp., 19, 326–336, 2005.

Daum, K. A., Watrous, M. G., and Neptune, M. D.: Data for First Responder Use of Photoionization Detectors for Vapor chemical Constituents, Idaho National Laboratory, Idaho Falls, Idaho, USA, 2006.

Eiceman, G. A. and Karpas, Z.: Ion Mobility Spectrometry, 3rd Edn., Taylor & Francis Group, New York, USA, 2013.

Freedman, A. N.: The photoionization detector: Theory, performance and application as a low-level monitor of oil vapour, J Chromatogr. A, 190, 263–273, 1980.

Kallis, K. T., Dietz, D., Subasi, E., Müller, M. R., Kontis, C., and Zimmer, C. M.: Design, simulation, fabrication and characterization of nano-scaled acceleration grids, Microelectronic Engineering, 121, 118–121, 2014.

Nishitani, R., Aono, M., and Tanaka, T.: Surface structures and work functions of the LaB_6 (100), (110) and (111) clean surfaces, Surf. Sci., 93, 535–549, 1980.

RAE Systems Inc.: The PID Handbook: Theory and applications of direct reading photoionization detectors (3rd Edn.), ISBN 0-9768162-1-0, 2013.

Seah, M. P. and Dench, W. A.: Quantitative electron spectroscopy of surfaces: A standard data base for electron inelastic mean free paths in solids, Surf. Interf. Anal., 1, 2–11, 1979.

Tarraf, A., Daleiden, J., Irmer, S., Prasai, D., and Hillmer, H.: Stress investigation of PECVD dielectric layers for advanced optical MEMS, J. Micromech. Microeng., 14, 317–323, 2004.

Yamauchi, H., Takagi, K., Yuito, I., and Kawabe, U.: Work function of LaB_6, Appl. Phys. Lett., 29, 638, 1976.

Zimmer, C. M., Kieschnick, M., Kallis, K. T., Schubert, J., Kunze, U., and Doll, T.: Nano photoelectron ioniser chip using LaB_6 for ambient pressure trace gas detection, Microelectron. Eng., 98, 472–476, 2012.

Zimmer, C. M., Asbeck, C., Lützenkirchen-Hecht, D., Glösekötter, P., and Kallis, K. T.: Backside illumination of an electronic photo ionization detector realized by UV transparent thin films, J. Nano Res., 25, 55–60, 2013.

Zimmer, C. M., Yoganathan, K., Giebel, F. J., Lützenkirchen-Hecht, D., Glösekötter, P., and Kallis, K. T.: Photoemission properties of LaB_6 thin films for the use in PIDs, Proceedings of the 14th IEEE International Conference on Nanotechnology, Toronto, Canada, 18–21 August 2014, 877–881, 2014.

Influence of the substrate on the overall sensor impedance of planar H$_2$ sensors involving TiO$_2$–SnO$_2$ interfaces

L. Ebersberger and G. Fischerauer

Chair of Metrology and Control Engineering, Universität Bayreuth, Bayreuth, Germany

Correspondence to: L. Ebersberger (mrt@uni-bayreuth.de)

Abstract. To date, very little has been written about the influence of the substrate layer on the overall sensor impedance of single- and multilayer planar sensors (e.g., metal-oxide sensors). However, the substrate is an elementary part of the sensor element. Through the selection of a substrate, the sensor performance can be manipulated. The current contribution reports on the substrate influence in multilayer metal-oxide chemical sensors. Measurements of the impedance are used to discuss the sensor performance with quartz substrates, (laboratory) glass substrates and substrates covered by silicon-dioxide insulating layers. Numerical experiments based on previous measurement results show that inexpensive glass substrates contribute up to 97 % to the overall sensor responses. With an isolating layer of 200 nm SiO$_2$, the glass substrate contribution is reduced to about 25 %.

1 Introduction

A metal-oxide sensor consists of a substrate (typically aluminum oxide, which, however, cannot be used for thin-film sensors owing to its high surface roughness), electrodes (typically aluminum or gold) and an active layer (differs in composition by application, mostly SnO$_2$ doped in one way or another). In the open literature, the substrate is listed as part of the sensor element (Barsan et al., 2007; Barsan and Weimar, 2001), but the research focusses on the active layer in the majority of cases and sometimes also on the electrodes. If at all, the substrate is only treated as a thermally important component (Simon et al., 2001). Rarely is it discussed with a focus on other aspects, like the influence of the substrate on the growth of the crystalline phases (Mardare et al., 2008). However, the substrate could play a decisive role in the overall impedance of the sensor system (Fischerauer et al., 2009). This contribution focusses on the influence of the substrate on the overall sensor impedance of planar (thin-film) hydrogen (H$_2$) sensors involving titanium-dioxide (TiO$_2$) stannic-dioxide (SnO$_2$) interfaces.

2 Sensor geometry

We have investigated three different types of sensor structures, called types A, B and C. All of them consist of planar interdigital electrodes (IDE) made of 200 nm ion-beam-deposited aluminum and respectively featuring 55 and 56 electrode fingers with a finger width of 15 μm and an interfinger space of 25 μm (Fig. 1). The finger length was 3450 μm. The substrate consisted of glass without an insulating layer (type A) or glass coated with an insulating layer of 200 nm reactive ion-beam-sputtered silicon dioxide (SiO$_2$, type B) or quartz (type C).

The 80 nm thick TiO$_2$ layer covers the top of all electrode fingers and 5 μm to their left and their right. The SnO$_2$ layer covers all electrodes, the TiO$_2$ layer and the space between the fingers (Fig. 2). A wide area of the busbar (see Fig. 2, left) is left blank for wire bonding.

The metal oxides were made by ion-beam deposition of the metals (titanium respectively tin) and thermal oxidation immediately afterwards. The oxidation was performed in a glass furnace which was heated to 500 °C in ambient atmosphere.

Table 1. Comparison between sensor types.

	Type A	Type B	Type C
Active layer		TiO_2-SnO_2-multilayer	
Electrodes		aluminum IDE	
Number of fingers		55/56	
Finger length		$3450 \, \mu m$	
Substrate	glass	glass + 200 nm SiO_2	quartz

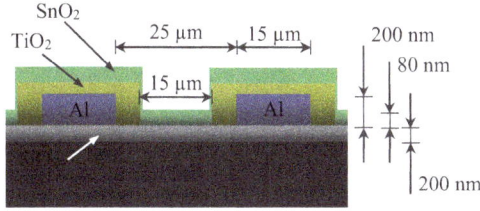

Figure 1. Schematic cross section of the sensor structure (type B).

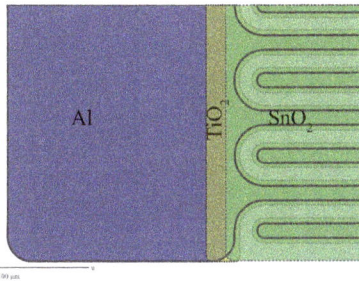

Figure 2. Schematic detail of the busbar area (blue: Al) and covering metal oxides (olive green: TiO_2, green: SnO_2) in a view from above. The darker parts shining through the thin films on the right-hand side are the electrode fingers.

The thermal postprocessing could lead to a doping of the active layers with aluminium from the electrodes due to the absence of a diffusion barrier layer. Furthermore, a diffusion of titanium to the tin-oxide layer (and reverse) could not be prevented. But as similar measurement results have been obtained with thermally oxidized Ti and with reactively ion-beam-sputtered TiO_2 layers, such a diffusion process appears to be insignificant.

As to the surface roughness of the metal oxides, it is caused completely by the thermal treatment of the sensors. Straight after the sputtering process the roughness was negligible and increased by the thermal oxidization.

3 Measurement setup

To characterize the sensors, they were mounted on the lower part of a DIL-14 metal case (with six grounded pins) and placed inside a measurement chamber, which in turn was placed in a temperature-controlled oven (Fig. 3). The chamber itself consisted of stainless steel and contained a base

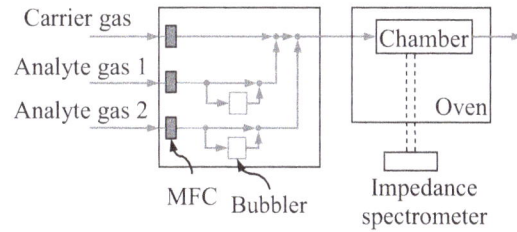

Figure 3. Schematic drawing of the test bed used to characterize the sensors.

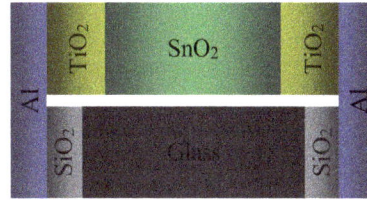

Figure 4. One-dimensional approximation of basis cell of an IDE-based gas sensor (type B) (Fischerauer et al., 2011).

out of Macor (Corning Inc.) which holds the metal case and contacts the pins of the sensor package. The atmosphere in the measurement chamber could be controlled via custom-specific gas-mixing equipment involving mass-flow controllers. The chamber was first heated to 270 °C in a nitrogen (N_2) atmosphere. When the steady state was reached, selected amounts of hydrogen and water were added to the nitrogen carrier gas. The temperature of 270 °C was chosen, because it is known that at least TiO_2 nanotubes are sensitive to H_2 in this temperature region (Varghese et al., 2003). The overall gas flow was held constant at $500 \, cm^3 \, min^{-1}$. For lower concentration levels, the analytes were partly pre-thinned to 5 % in nitrogen carrier gas. A bubbler was used for the humidification of the gas.

The sensor response to concentration steps in hydrogen or H_2O was observed via the electrical terminals of the sensors considered as two-terminal devices. These terminal characteristics were measured by impedance spectrometers (Meodat Impspec LF HF or Agilent E4980A) in the frequency range from at least 20 Hz to 10 kHz. The frequencies were logarithmically equidistantly spaced (typically 30 points per decade).

To model the sensor element, a one-dimensional approximation with a geometry comprising two parallel tracks was used as described in Fischerauer et al. (2011) and illustrated in Fig. 4.

The overall impedance $\underline{Z}(f)$ is then calculated as

$$\underline{Z}(f) = \frac{\underline{Z}_a(f) \cdot \underline{Z}_s(f)}{\underline{Z}_a(f) + \underline{Z}_s(f)}, \tag{1}$$

where $\underline{Z}_a(f)$ denotes the impedance of the active layer and $\underline{Z}_s(f)$ is the impedance of the substrate (including insulating layers).

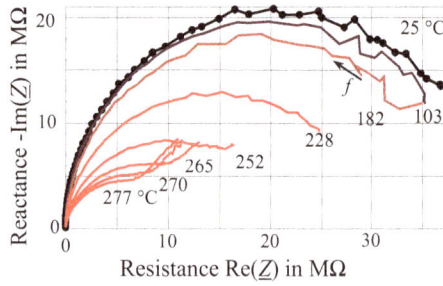

Figure 5. Nyquist plot of the measured impedance of an unbonded sensor in its test cell at various temperatures. The frequency was swept from 10.3 Hz to 10 kHz. For the temperature of 25 °C, the measurement points are also plotted. At the highest temperature, different levels of H_2 were added to the nitrogen.

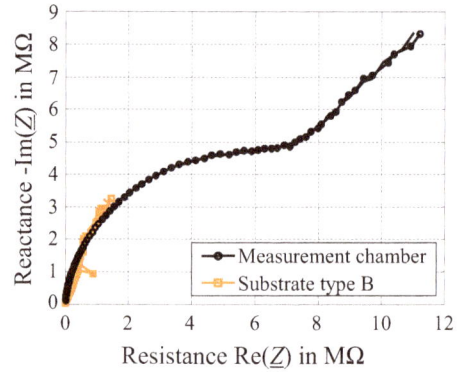

Figure 6. Nyquist plot of the measured impedance of of a type-A sensor element without active layers (pure IDE on glass substrate). Again, the frequency was swept from 10.3 Hz to 10 kHz. At the peak temperature of 247 °C H_2O was added, but produced no effect. For the 218 °C curve, the inflexion frequency f_b has been marked; it will be discussed later.

It turned out, however, that the test environment also contributed to the device response. To include this influence (as a basis for an eventual de-embedding procedure), Eq. (1) is modified to

$$\underline{Z}(f) = \qquad\qquad\qquad\qquad\qquad\qquad (2)$$
$$\frac{\underline{Z}_a(f) \cdot \underline{Z}_s(f) \cdot \underline{Z}_T(f)}{\underline{Z}_a(f) \cdot \underline{Z}_s(f) + \underline{Z}_s(f) \cdot \underline{Z}_T(f) + \underline{Z}_a(f) \cdot \underline{Z}_T(f)},$$

where $\underline{Z}_T(f)$ denotes the impedance contribution of the test environment (including cable capacitances and insulator resistances). This model is solely based on the parallel connection of lumped elements and as such is quite crude. The limitations of the simple network are made up for by allowing all network elements to be frequency-dependent. This procedure is common practice in impedance spectroscopy. A more detailed mechanistic model of the test environment that would improve the correction even further is out of focus for this paper.

4 Measurement results

The measurement results will be presented in three steps. First, in Sect. 4.1, the characteristics of the bare substrate

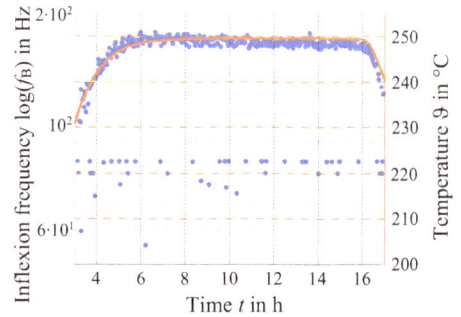

Figure 7. Comparison of type-B sensor impedance (element without any active layer) $\underline{Z}(\underline{Z}_T, \underline{Z}_{S,B})$ and the measured impedance of the empty test chamber $\underline{Z}(\underline{Z}_T)$ at an elevated temperature ($\vartheta = 280$ °C).

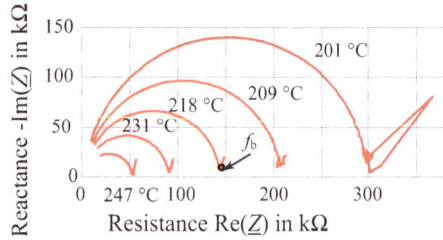

Figure 8. Time series of the inflexion frequency f_B and the temperature as measured with a thermocouple inside the measurement chamber.

will be discussed. Next, Sect. 4.2 focusses on the influence of humidity on the sensor signal. Finally, the key aspect of Sect. 4.3 is the influence of insulating layers or substrates on the sensor response.

4.1 Bare substrate

To estimate the influence of the test environment, unbonded sensors were characterized. In this manner, all aspects of the test environment (contact resistances, cable capacitances, etc.) can be quantified for later de-embedding of measurement data obtained with sensors.

As one can see in Fig. 5, even at elevated temperatures, the insulation of the chamber and the package is very high. As expected, it does not respond to H_2, which means that neither Macor nor other insulation materials are sensitive to it. In addition, neither the materials (Macor, insulating materials) nor the substrate is sensitive to H_2O. The fact, that no effect of the water is visible in Fig. 6, leads to the conclusion that none of the materials (Macor, insulation, substrate) causes an H_2O sensitivity of the sensors.

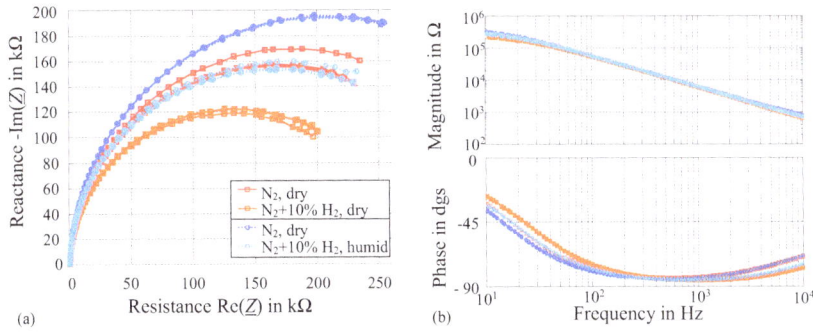

Figure 9. Measured type-B sensor impedance $\underline{Z}(\underline{Z}_T, \underline{Z}_{S,B}, \underline{Z}_a)$ in various atmospheres. All curves have been obtained with the same sensor specimen measured at various times. Case 1: N$_2$ with or without dry H$_2$ (orange and red solid lines, respectively). Case 2: N$_2$ with or without humid H$_2$ (light blue and dark blue dashed lines, respectively). Frequency from 10.3 Hz to 10 kHz. **(a)** Nyquist diagram and **(b)** Bode diagram.

As revealed by Figs. 5 and 6 , the substrate (glass) conductivity of a type-A sensor is about 100 times higher than that of the measurement environment at the same temperatures. Figure 7 demonstrates that the type-B substrate (insulated glass) leads to a slightly higher shunt conductance (parallel to the coated IDT) than the test bed alone. This means that the test bed influence can be neglected for "high-conductivity" substrates as in type-A sensors, but must be taken into account with "low-conductivity" substrates as in type-B sensors.

Apart from this, the results clearly show that a layer thickness of about 200 nm (atomic force microscopy (AFM) measurements yielded a layer thickness, for the given sensor, of 175 ± 5 nm) increases the resistivity of the substrate sufficiently.

To estimate the magnitude of the influence, we performed a numerical experiment based on measurement data for the test environment with a type-A substrate (yielding $\underline{Z}(\underline{Z}_T, \underline{Z}_{S,A})$), a type-B substrate (yielding $\underline{Z}(\underline{Z}_T, \underline{Z}_{S,B})$), and a type-B sensor (yielding $\underline{Z}(\underline{Z}_T, \underline{Z}_{S,B}, \underline{Z}_a)$) at 240 °C. The contributions of the active layer $\underline{Z}_a(f)$, the substrate A $\underline{Z}_{S,A}(f)$, the substrate B $\underline{Z}_{S,B}(f)$, and the test environment $\underline{Z}_T(f)$ were calculated by solving Eq. (2). To simulate a test gas, the impedance of the active layer $\underline{Z}_a(f)$ was increased by 100 %, leading to $\underline{Z}_{a,g}(f) = 2 \cdot \underline{Z}_a(f)$. The computations revealed that a sensor of type A would respond to this by an impedance increase of 3 % at low frequencies (10.3 Hz) and up to 92 % at high frequencies (10 kHz). In contrast, a type-B sensor would increase its impedance by 75 % at low frequencies (10.3 Hz) and 92 % at high frequencies (10 kHz). Type-C sensors would respond by even higher impedance increases. However, their low-conductivity substrate could not be measured reliably. The type-C sensors are further discussed in Sect. 4.3.

As introduced in Fischerauer and Fischerauer (2011), the inflexion frequency in the Nyquist plot of the sensor impedance (cf. Fig. 5) strongly depends on temperature. Because the dependency is non-linear, this could be used to determine the temperature of the sensor element (cf. Fig. 8).

The outliers in this plot are due to noise which perturbs the reconstruction algorithm for the inflexion frequency.

4.2 Influence of humidity

Figure 9 shows measurement results obtained for type-B sensors in various atmospheres (N$_2$, with dry H$_2$, with H$_2$ plus water). At first sight, one is inclined to think that the sensor resistance increases with humidity. In fact, however, humidity changes the sensor response to H$_2$ only very little. The seeming humidity influence is due to the baseline drift inherent to the present sensors and resulting in time-variant impedance levels. We attribute the baseline drift to aging, relaxation, poisoning or similar effects during the time interval between the measurements of cases 1 and 2 (about 6 h). Temperature fluctuations can be ruled out as source of the drift because they were held below 0.5 °C.

The assertation that humidity does not change the sensor response is corroborated by the fact that the *shape* of the Nyquist plot does not change with humidity and that characteristic quantities such as the sensor resistance, at say 10 Hz, vary with H$_2$ content, but not with humidity (said resistance decreases by 15 % when 10 % H$_2$ is added to the N$_2$ atmosphere).

4.3 Influence of insulating substrates

To estimate the influence of the substrate, different sensors were characterized. Figure 10 shows the results for a type-B sensor (insulated glass substrate) as compared to a type-C sensor (quartz substrate). Although the impedance level of the former is lower by about 30 %, it still is in the same range as that of the latter. The variation visible in the figure is caused by noise, not by drift. No recognizable drift could be identified during the illustrated detail, because the repeatedly measured spectra did not "move" in a specific direction.

Thus, even a thin layer of SiO$_2$ may suffice to make an inherently unsuitable, but inexpensive substrate (glass) suitable

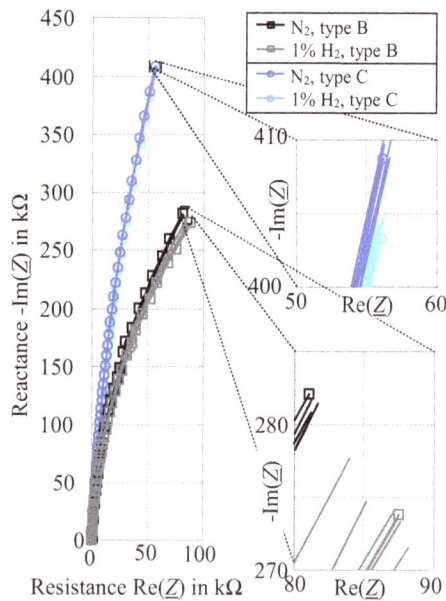

Figure 10. Measured impedances of a type-B sensor (squares) and a type-C sensor (circles). The dark symbols refer to measurement in pure N_2 repeated 7 times for each sensor type. The light symbols to measurement in N_2 plus 1 % H_2 (also repeated 7 times). Frequency from 20 Hz to 10 kHz. Time interval between the repeated measurements was 2 min.

for the present purpose and a serious competitor of a superior, but more expensive substrate such as quartz. Anyway, the impedance level of a type-B sensor, described by, e.g., $\left|\underline{Z}(f = 20\,\mathrm{Hz})\right|$, exceeds that of its type-A sensor counterpart by a factor of about 5.

Besides this, it can also be concluded from the curves in Fig. 10 that the response of the type-B and type-C sensors to hydrogen is similar. The respective reactances decrease by about 2 and 1 % in the two cases. We would not claim that type-B sensors systematically respond to H_2 more strongly than type-C sensors. Aging and other effects like fabrication tolerances of the layer thicknesses could have a non-negligible impact on the sensor response.

We assume that the influence of the more conductive substrates (type-A) are caused by the ionic conductivity of the glass at elevated temperatures. For type-B sensors the amount of insulating SiO_2 suffice to disrupt this type of conductivity through the substrate. Quartz (type-C) is considered to be a nearly ideal insulator. We assume that the small differences between type-B and type-C substrates are caused by impurities in the insulating SiO_2 layer which enables a minimum of ionic conductivity through the glass substrate.

5 Conclusion

We have demonstrated experimentally that the substrate plays an essential role for the overall performance of planar H_2 sensors involving TiO_2 / SnO_2 thin films. Aspects of

the impedance spectrum (the inflexion frequency) of high-conductivity substrates (like glass, an ion conductor at elevated temperatures) strongly depend on temperature and could therefore be used to recognize even small variations of the temperature.

Numerical experiments showed that high-conductivity substrates like glass can dominate the overall sensor response (impedance change) by 97 % at a frequency of 10.3 Hz. But a thin layer of SiO_2 (200 nm) suffices to increase the impedance level. Thereby just about 25 % of the impedance increase is hidden by the substrate influence. Only sensors based on a highly insulating substrate like quartz can excel this.

Thin-film gas sensors do not need special properties of the substrate (like crystallinity) except the compatibility with thin-film technology and mechanical stability. Hence, there is no need to use expensive substrates such as quartz. By depositing silicon dioxide on top of the substrate, even inexpensive substrates like glass can be made to work.

Sensors with both conducting substrate plus insulating thin film and insulating substrate respond to H_2 very similarly. Therefore, it can be excluded that the tested substrates influence the sensing mechanism in an unacceptable manner. Furthermore, neither the test bed nor the sensors showed a cross-sensitivity to humidity.

Acknowledgements. The authors gratefully acknowledge support by the German Research Foundation (DFG) under contract number Fi 956/4-1.

References

Barsan, N. and Weimar, U.: Conduction Model of Metal Oxide Gas Sensors, J. Electroceram., 7, 143–167, doi:10.1023/A:1014405811371, 2001.

Barsan, N., Koziej, D., and Weimar, U.: Metal oxide-based gas sensor research: How to?, Sensor. Actuat. B-Chem., 121, 18–35, doi:10.1016/j.snb.2006.09.047, 2007.

Fischerauer, A. and Fischerauer, G.: Physikalisches Modell der Materialparameterabhängigkeit des Impedanzspektrums planarer Chemosensoren in Mehrschichtbauweise, tm – Technisches Messen, 78, 15–22, doi:10.1524/teme.2011.0074, 2011.

Fischerauer, A., Schwarzmüller, Ch., and Fischerauer, G.: Substrate influence on the characteristics of interdigital-electrode gas sensors, Proc. Sixth Int'l Multi-Conf. on Systems, Signals & Devices (SSD'09), Djerba, 5 pp., doi:10.1109/ssd.2009.4956738, 23–26 March, 2009.

Fischerauer, A., Fischerauer, G., Hagen, G., and Moos, R.: Integrated impedance based hydro-carbon gas sensors with Na-zeolite/Cr2O3 thin-film interfaces: From physical modeling to devices, Phys. Status Solidi A, 208, 404–415, doi:10.1002/pssa.201026606, 2011.

Mardare, D., Iftime, N., and Luca, D.: TiO$_2$ thin films as sensing gas materials, J. Non-Cryst. Solids, 354, 4396–4400, doi:10.1016/j.jnoncrysol.2008.06.058, 2008.

Simon, I., Bârsan, N., Bauer, M., and Weimar, U.: Micromachined metal oxide gas sensors: opportunities to improve sensor performance, Sensor. Actuat. B-Chem., 73, 1–26, doi:10.1016/S0925-4005(00)00639-0, 2001.

Varghese, O. K., Gong, D., Paulose, M., Ong, K. G., and Grimes, C. A.: Hydrogen sensing using titania nanotubes, Sensor. Actuat. B-Chem., 93, 338–344, doi:10.1016/S0925-4005(03)00222-3, 2003.

14

Investigations into packaging technology for membrane-based thermal flow sensors

G. Dumstorff, E. Brauns, and W. Lang

Institute of Microsensors, -actuators, and -systems (IMSAS), Microsystems Center Bremen, University of Bremen, Bremen, Germany

Correspondence to: G. Dumstorff (gdumstorff@imsas.uni-bremen.de)

Abstract. A new packaging method to mount a membrane-based thermal flow sensor, flush with the surface, is presented. Therefore, a specific design for the housing is shown, which is also adaptable to other conditions. It has been experimentally shown that it is important to mount the sensor flush with the surface. In addition, the experimental results are discussed. If a membrane-based thermal flow sensor is not mounted flush with the surface, vortices can occur (depending on velocity and fluid properties) or the reduction in the channel cross section plus a decrease in sensitivity have to be taken into account.

1 Introduction

Over the last decades, many different thermal flow sensors have been developed for the accurate measurement of gas flows (Van Putten and Middelhoek, 1974; Van Herwaarden and Sarro, 1986; Johnson and Higashi, 1987; Ashauer et al., 1999; Mailly et al., 2001; Ito et al., 2010; Adamec and Thiel, 2010; Ma et al., 2009). A wide range of flow sensors have been developed for different applications, such as automotive, industrial, and medical. Commercial flow measurement systems using MEMS (micro-electro-mechanical systems) devices are also available (sen, http://www.sensirion.com/; axe, http://www.axetris.com/; ist, http://www.ist-ag.com/; hon, https://www.honeywellprocess.com).

Besides sensor fabrication, the packaging technology is equally essential for building a flow measurement system. Here are some reasons underpinning the importance of flow sensor packaging.

- The chip has to be placed in a fluidic channel. Turbulence caused by, e.g., sharp edges of the chip, should be avoided. Eberhardt et al. (2003) showed that there is a point of discontinuity in the characteristic curve, because of turbulence at the edge of the flow sensor (Eberhardt et al., 2003). Due to a misalignment of the flow sensor in the channel, the sensitivity can significantly decrease.

- The membrane should be pressure equalized. If not, the membrane may be deflected and the characteristic curve may change. In addition, the heater resistance increases because of the piezoresistive effect. This can have an influence on the heater regulation.

- The flow system has to be sealed. Otherwise, the measurement results will be biased.

For the building up of a flow measurement system, different approaches have been developed over the years. A common approach relies on a monolithically integrated channel implemented at wafer level. Different types of these channels have been developed in the past years (Liu et al., 2012; Dijkstra et al., 2008; Buchner et al., 2007; Billat et al., 2008). The main disadvantages of such monolithically integrated microchannels are the need for an expensive wafer area, less material choice, and the fact that, due to the microchannels, large flow rates can only be measured by means of the bypass technique. In addition, packaging is still necessary to connect the tiny fluidic inlets and outlets, which could be a tricky task. Another approach to building a sensor system is in the post-processing, where a flow sensor is integrated into a housing. This is what we focus on in this paper. In most publications, post-processing is only superficially treated, and sensor design, technology and performance are discussed in detail (Dillner et al., 1997; Sabaté et al., 2004; Bruschi et al., 2005). Different concepts for a packaging technology have already

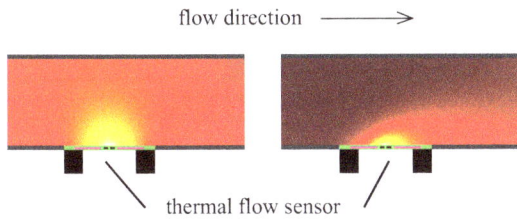

Figure 1. Illustration of the thermal field of a flow sensor. Left: $\dot{V} = 0$; right: $\dot{V} \neq 0$

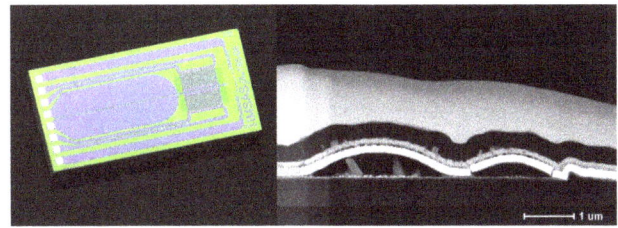

Figure 2. Left: the IMSAS flow sensor; right: the tungsten nitride layer is spalling because of stress and thermal shock in the LPCVD passivation process.

been presented, e.g., by Bruschi et al. (2009) or Kaltsas et al. (2002), but detailed information about housing design and the packaging process can not be found. Furthermore, there are no investigations on how misarrangement of the sensor can affect the characteristic curve. Of course, no detailed information on commercial flow measurement systems like sen, axe, ist and hon can be found (see the Introduction, first paragraph, p. 1).

For these reasons, we will first focus on the technology of packaging and housing for a precise assembly of a flow sensor chip in a channel. Afterwards we will demonstrate and discuss the importance of mounting the flow sensor flush with the surface. Therefore, we assembled flow sensors in a channel with a known misalignment.

2 IMSAS flow sensor

2.1 Theory

The IMSAS flow sensor (Fig. 2) is based on the calorimetric principle. It consists of a silicon nitride membrane with two thermal sensing elements and a heater in between. A membrane is essential for a high-performance thermal flow sensor because of the thermal decoupling of the functional elements. This in combination with a small thermal mass is the reason for a high sensor performance. When the heater is switched on and the volume flow $\dot{V}$ is equal to zero, a symmetric thermal field is generated in the flow channel (see Fig. 1, left). The temperature difference between the thermal sensing elements is zero. By applying a pressure difference between the channel inlet and the channel outlet, the thermal field becomes asymmetric and a temperature difference is measured (see Fig. 1, right). The temperature measurement is done by thermopiles based on the SEEBECK effect. While the hot junction of the thermopile is placed at a defined distance d_{HT} from the heater, the cold junction is placed on the bulk silicon, which acts as a heat sink due to its high thermal conductivity. A mathematical model of the IMSAS flow sensor can be found in Sosna et al. (2008). To get higher output voltages, several thermopiles are connected in series.

2.2 Fabrication

An IMSAS flow sensor and its fabrication were first presented in Buchner et al. (2006). The base of this sensor is a 600 nm thick membrane (800 μm long and 600 μm wide in the flow direction) made of low-stress silicon-rich LPCVD-Silicon-Nitride (LPCVD-SiN). All functional structures are passivated by this silicon nitride membrane. The thermocouples are made of boron-doped poly-silicon and tungsten–titanium as well as a heater and a thermistor. The SEEBECK coefficient of one thermopile is 0.24 mV K^{-1}; 15 thermopiles are connected in series. Connecting thermopiles in series has an impact on the accuracy: while the output voltage increases with n (the number of thermopiles), the Nyquist noise increases with $\sqrt{n}$. The single chips have a length of 3.5 mm and a width of 1.8 mm and consist of eight bond pads. Over the past years, new developments have been introduced into the process flow. The most important innovation is a high-temperature stable diffusion barrier at the junction between the metal and the semiconductor, made of tungsten nitride. Without the tungsten nitride layer, it would not be possible to passivate the functional structures in the LPCVD process because diffusion would destroy the junction, which can be seen in Fig. 2. The fabrication processes are described and illustrated in Fig. 3.

3 Packaging technology

3.1 Housing

The housing, shown in Fig. 4, consists of two parts: A is the base part and B is the flow channel part. In the base part, a 500 μm deep rectangular cavity (A1, Fig. 4) is integrated to accommodate the flow sensor (C, Fig. 4). The cavity is made 400 μm wider than the sensor. No influence on the flow, e.g., turbulence, is expected. A further cylindrical cavity (500 μm deep, 1 μm in diameter) is placed inside the rectangular cavity in the membrane position (A2, Fig. 4). This cylindrical cavity is important for the pressure equalization. It should prevent adhesive flows through capillary forces under the membrane from sealing the back volume of the flow sensor. Of course, due to this setup, there is a bypass under the membrane. From simulations of the geometry, it was found that

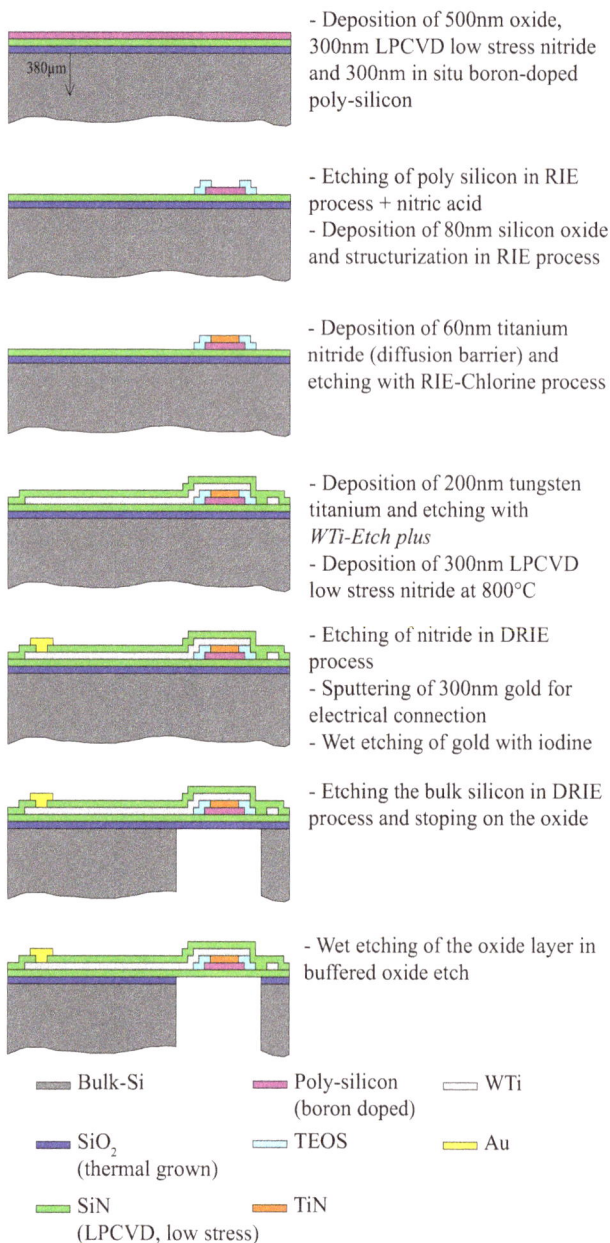

- Deposition of 500nm oxide, 300nm LPCVD low stress nitride and 300nm in situ boron-doped poly-silicon

- Etching of poly silicon in RIE process + nitric acid
- Deposition of 80nm silicon oxide and structurization in RIE process

- Deposition of 60nm titanium nitride (diffusion barrier) and etching with RIE-Chlorine process

- Deposition of 200nm tungsten titanium and etching with *WTi-Etch plus*
- Deposition of 300nm LPCVD low stress nitride at 800°C

- Etching of nitride in DRIE process
- Sputtering of 300nm gold for electrical connection
- Wet etching of gold with iodine

- Etching the bulk silicon in DRIE process and stoping on the oxide

- Wet etching of the oxide layer in buffered oxide etch

Figure 3. Process flow of the IMSAS flow sensor.

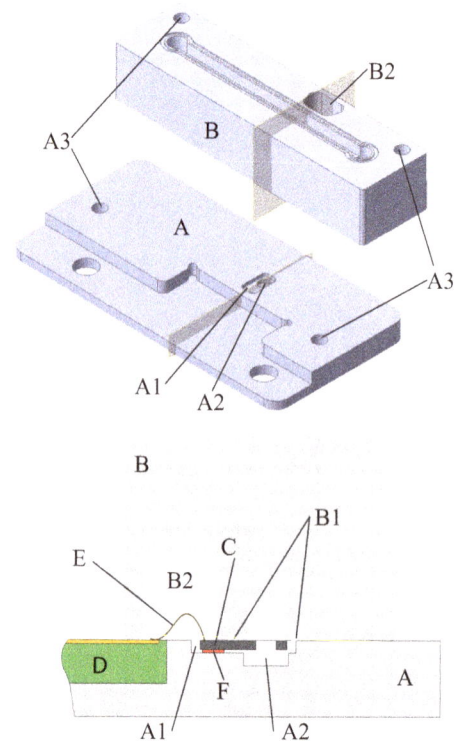

Figure 4. Top: CAD drawing of the base part (top view) and flow channel part (bottom view); bottom: cross section through the assembled parts referring to the plane in the CAD drawing. A = base part, A1 = cavity for the sensor, A2 = cavity for pressure equalization, A3 = drill holes for assembling both parts with aligning pins, B = flow channel part, B1 = capillary stop for adhesive, B2 = cavity to protect bond wires, C = sensor, D = PCB, E = bond wire, and F = adhesive.

the flow thorough the bypass is less than 0.2 % at 1 sLm for air as fluid. Furthermore, the base part has an accommodation for a PCB board, which can be placed next to the base part for electrical connection by wire bonding. The hollow is designed in such a way that the PCB board is mounted flush with the surface (D, Fig. 4). The flow channel part contains the flow channel, whose geometry can be adapted by varying the cross section. In our setup, we set the width of the channel to $a = 1.5$ mm and the channel height to 1.75 mm. It should be mentioned that the possibilities of changing the width of the channel are limited because of the dimensions of the sensor. The membrane should be placed in the middle

of the flow channel. Otherwise, there is a reduction in sensitivity. The flow channel is surrounded by a smaller channel integrated as a capillary stop for the adhesive when the flow channel part and the base part are fixed (B1, Fig. 4). To protect the bond wires, a cavity is built into the flow channel part (B2, Fig. 4). If electrical conductive fluids are used, the bond wires have to be sealed by a glob top. The measurements in this paper are only done with air; thus, we did not use any protection for the bond wires. Two drill holes in the base and flow channel part (A3, Fig. 4) are integrated for assembling both parts with aligning pins. Fabrication was done by milling the parts out of polycarbonate because tests were only performed with air.

3.2 Sensor integration in the housing

Sensor integration in the base part is done with a microsystem assembling machine from *finetech*. It is especially made for flip chip processes, and it consists of several tools, e.g., for the targeted inclusion of chips in a housing. The principle structure of the machine can be seen in Fig. 5. To integrate the flow sensor into the cavity of the base part, a thermode was developed (see Fig. 5). It can be placed in a special mechanical receiver integrated into the machine. The thermode must satisfy two requirements:

– picking the sensor without destroying the membrane; and

– mounting the sensor in the base part flush with the surface to avoid sharp edges provoking turbulence.

To satisfy the first requirement, the thermode picks the sensor outside of the membrane. Through a hole in the thermode, the sensor can be held by a vacuum. For the second requirement, the thermode is wider than the cavity. When the sensor is placed in the cavity, the thermode rests on the surface (Fig. 6d). To cure the adhesive, the thermode can be heated by a temperature controller.

The process of sensor integration is illustrated in Fig. 6. In the beginning of the process, the base part is put on the swivel table. Using the dispensing system, a drop of a fast curing adhesive (*DELOMONOPOX® NU355*) is placed in its cavity. Then, the sensor is picked by the thermode (Fig. 6a). Curing the adhesive thermode with the sensor is aligned with the cavity of the base part. As seen in Fig. 6c, the top camera with the beam splitter shows the sensor held by the thermode and the base part with the cavity where the drop of adhesive lies. By pivoting the swivel head forward, the sensor is placed in the cavity flush with the surface of the base part (Fig. 6d). The thermode is heated to $T = 140\,°C$ and the adhesive is cured in a few seconds. The vacuum for holding the sensor is switched off and the thermode is lifted up.

To contact the flow sensor, Al bond wire with a wedge–wedge process was used (see Fig. 6). The bond wires are not sealed and thus it is only possible to measure air or other electrical non-conductive fluids. If water or other fluids are used, the bond wires have to be protected by a glob top.

In the last step, the base part and the flow channel part are combined and fixed with a common two-component adhesive. By optical inspection it was found out that the capillary stop in the flow channel part worked very well: the epoxy resin stopped exactly at the capillary stop. For flow connection, two fittings are screwed into the flow channel part: at the inlet and the outlet (see Fig. 7).

4 Characterization

4.1 Flow measurement

As flow references, *MKS 1179* mass flow controllers (pressure at inlet 2 bar, resolution 0.002 sLm) have been used. The data acquisition from the sensor has been carried out with a USB-6212 card from National Instruments in combination with LabView. A sampling rate for data acquisition of 1 kHz has been chosen. To control the heater, three different principles can be used:

– constant voltage;

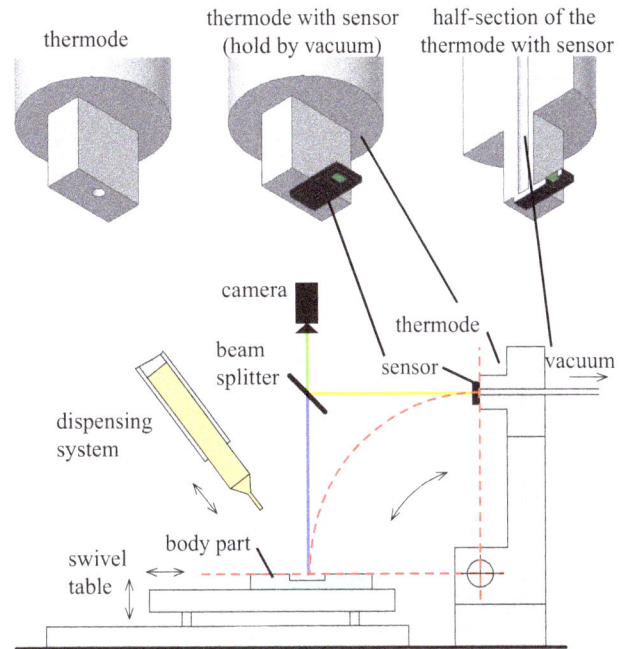

Figure 5. Principle illustration of the FINEPLACER® from *finetech*.

– constant temperature; and

– constant power.

Controlling systems for constant power and constant temperature can be found in Sosna et al. (2010) and Sturm et al. (2012). For our measurements, we used a constant voltage of 3.2 V for the heater. The deviation of the heater resistance in our systems was less than 1 %. Thus, the temperature of the heater at 3.2 V is nearly the same in all systems, and the measurements are comparable. We also compared constant voltage operation with constant power operation – however, we could not see any difference between the modes, because the resistance of the heater varies only slightly with the temperature.

4.2 Influence of the sensor on the step height

A flow sensor can be mounted in a channel in three different ways illustrated in Fig. 8: no step, step up, and step down. To analyze the influence of the step height on the characteristic curve, sensor systems of these three different types were built. Mounting the sensor step up was done with the help of a 100 μm thick stainless steel sheet with a hole (according to the geometry of the cavity) that was laid on the body part, and the sensor was then mounted as described in Sect. 3. The sensor mounted step down was fixed with adhesive on the bottom of the sensor cavity (see A1 in Fig. 4). Both step heights were around $100 \pm 2\,\mu m$, which was tactile measured (*Ambios XP-2*). Of course, the system with no step was mounted in the way described in Sect. 3. The result

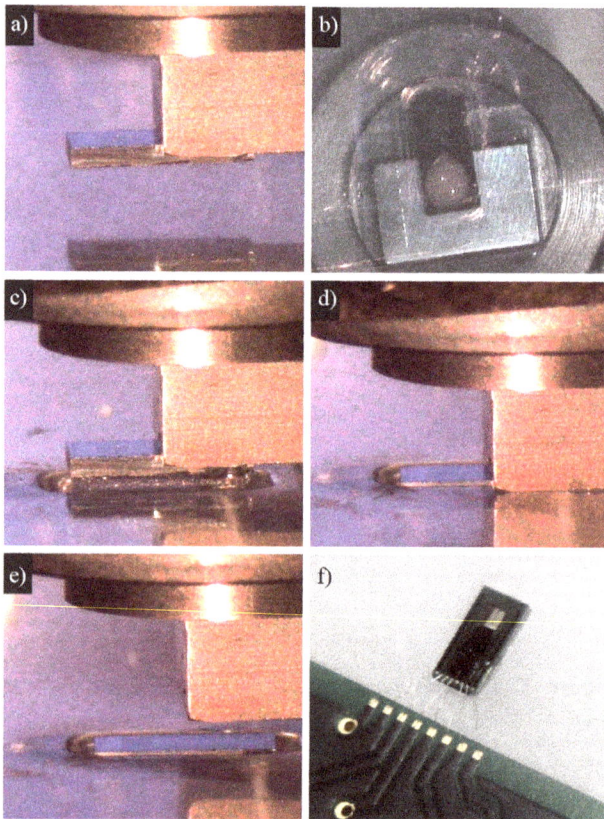

Figure 6. Flush mounting of the sensor in the base part. (**a**) Picking of the sensor outside of the membrane (**b**) because of the camera with a beam splitter sensor and cavity can been seen and aligned; (**c**) the sensor is set in the cavity; (**d**) the thermode is heated; (**e**) the thermode is lifted, and the sensor is mounted flush with the surface; (**f**) IMSAS flow sensor integrated into the base part and electrically connected by Al bond wires.

Figure 7. IMSAS flow sensor system.

of the measurement can be seen in Fig. 9. When the sensor is mounted step down, there is a point of discontinuity in the characteristic curve, comparable with the results of Eberhardt et al. (2003). Both thermopiles show this point of discontinu-

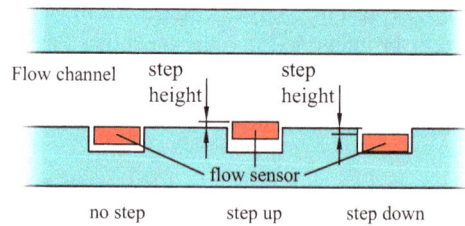

Figure 8. Flow sensor mounted in a channel in three different ways. Left: no step; middle: step up; right: step down.

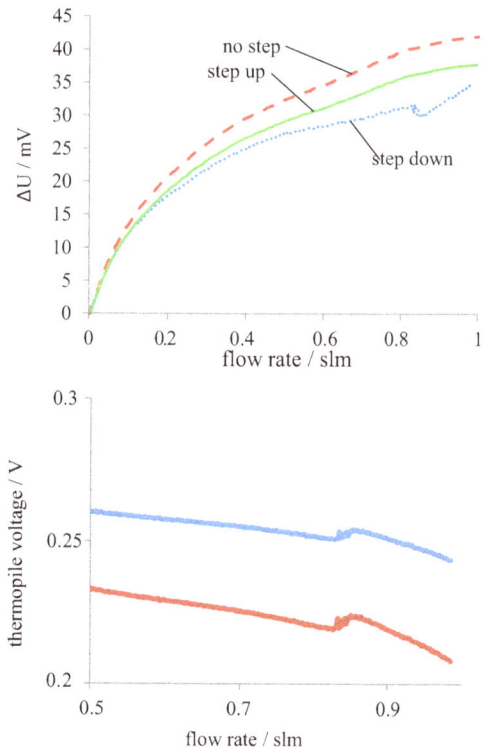

Figure 9. Top: voltage difference between the thermopiles ΔU for different step heights between the sensor and the channel wall in dependence on the flow rate Q: dashed line no step, solid line step up, dotted line step down. Bottom: thermopile voltage in dependence on the flow rate for the sensor mounted step down.

ity. In contrast to the sensor mounted step down, no point of discontinuity appears in the characteristic curve of the sensor mounted step up or flush with the surface.

5   Discussion

Regarding the previous experimental investigations, the reason for the point of discontinuity on flow sensors mounted step down could be the change from laminar to turbulent flow. The Reynolds number for our setup is $Re = 630$ for $Q = 1\,\text{sLm}$. In addition, we calculated the standard deviation of the samples at this point to quantify the noise and thus turbulences, but the standard deviation did not change

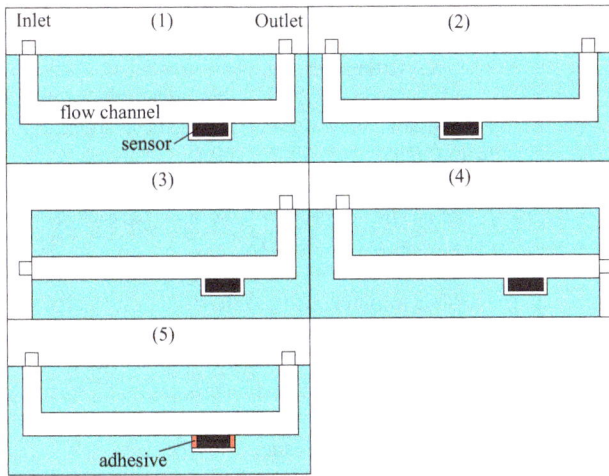

Figure 10. Different geometries of the channel: (1) rectangular inlet and outlet + sensor close to the outlet, (2) rectangular inlet and outlet + sensor in the middle of the channel, (3) straight inlet and rectangular outlet + sensor close to the outlet, (4) rectangular inlet and straight outlet + sensor close to the outlet, and (5) no pressure equalizing, fully sealing of the sensor. The sensor was always mounted flush with the surface. All measurements were also performed by changing the inlet and outlet. No points of discontinuity in the characteristic curve were measured.

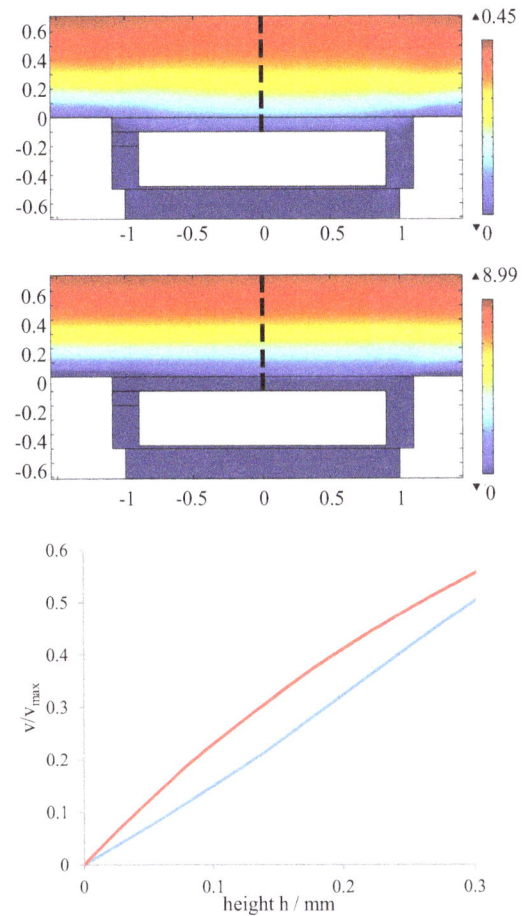

Figure 11. Top and middle: simulation of the flow velocity in the channel for flow rates 0.05 and 0.8 sLm (COMSOL MULTIPHYSICS, laminar flow study with compressible flow). Boundary conditions: left side *inlet with laminar inflow flow rate*, right side $p = 0$, wall $u = 0$. Bottom: normalized flow velocity v/v^{max} for 0.05 sLm (red) and 0.8 sLm (blue) in dependence on the height of the dashed black line above the sensor.

in comparison to the standard deviation at lower flow rates. Thus, we can state that there is still a laminar flow. To exclude other geometry influences, we also performed different channel inlet and outlet setups as well as a fully sealed flow sensor without pressure equalizing. The measurements did not show any point of discontinuity, while they show a point of discontinuity when the sensor is mounted step down.

To investigate the flow behavior in the channel to see any causes of the point of discontinuity, we did a laminar flow simulation of the channel. The results for low and high flow rates are shown in Fig. 11. When the flow rate is low, there is flow in the cavity formed by the sensor. While increasing the flow rate, the flow in this cavity gets less, and the laminar flow profile also changes, as seen in Fig. 11 (bottom). But, a sudden change in the characteristic curve can not be explained with simple laminar flow conditions.

It is widely known that vortices can appear, although the flow is not turbulent (Tabeling, 2005; Abu-Mulaweh, 2003; Kang et al., 1999). These vortices can appear in laminar flows when there are obstacles or sudden changes in the geometry. An explanation that there is a discontinuity in the characteristic curve for a flow sensor mounted step down but not for the one mounted step up can be given by having a look at Fig. 12. At low flow rates, no vortices appear in the cavity, as shown in the simulation in Fig. 11. At higher flow rates, vortices appear in the cavity above the sensor, influencing the thermal field and thus changing the characteristic curve, which is shown in the measurement by the point of discontinuity. For the sensor mounted step up, vortices appear before and after

the step (sensor) but not above the sensor. Therefore, we expect a discontinuity in the characteristic curve for step down, but not for step up. This is actually observed (Fig. 12) in different scientific papers that study a fundamental Computational Fluid Dynamics (CFD) analysis for similar geometries (Shankar, 1993; Cheng and Hung, 2006; van Dijk and de Lange, 2007). Making a CFD analysis to give a fundamental theoretical answer to what is happening in our setup when the sensor is mounted step down or step up would be rather complicated and very extensive work. It is rather complicated to simulate vortex flows, because the occurrence of vortices is dependent on the geometry, investigations into the turbulent length have to be made, and the change from laminar to turbulent flow is still part of current research and scientific papers (Cheng and Hung, 2006; Pan and Acrivos, 1967; Imberger, 1974; Shen and Floryan, 1985; Pey et al., 2014). For a flow sensor, the influence of the thermal field above the

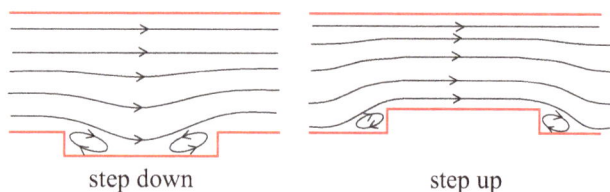

step down   step up

Figure 12. Explanation for the discontinuity in the characteristic curve: vortices appear in the cavity and thus above the sensor mounted step down, influencing the thermal field and the characteristic curve, while vortices appear before and after but not above the sensor mounted step up.

sensor has to be taken into account too, which makes it more complicated.

So, it could be determined that a flow sensor should be mounted flush with the surface. Otherwise, a point of discontinuity in the characteristic curve can appear when the sensor is mounted step down. Mounting a sensor step up has no point of discontinuity in the characteristic curve, but it has two disadvantages: on the one hand, the cross section of the channel decreases, and on the other hand, the flow under the sensor (bypass) increases around 1.7 % at 1 sLm, which results in a lower sensitivity.

6   Conclusions

A new packaging technology for building flow measurement systems has been presented. In the beginning, an introduction to the functional principle of flow sensors is given, and the fabrication process of our flow sensor is explained. The new packaging method was presented by describing the housing and the entire process from integrating the sensor into the bottom part to the sealing of the system. By placing a flow sensor in a channel in different step arrangements, the importance of mounting the sensor flush with the surface, in order to avoid turbulence or a reduction in the channel cross section or sensitivity, has been demonstrated. The main result of the characterization is a point of discontinuity in the characteristic curve when the sensor is mounted step down. This was analyzed by the help of theoretical investigations and scientific papers.

References

Abu-Mulaweh, H.: A review of research on laminar mixed convection flow over backward-and forward-facing steps, Int. J. Therm. Sci., 42, 897–909, 2003.

Adamec, R. J. and Thiel, D. V.: Self heated thermo-resistive element hot wire anemometer, IEEE Sens. J., 10, 847–848, 2010.

Ashauer, M., Glosch, H., Hedrich, F., Hey, N., Sandmaier, H., and Lang, W.: Thermal flow sensor for liquids and gases based on

combinations of two principles, Sensor. Actuat. A-Phys., 73, 7–13, 1999.

Billat, S., Kliche, K., Gronmaier, R., Nommensen, P., Auber, J., Hedrich, F., and Zengerle, R.: Monolithic integration of microchannel on disposable flow sensors for medical applications, Sensor. Actuat. A-Phys., 145, 66–74, 2008.

Bruschi, P., Diligenti, A., Navarrini, D., and Piotto, M.: A double heater integrated gas flow sensor with thermal feedback, Sensor. Actuat. A-Phys., 123, 210–215, 2005.

Bruschi, P., Piotto, M., and Bacci, N.: Postprocessing, readout and packaging methods for integrated gas flow sensors, Microelectr. J., 40, 1300–1307, doi:10.1016/j.mejo.2008.08.009, 2009.

Buchner, R., Sosna, C., Maiwald, M., Benecke, W., and Lang, W.: A high-temperature thermopile fabrication process for thermal flow sensors, Sensor. Actuat. A-Phys., 130, 262–266, 2006.

Buchner, R., Bhargava, P., Sosna, C., Benecke, W., and Lang, W.: Thermoelectric Flow Sensors with Monolithically Integrated Channel Structures for Measurements of Very Small Flow Rates, in: IEEE Sensors, 828–831, doi:10.1109/ICSENS.2007.4388529, 2007.

Cheng, M. and Hung, K.: Vortex structure of steady flow in a rectangular cavity, Comput. Fluids, 35, 1046–1062, 2006.

Dijkstra, M., de Boer, M., Berenschot, J., Lammerink, T., Wiegerink, R., and Elwenspoek, M.: Miniaturized thermal flow sensor with planar-integrated sensor structures on semicircular surface channels, Sensor. Actuat. A-Phys., 143, 1–6, doi:10.1016/j.sna.2007.12.005, 2008.

Dillner, U., Kessler, E., Poser, S., Baier, V., and Müller, J.: Low power consumption thermal gas-flow sensor based on thermopiles of highly effective thermoelectric materials, Sensor. Actuat. A-Phys., 60, 1–4, doi:10.1016/S0924-4247(96)01409-4, 1997.

Eberhardt, W., Kück, H., Münch, M., Schilling, P., Ashauer, M., and Briegel, R.: MID-Gehäuse für ein Durchfluss-Sensorsystem, Kunststoffe, 93, 51–53, 2003.

Imberger, J.: Natural convection in a shallow cavity with differentially heated end walls. Part 3. Experimental results, J. Fluid Mech., 65, 247–260, 1974.

Ito, Y., Higuchi, T., and Takahashi, K.: Submicroscale flow sensor employing suspended hot film with carbon nanotube fins, J. Therm. Sci. Tech., 5, 51–60, 2010.

Johnson, R. and Higashi, R.: A highly sensitive silicon chip microtransducer for air flow and differential pressure sensing applications, Sensor. Actuator., 11, 63–72, doi:10.1016/0250-6874(87)85005-9, 1987.

Kaltsas, G., Nassiopoulos, A., and Nassiopoulou, A.: Characterization of a silicon thermal gas-flow sensor with porous silicon thermal isolation, IEEE Sens. J., 2, 463–475, doi:10.1109/JSEN.2002.806209, 2002.

Kang, S., Choi, H., and Lee, S.: Laminar flow past a rotating circular cylinder, Phys. Fluids, 11, 3312, doi:10.1063/1.870190, 1999.

Liu, J., Wang, J., and Li, X.: Fully front-side bulk-micromachined single-chip micro flow sensors for bare-chip SMT (surface mounting technology) packaging, J. Micromech. Microeng., 22, 035020, doi:10.1088/0960-1317/22/3/035020, 2012.

Ma, R.-H., Wang, D.-A., Hsueh, T.-H., and Lee, C.-Y.: A MEMS-based flow rate and flow direction sensing platform with integrated temperature compensation scheme, Sensors, 9, 5460–5476, 2009.

Mailly, F., Giani, A., Bonnot, R., Temple-Boyer, P., Pascal-Delannoy, F., Foucaran, A., and Boyer, A.: Anemometer with hot platinum thin film, Sensor. Actuat. A-Phys., 94, 32–38, 2001.

Pan, F. and Acrivos, A.: Steady flows in rectangular cavities, J. Fluid Mech., 28, 643–655, 1967.

Pey, Y. Y., Chua, L. P., and Siauw, W. L.: Effect of trailing edge ramp on cavity flow structures and pressure drag, Int. J. Heat Fluid Fl., 45, 53–71, 2014.

Sabaté, N., Santander, J., Fonseca, L., Gràcia, I., and Cané, C.: Multi-range silicon micromachined flow sensor, Sensor. Actuat. A-Phys., 110, 282–288, doi:10.1016/j.sna.2003.10.068, 2004.

Shankar, P.: The eddy structure in Stokes flow in a cavity, J. Fluid Mech., 250, 371–383, 1993.

Shen, C. and Floryan, J.: Low Reynolds number flow over cavities, Phys. Fluid., 28, 3191, doi:10.1063/1.865366, 1985.

Sosna, C., Buchner, R., Benecke, W., and Lang, W.: A Simple One-Dimensional Analytical Model for Calculation of the Output Signal of Thermal Flow Sensors, in: Eurosensors XXII, Dresden, Germany, 129–132, 2008.

Sosna, C., Buchner, R., and Lang, W.: A temperature compensation circuit for thermal flow sensors operated in constant-temperature-difference mode, IEEE T. Instrum. Meas., 59, 1715–1721, 2010.

Sturm, H., Dumstorff, G., Busche, P., Westermann, D., and Lang, W.: Boundary Layer Separation and Reattachment Detection on Airfoils by Thermal Flow Sensors, Sensors, 12, 14292–14306, 2012.

Tabeling, P.: Introduction to Microfluidics, Oxford University Press, 2005.

van Dijk, A. and de Lange, H.: Compressible laminar flow around a wall-mounted cubic obstacle, Comput. Fluids, 36, 949–960, doi:10.1016/j.compfluid.2006.05.003, 2007.

Van Herwaarden, A. and Sarro, P.: Thermal sensors based on the Seebeck effect, Sensor. Actuator., 10, 321–346, 1986.

Van Putten, A. and Middelhoek, S.: Integrated silicon anemometer, Electron. Lett., 10, 425–426, 1974.

15

Partially integrated cantilever-based airborne nanoparticle detector for continuous carbon aerosol mass concentration monitoring

H. S. Wasisto[1,2], **S. Merzsch**[1,*], **E. Uhde**[3], **A. Waag**[1,2], **and E. Peiner**[1,2]

[1]Institut für Halbleitertechnik (IHT), Technische Universität Braunschweig, Braunschweig, Germany
[2]Laboratory for Emerging Nanometrology (LENA), Braunschweig, Germany
[3]Material Analysis and Indoor Chemistry Department (MAIC), Fraunhofer-WKI, Braunschweig, Germany
[*]now at: Infineon Technologies AG, Munich, Germany

Correspondence to: H. S. Wasisto (h.wasisto@tu-bs.de)

Abstract. The performance of a low-cost partially integrated *can*tilever-based airborne nanoparticle (NP) detec*tor* (CANTOR-1) is evaluated in terms of its real-time measurement and robustness. The device is used for direct reading of exposure to airborne carbon engineered nanoparticles (ENPs) in indoor workplaces. As the main components, a miniaturized electrostatic aerosol sampler and a piezoresistive resonant silicon cantilever mass sensor are employed to collect the ENPs from the air stream to the cantilever surfaces and to measure their mass concentration, respectively. Moreover, to realize a real-time measurement, a frequency tracking system based on a phase-locked loop (PLL) is built and integrated into the device. Long-term ENP exposure and a wet ultrasonic cleaning method are demonstrated to estimate the limitation and extend the operating lifetime of the developed device, respectively. By means of the device calibrations performed with a standard ENP monitoring instrument of a fast mobility particle sizer (FMPS, TSI 3091), a measurement precision of ENP mass concentrations of $< 55\%$ and a limit of detection (LOD) of $< 25\,\mu\mathrm{g\,m^{-3}}$ are obtained.

1 Introduction

Over the last few decades, nanotechnology, which covers a compilation of technologies and methods for manipulating material on the nanoscale (i.e., nanomaterial or nanoparticle (NP)), has been attracting immense attention in society and has been hailed by some scientists as the next industrial revolution. The possible interests in nanotechnology originate mainly from the novel properties and characteristics of the nanomaterials, which are not the same as bulk materials and may be unpredictable and unimagined as scale effects (Maynard, 2007; Hullman, 2007). Thus, this technology has rapidly been developed and used across a variety of industries (e.g., electronics, medicine, cosmetics, pharmaceuticals, food packaging, household appliances, and national defense), leading to increased economic growth and new job vacancies (Bekker et al., 2013).

Although nanotechnology provides society with enormous feasibilities, questions have also been raised about uncertainties concerning the risks to and potential health effects of released NPs on the environment. The formation and release of the NPs into indoor environments and workplaces can occur through both incidental (i.e., unintentional NPs) and planned manufacturing processes (i.e., engineered NPs (ENPs)) (Balbus et al., 2007). Moreover, much higher awareness should be given to the workers, who manufacture and handle NPs directly in large quantities during the line productions (Brouwer, 2010). Thus, a direct-reading airborne NP mass concentration detector is very useful for the assessment of personal- and location-dependent monitoring in workplaces and indoor environments.

For individual NP mass monitoring, the complete system relies on mini portable devices that can be held and carried easily by the workers. Currently, the already developed NP mass sensors based on microelectromechanical systems

(MEMS) need to be operated at high flow velocity (50–150 m s^{-1}) using an external pump (Schmid et al., 2013) or partial vacuum requiring bulky, heavy, and expensive vacuum tools (Hajjam et al., 2011). In contrast, our first generation of the *can*tilever-based airborne NP detec*tor* (CANTOR-1) evaluated in this work has a small size, a low weight, an appropriate flow rate, an atmospheric pressure working condition, and low-cost components. This device had been successfully tested to prove the NP mass sensing principle with detected ENPs down to ~ 20 nm in diameter (Wasisto et al., 2013a). However, these measurements were done offline, where the ENP sampling and the sensor characterization were performed separately and controlled manually. Thus, some supporting components were needed to be added to the device. Recently, we reported on a frequency tracking system based on a phase-locked loop (PLL) circuit and a high voltage (HV) module based on a DC amplifier for nanoparticle mass concentration monitoring in real time (Wasisto et al., 2014). This new device (CANTOR-1) was successfully tested in normal indoor ambience, showing the ability to detect, e.g., cigarette smoke. However, the quantitative detection of NPs was impeded by coarse particulate matter, which was also deposited on the cantilever. To ensure that CANTOR-1 monitors only NPs, coarse particles have to be removed from the air passing through the NP sampler. Furthermore, to give a quantitative NP concentration read-out, CANTOR-1 had to be calibrated under defined ambient conditions.

Therefore, in this paper, we address a first quantitative test and evaluation of CANTOR-1 in real-time ENP measurements. For this purpose, measurements were performed under a defined ENP exposition in a sealed chamber and with standard stationary aerosol monitoring equipment; i.e., a fast mobility particle sizer (FMPS) was used to provide reference data for the time-dependent ENP concentration. Membrane filters and impactors were described to prevent interfering coarse particulate matter from entering the sampler. A recycling process of ultrasonic wet cleaning is also performed to regenerate the sensor after being fully loaded by ENPs during long-term exposure.

2 Direct-reading cantilever-based airborne nanoparticle detector

2.1 Main components

Figure 1a shows the principle of the direct-reading CANTOR-1. Particle-laden air is flown through a filter, which removes coarse particulate matter, towards a cantilever biased to a negative potential (V_{EP}) by which the positively charged fraction of the airborne nanoparticles is attracted. The mass added to the cantilever by the attached nanoparticles lowers the resonant frequency of the cantilever, which is driven by a piezostack (V_{dr}) and read-out (ΔV) using an integrated piezoresistive strain gauge supplied by V_0.

In principle, the CANTOR-1 system comprises two main components (i.e., a miniaturized electrostatic aerosol sampler and a piezoresistive resonant silicon cantilever mass sensor). The first main module is a cylindrical miniaturized electrostatic NP sampler made of aluminum material and designed as a tube with a diameter of 20 mm and a length of 45 mm (Fig. 1b). The tube consists of three single aluminum parts joined by threads denoted as inlet part, middle part, and outlet part, respectively. Aiming to direct the aerosol flow against the cantilever positioned on the tube axis, the inlet part was conically shaped with an inclination of 15° to the tube axis (Fig. 1c). In the middle part of the tube, the second module of a silicon cantilever resonator was mounted by double-sided tape on a small ceramic printed circuit board (PCB) to integrate it with the piezoelectric stack actuator (Fig. 1d). Gold bonding microwires of 30 µm were used for getting electrical contact between cantilever and board. To generate stable aerosol flow of 0.68 L min^{-1}, a small fan (MF10A03A, SEPA Europe GmbH) was mounted at the outlet part of the tube. To operate the NP sampler, a voltage of 3.5 V from either a battery pack or a power supply (HP E3631A) was used. This voltage, however, needed to be amplified first up to 0.5 kV using a DC voltage amplifier (i.e., a factor of ~ 140 higher).

The employed second module is a silicon cantilever ($V_{cant} = 2750 \times 100 \times 50 \, \mu m^3$, $m_{cant} = 32.04 \, \mu g$, $f_{cant} = 9.4$ kHz), which has a full square Wheatstone bridge on its clamped end as a piezoresistive element and works based on the strain-to-resistivity change to read the sensor signal output (Fig. 1e) (Wasisto et al., 2013b). The cantilevers are fabricated by utilizing silicon bulk micromachining processes (i.e., photolithography, thermal oxidation, dopant diffusion/implantation, and inductively coupled plasma (ICP) cryogenic deep reactive ion etching (cryo-DRIE)). Figure 2 depicts the fabrication process steps of the self-sensing piezoresistive silicon cantilever resonators, which can be described in detail as follows.

1. The fabrication was started with a preparation of (100)-oriented *n*-type bulk silicon wafers (3–5 Ωcm) with a thickness of 300 µm. In contrast to piezoelectric AlN/Si cantilever resonators (Sökmen et al., 2010), which use SOI for the device fabrication, bulk silicon wafers are more preferable for the current sensor fabrication considering their lower price, high stability, high mechanical Q factor, and high degree of freedom for the geometrical resonant cantilever design. At the beginning, a wafer was placed in a furnace to grow a thermal silicon dioxide (SiO_2) thin layer on the silicon surface. A 400 nm SiO_2 layer was obtained within 1 h at a temperature of 1100 °C.

2. Subsequently, the oxidized wafer coated with a photoresist mask was patterned utilizing UV lithography. Buffered hydrofluoric acid solution was then used to open the areas for the definition of *p*-type piezoresis-

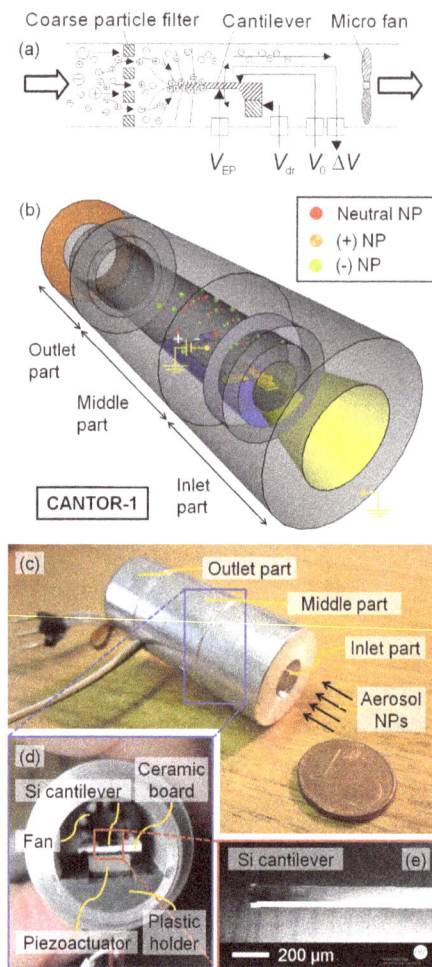

Figure 1. (**a**) 2-D and (**b**) 3-D schematics, and (**c**) photographs of the handheld CANTOR-1 showing (**d**) the main components of its middle part. (**e**) SEM image of the silicon piezoresistive cantilever resonator.

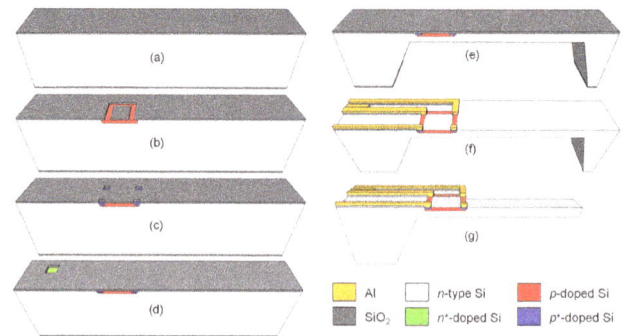

Figure 2. The fabrication process flow of the self-sensing piezoresistive silicon cantilever resonators showing (**a**) thermal oxidation, (**b**) p-type piezoresistor creation, (**c**) p^+-type contact formation, (**d**) phosphorus diffusion, (**e**) backside membrane wet etching, (**f**) metallization, and (**g**) cantilever free release dry etching.

4. Since an ohmic contact was needed for electrostatic NP collection, a phosphorus diffusion/implantation was then carried out and placed in an n^+-type well close to one of the p^+-type piezoresistor contacts.

5. Next, to create the backside membranes on the bottom side of the wafer, the selected bottom oxide layers were patterned by UV lithography and buffered Hydrofluoric acid (HF) etching. Through the oxide openings, a potassium hydroxide solution (KOH) at elevated temperatures or cryo-DRIE was introduced to etch the silicon down to a residual thickness of 25–50 μm. This process is very critical because it determines the cantilever thickness and thus its operating resonant frequency. From the experiments, a tolerance of $\pm 0.5\,\mu$m (i.e., 1 to 2 %) was normally obtained using KOH etching.

6. Subsequent to the opening of the SiO_2 contact holes, a top-side metallization was then deposited by 300 nm aluminum electron-beam evaporation. Furthermore, to provide large-enough landing areas for non-permanent electrical connections using spring-loaded contact pins, large contact pads were provided (i.e., designed as a 0.75×1 mm^2 large area).

7. To finish the fabrication process, the cantilevers were lithographically patterned and released by cryo-DRIE from the front side of the samples. A dry etcher (SI 500 C, SENTECH Instruments GmbH, Berlin, Germany) was used along with a photoresist serving as the etching mask.

2.2 Microparticle filtration components

In addition, to filter out the undesired microparticles, which could possibly approach the cantilever, two filtration stages

tors. Afterwards, to create a full piezoresistive Wheatstone bridge on the Si wafer, boron implantation or diffusion was performed.

3. Furthermore, to form p^+ feed lines to the Wheatstone bridge and improve contact formation, additional boron diffusion or implantation was used. For the implanted and diffused wafers, the standard deviations of the measured resistivity from the desired value were ~ 0.6 and ~ 4.1 % in proportion to their lateral doping distribution, respectively. Typical values of the junction depth and the surface doping concentration amounted to $\sim 4.5\,\mu$m and 1.5–3.0×10^{18} cm^{-3}, respectively, representing a tradeoff between a large piezoresistive coefficient $\pi_{44} \approx 1$ GPa^{-1} and a low temperature coefficient around $-3 \times 10^{-3}\,°$C^{-1} (Cho et al., 2008; Peiner et al., 2008).

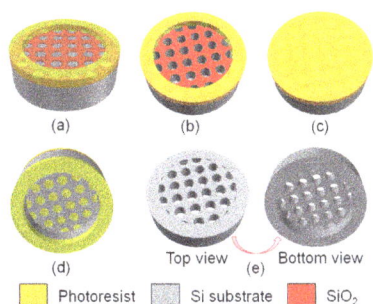

Figure 3. Fabrication process for the Si microfilter, including (**a**) SiO$_2$ masking, (**b**) ICP-RIE, (**c**) protective coating, (**d**) back side etching, and (**e**) cleaning using an HF dip.

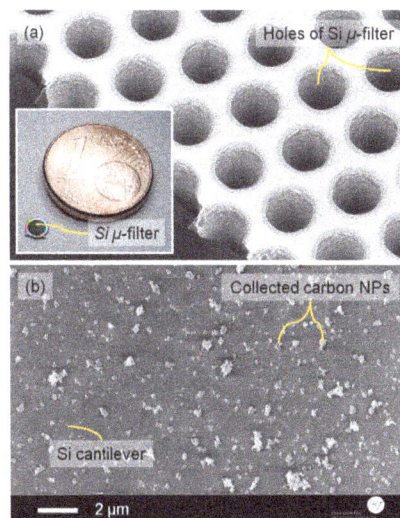

Figure 4. SEM images of the (**a**) Si microfilter used in CANTOR-1 and (**b**) the collected carbon ENPs on the cantilever that could pass through the filter holes.

were also developed and integrated on the NP sampler head of CANTOR-1. From the market, commercially available grids for transmission electron microscopy (TEM) might be a solution for cavity diameters down to $0.6\,\mu$m. However, those grids were very fragile, leading to difficult mounting and use in our NP sampling system. Therefore, the first stage of microparticle filtration was to use a home-built silicon microfilter with a grid diameter of $2.5\,\mu$m. Its manufacturing processes are illustrated in Fig. 3, which can be described as follows.

1. Silicon dioxide (SiO$_2$) was thermally grown on a silicon wafer to be employed as a grid mask by subsequent lithography and HF dips. A second lithography is done to prepare the rim on the front side of the filter.

2. By using ICP cryo-DRIE, the holes with a diameter of $2.5\,\mu$m were transferred into the silicon wafer.

3. Before structuring the back side of the wafer, its front side had to be firstly protected by a resist layer to prevent mechanical deformation because of harsh contact with the aluminum plate carrier. Moreover, this process flow had been selected to prevent grid artifacts (i.e., a mask undercut that was expected due to the impaired release of the etching-induced heat through a membrane).

4. The backside etching was performed to reduce the membrane thickness down to $\sim 4\,\mu$m. This step is important to determine the ability of the aerosol to pass through the filter channels, because creating too thick a membrane will provide a lower aerosol flow rate, which is a disadvantageous effect for the NP sampling system.

5. After stripping off the photoresist and SiO$_2$ layers, the silicon microfilter had been finally realized.

To check on the utility of the fabricated silicon microfilter shown in Fig. 4a, a preliminary simple test was carried out in the ENP exposure assessment prior to the real-time monitoring of carbon aerosol ENPs (Sigma-Aldrich Chemie GmbH). From scanning electron microscope (SEM) investigation, it

is clearly shown that, by placing the microfilter at the opening inlet of the CANTOR-1 sampler head, the microparticles and their agglomerates ($d_P > 2.5\,\mu$m) had been rejected. Thus, only the ENPs ($d_P \leq 2.5\,\mu$m) could pass through the filter holes and subsequently reach the cantilever surfaces (Fig. 4b).

To further reject the smaller microparticles ($d_P > 1\,\mu$m), an impactor made of aluminum was also made and placed in between the microfilter and the cantilever sensor, acting as the second large particle filtration stage in the system (Fig. 5a). In other studies, this impaction technology had also been used for various devices and described in detail (Mehdizadeh et al., 2013; Schmid et al., 2013). For our Al impactor, its location is adjustable, which influences the velocity of the aerosol flow. By integrating this impactor, microparticles of diameters in the range of $1\,\mu m < d_P \leq 2.5\,\mu$m are expected to be impacted, and no longer in the aerosol flow stream. Figure 5b shows as proof microparticles captured on the impactor during a typical particle exposure experiment. Nevertheless, the maximum aerosol velocity in the CANTOR-1 system was set to 8–20 m s^{-1} to avoid any disturbances to the operating resonators during the aerosol NP detection. The airflow through the sampler modeled using COMSOL Multiphysics 4.3b is depicted in Fig. 5c. The used simulation module is particle tracing.

2.3 System integration

Besides the main components (i.e., an NP sampler with its mounted cantilever resonator) and the microparticle filtration stages (i.e., a silicon microfilter and aluminum impactor), there are three other supporting modules (i.e., electronic circuit, power supply, and data acquisition control system),

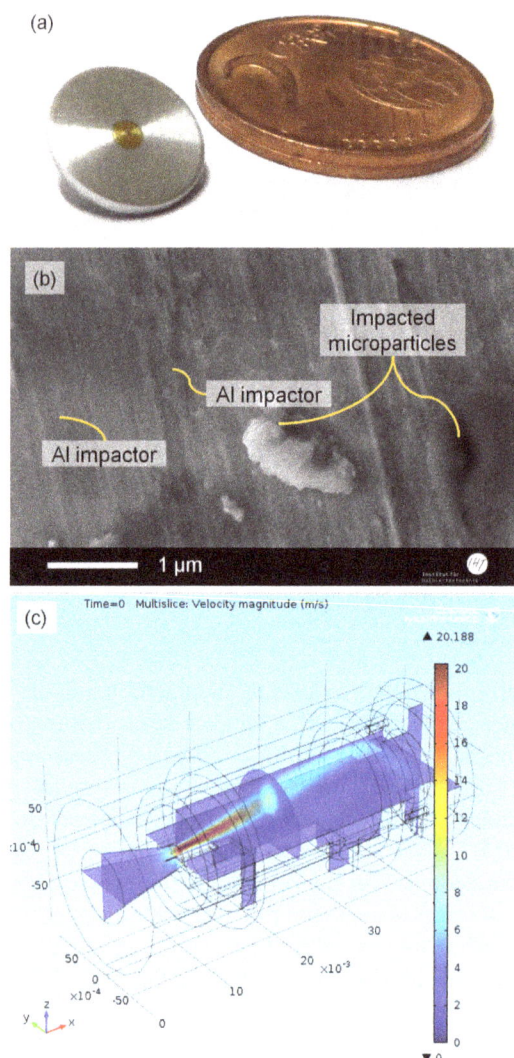

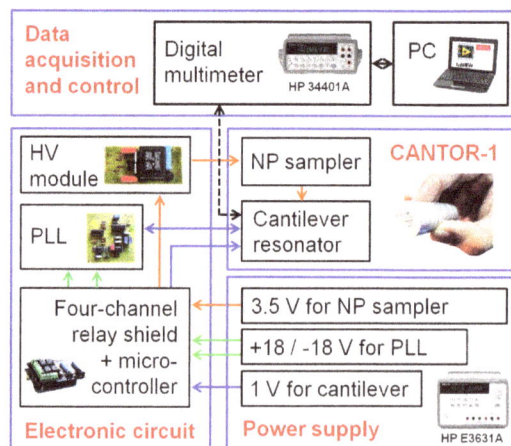

Figure 6. Schematic of the partially integrated CANTOR-1 showing all its components (i.e., a silicon cantilever resonator mounted in an NP sampler, power supply, electronic circuits, and data acquisition control system).

Figure 5. (**a**) Homebuilt aluminum impactor and (**b**) impacted microparticles on its surface. (**c**) 3-D COMSOL simulation result showing the aerosol flow velocity in a color gradient representation ranging from 0 (blue) to 20 m s^{-1} (red) inside the sampler.

which need to be integrated to build a complete CANTOR-1 system (Fig. 6). In the electronic circuit module, the frequency tracking system based on PLL and HV modules based on a DC amplifier are realized and involved to track the resonant frequency of the silicon cantilever in real time and amplify the input voltage from 3.5 V to 0.5 kV in the NP sampler, respectively (Wasisto et al., 2014). During particle sampling, a negative high voltage of −0.5 kV is set to one of the cantilever electrodes, which is connected with the bulk contact of the silicon cantilever to generate an electric field over the sensor. Hence, the positively charged and uncharged NPs will be attracted and polarized to the cantilever surfaces by electrophoresis and dielectrophoresis mechanisms, respectively. The detailed mechanism of the airborne NP de-

position in an electrostatic sampler had been described by Krinke et al. (2002), who had modeled the NP trajectory as a force balance on a single NP approaching the substrate.

Additionally, a combination of two small electronic components (i.e., an Arduino UNO microcontroller and a Seeedstudio relay shield) was also used to automatically control the periodic switching between NP sampling and resonant frequency measurement phases. The yielded response time of this relay shield was in milliseconds. This fast switching control had overcome the previously raised issue in the software development (i.e., a time delay of about 6–7 s due to the communication overhead between measuring instruments and the notebook computer) (Wasisto et al., 2014). The power sources used in CANTOR-1 originate from the two power supplies (HP 3631A) to supply the different power requirements of the NP sampler (i.e., 3.5 V), the cantilever resonator Wheatstone bridge (i.e., 1 V), and the PLL circuit (i.e., ±18 V).

In general, the operating process of CANTOR-1 can be divided into two phases, which are run alternately. First, the resonant frequency of CANTOR-1 will be tracked with a time resolution of 1 s under the condition when the PLL integrated with the cantilever sensor is switched on and the NP sampler is switched off by relay. Next, after the PLL integrated with the cantilever sensor has been turned off and the sampler head is subsequently turned on, NP sampling begins to be performed. It should be noted that, during the entire switching, a digital multimeter (HP 34401A) is always kept in reading status to continuously observe the resonant frequency change, which will be further converted into detected NP mass concentration in a LabVIEW-installed PC.

In its present configuration, CANTOR-1 consists of all components for real-time ENP monitoring except those for data acquisition control, logging, and display, for which a laptop PC is used. Thus, CANTOR-1 is more than a lab demonstrator, but can also be used as a portable indoor ENP monitor comparable in weight (less than 1 kg, including the PC) with state-of-the-art instrumentation (e.g., an Aerasense nanotracer of 750 g, Buonanno et al., 2014, or a Testo mini-Disc of 670 g, Fierz et al., 2011). A fully integrated system, which will include all necessary components in one base unit of a total weight of < 400 g, will be developed as the second version (CANTOR-2).

3 Detector performance

3.1 Test aerosol nanoparticle generation

After integrating all components into a complete system, CANTOR-1 was then assessed in carbon ENP exposures (Sigma-Aldrich Chemie GmbH), which were performed in a test chamber under typical workplace conditions (i.e., $V = 1$ m^3, $T = 23$ °C, RH $= 30$ %, and $p = 1$ atm), as depicted in Fig. 7a. The generation of stable test aerosols was started by nebulizing a suspension of ENPs in a solution of water/ethanol or water/isobutanol using a constant output atomizer (TSI 3076, TSI Inc.). Inside the atomizer, the incoming compressed air swelled through an orifice forming a high-velocity jet. The liquid solution was then pulled into the atomizing part through a vertical channel and subsequently atomized by the jet. Large droplets were removed by impaction on the barrier opposite to the jet. Meanwhile, the excess liquid was drained at the bottom part of the atomizer to the closed reservoir. As a result, fine spray could leave the atomizer through a channel at the top. By using this technique, the average particle size of the produced aerosol could be varied from 20 to 300 nm.

However, it should be noted that the produced aerosols from the atomizer were still wet. Thus, they had to be dried in a diffusion dryer. In this device, a water trap was incorporated into its inlet to collect coarse water droplets. Moreover, this device consists of two concentric cylinders as the main parts (i.e., inner and outer parts) formed by wire screen and acrylic cylinders, respectively. The space between those two cylinders was filled with a volume of round-shaped silica gel, which acted as a desiccant for having a strong affinity with water molecules and maintaining a dry atmosphere at the tube. During the flowing of the wet aerosol through the inner cylinder, water vapor would diffuse into the silica gel through the wire screen. At the outlet of the dryer, the dried aerosol ENPs can be obtained, which were then flown into a sealed glass chamber (cf. Fig. 7a). To circulate the aerosol inside the chamber (i.e., dynamic condition), a fan was used under a clean air supply of 12 L min^{-1} and put close to one of the inner chamber walls. Furthermore, in situ monitoring of ENP number concentration and size distribution was realized by a fast mobility particle sizer (FMPS, TSI 3091, TSI Inc.) with a time resolution of 1 s (Fig. 7b). The FMPS worked principally based on the elec-

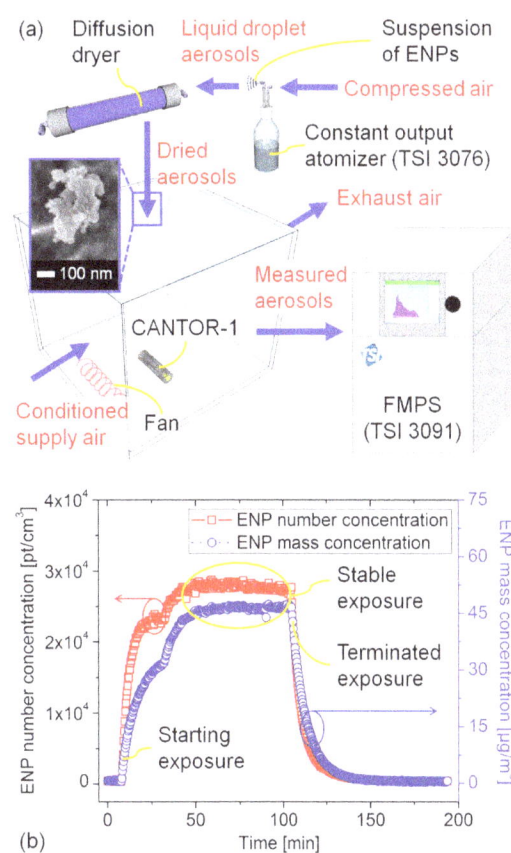

Figure 7. (a) Schematic of the aerosol ENP generation setup in a test chamber, involving an atomizer, a diffusion dryer, a fan, and an FMPS. (b) Typical generated particle number and mass concentrations for carbon airborne ENPs measured by FMPS as a function of time.

trical aerosol spectrometer (EAS) (Tammet et al., 2002). The incoming ENPs, which were firstly charged in a unipolar corona charger, were classified and measured with an array of 22 electrometers, where each channel corresponds to a defined electrical mobility bandwidth. Thus, by knowing the ENP charge levels, the measured currents of the electrometers can be extracted into ENP number size distributions (6–523 nm). Summing all measured ENP numbers will yield the total concentration of the flowing aerosols inside the chamber. Moreover, by assuming the ENPs are in a spherical shape, their mass concentration can be calculated. The typical stable generated carbon aerosols comprise two size modes of NP diameters of ~ 20 and ~ 120 nm, respectively, at total number concentrations on the order of 10^4 pt cm^{-3}.

3.2 Real-time engineered nanoparticle exposure assessments

The performance of the complete system of CANTOR-1 was evaluated in two real-time ENP exposure tests with different purposes (i.e., a minimum NP sampling investigation and a device continuation, respectively). In the first test, the setup of ENP detection was performed according to the specified time (i.e., three cycles comprising 5 min measuring states and varied sampling states (10, 8, 6, 4, 2, 1, and 0.5 min, respectively) for each cycle). CANTOR-1 was automatically switched to the measuring status to track its resonant frequency, whenever the time of the ENP sampling had expired (Fig. 8a).

After experiencing three cycles of repeated exposure tests of carbon aerosol ENPs with a stable concentration of $4.9 \pm 1.2 \times 10^4$ pt cm^{-3} in a sequence of periodic states (i.e., consecutive sampling and measuring), a real-time behavior of CANTOR-1 can then be observed. The increasing mass loading of the deposited ENPs had caused a linear drop in cantilever resonant frequency during a total sampling time of 94.5 min (Fig. 8b). From the analyzed results, the collected ENP mass amounted to $\Delta m_{ENP} = 2 \times m_{cant} \times \Delta f_{ENP} / f_{cant} = 26.88$ ng, which corresponded to a frequency shift of $\Delta f_{ENP} = -3.95$ Hz. Thus, a mass sensitivity of CANTOR-1 of $S_{ENP} = \Delta f_{ENP} / \Delta m_{ENP} = -0.15$ Hz ng^{-1} could be calculated under this condition. Furthermore, from this test, it was shown that there was a necessary waiting time (i.e., ~ 1–2 min) of the resonant frequency measurement before CANTOR-1 could operate under a stable condition, which was related to the PLL capturing time. Moreover, the ability of CANTOR-1 to detect the airborne ENP within only 1 min sampling had been confirmed, although some unstable data points were also present, which could be attributed to an unwanted lift-off of already deposited ENP fractions.

Next, differentiating from the first online test where the measurement was done in only a day, the second exposure assessment of CANTOR-1 was performed on two subsequent days to investigate the continuation of device use. Moreover, even though, in the first test of online measurement, CANTOR-1 could already detect carbon ENPs with only 1 min sampling time resolution, in the second test, CANTOR-1 was set at 5 min sampling time and 3 min measuring time to better interpret the data (i.e., a sufficient frequency shift). Thus, in total, CANTOR-1 will need 8 min to obtain a single data point of the ENP monitoring. This value is well below the 10–15 min period recommended for short-term aerosol exposure measurement (Duarte et al., 2014).

For the first day of the second online test, CANTOR-1 was operated only to measure the collected ENP mass in a stable carbon aerosol concentration of $2.1 \pm 0.3 \times 10^4$ pt cm^{-3} (i.e., $33 \pm 7 \mu$g m^{-3}). Therefore, the concentration inside the chamber after the ENP exposure termination (i.e., aerosol evacuating process) was not recorded. As expected, the similar effect to that obtained from the first online test was also clearly seen in this experiment (Fig. 9a). The linear decrease in the CANTOR-1 resonant frequency (i.e., -0.07 ± 0.02 Hz min^{-1}) was proportional to the increase in the collected ENP mass (i.e., 0.46 ± 0.11 ng min^{-1}), with an ENP sampling efficiency of 2.1 ± 0.5 %. In the last period of the measurement, the resonant frequency of CANTOR-1 be-

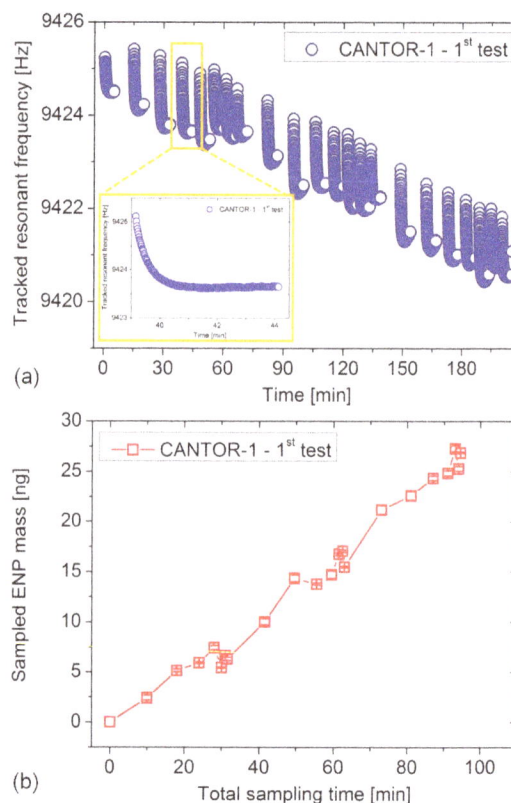

(a)

(b)

Figure 8. (a) The tracked resonant frequency and (b) its corresponding collected ENP mass monitored with CANTOR-1 during the first test of carbon ENP sampling.

fore being switched off was 9422.03 ± 0.04 Hz, which was also identical for the initial value on the next measurement day.

On the second day of the second online test, CANTOR-1 was kept operational to measure the ENP mass concentration from the early beginning of the exposure up to the end of the chamber evacuation of ENPs. However, at a point where the proof of stable ENP mass concentration has been sufficient, the flowing of the dispersed aerosols to the chamber was stopped. This action was used to validate the CANTOR-1 performance in terms of its sensitivity to the transitions of the different ENP concentrations from low to high exposure levels and in the opposite way. It can be obviously seen from Fig. 9b that the CANTOR-1 resonant frequency gradually decreased as the aerosol ENPs started to be injected into the chamber. This frequency reduction became linear (i.e., -0.05 ± 0.03 Hz min^{-1}, corresponding to a collected mass of 0.34 ± 0.09 ng min^{-1} with an ENP sampling efficiency of 1.1 ± 0.6 %) after the aerosols reached their stable condition of $2.7 \pm 0.2 \times 10^4$ pt cm^{-3} (i.e., $46 \pm 1 \mu$g m^{-3}). However, after exposure termination, the CANTOR-1 resonant frequency did not degrade into lower values anymore. Instead, it stayed at almost constant values of 9418.93 ± 0.13 Hz, meaning that there were no ENPs being sampled on the cantilever surfaces.

(a)

(b)

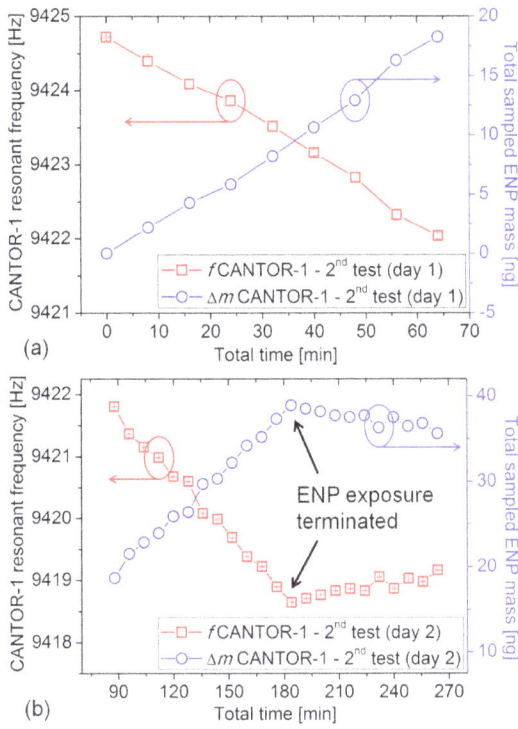

Figure 9. The tracked resonant frequency of CANTOR-1 during the second test of carbon ENP sampling and its corresponding collected ENP mass for **(a)** day 1 and **(b)** day 2. On day 2, CANTOR-1 was kept operational after ENP exposure termination.

There was even a slight decrease in total sampled ENP mass. It could be a condition where the already deposited ENPs or their agglomerates were being lifted off again as the cantilever sensor of CANTOR-1 kept vibrating.

3.3 Detector calibration

Although CANTOR-1 had already exhibited good performance in the two online ENP mass monitoring tests as a microbalance, it still became a necessity to calibrate this home-built device with a standard NP monitoring instrument in well-defined aerosols to recalculate the measured data in the standard unit for aerosol mass concentration (i.e., μg m^{-3}). Thus, during measurement on the second day of the second online test, the results obtained by CANTOR-1 are also calibrated with FMPS (TSI 3091) (cf. Figs. 7b and 9b). The calibrated ENP mass concentration ($C_{m\_\text{CANTOR-1}}$) in μg m^{-3} can be calculated using

$$C_{m\_\text{CANTOR-1}} \left(\mu\text{g m}^{-3} \right) = \text{CF}(T, \text{RH}, p) \qquad (1)$$
$$\times \frac{\Delta f}{\Delta t} \left(\text{Hz min}^{-1} \right),$$

where CF is the calibration factor in terms of the interferences from temperature, relative humidity, and pressure.

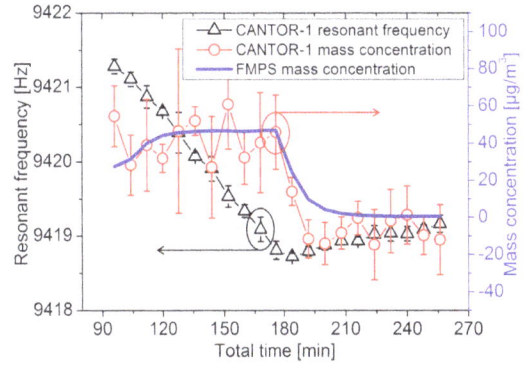

Figure 10. The tracked resonant frequency and its corresponding ENP mass concentration calibrated towards FMPS (TSI 3091) monitored with CANTOR-1 during the second day test of real-time carbon ENP sampling.

Meanwhile, $\Delta f / \Delta t$ is the resonant frequency shift per time unit of CANTOR-1 (i.e., Hz min^{-1}). However, prior to the determination of CF, the ENP number concentration measured by FMPS (i.e., C_{n_FMPS}) must first be converted into ENP mass concentration (i.e., C_{m_FMPS}) by assuming that the size-distributed particles are in a spherical shape. The CF value (i.e., μg min (m^3 Hz)$^{-1}$), in this case, is given by

$$\text{CF} = \frac{C_{m\_\text{FMPS\_avg}}}{X_{\Delta f / \Delta t\_\text{avg}}} \left(\mu\text{g min (m}^3 \text{ Hz)}^{-1} \right), \qquad (2)$$

where $C_{m\_\text{FMPS\_avg}}$ and $X_{\Delta f / \Delta t\_\text{avg}}$ are the averaged value of C_{m_FMPS} under stable conditions and its corresponding mean value of the CANTOR resonant frequency shift per time unit, respectively.

Figure 10 shows the calibrated ENP mass concentrations measured by CANTOR-1 in comparison with those measured by FMPS. In this case, the CF used for CANTOR-1 was 815 μg min (m^3 Hz)$^{-1}$. During 1 h assessment of a stable carbon ENP mass concentration of $46 \pm 1\,\mu$g m^{-3} measured by FMPS, the precision of the calibrated ENP concentrations measured by CANTOR-1 was found to be $< 55\,\%$, which was taken as the deviation of each measured data point. Moreover, by multiplying the minimum frequency shifts that can be resolved with CF, a limit of detection (LOD) of $< 25\,\mu$g m^{-3} was obtained for this detector. Regardless of some drawbacks of CANTOR-1 (i.e., time-consuming sensor preparation and a low ENP sampling efficiency of 1.07 %), the overall experimental results have verified the good sensitivity and working status of CANTOR-1, because in the real workplace, this detector aims to be used as a first alert for informing the workers in regards to the danger caused by the unexpected excessive concentration of the ENPs in air that can occur suddenly during their working shifts.

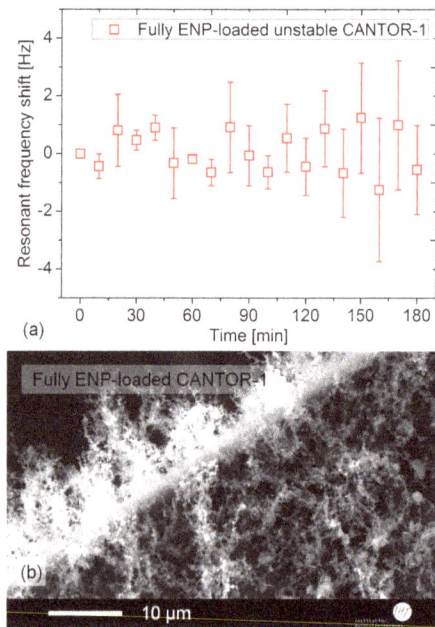

Figure 11. (a) Typical unstable resonant frequency signal and (b) surface condition of a fully ENP-loaded CANTOR-1.

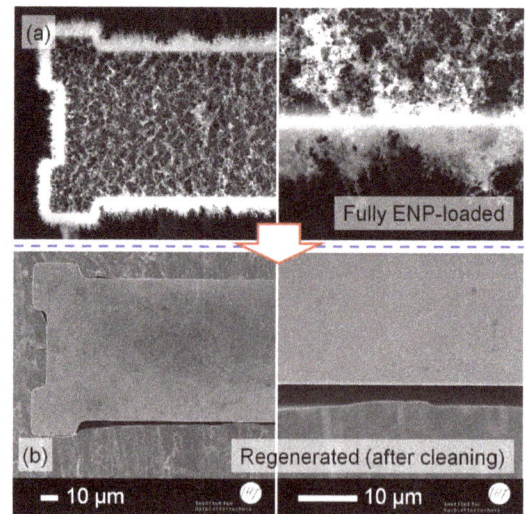

Figure 12. The surface conditions of fully ENP-loaded CANTOR-1 (a) before and (b) after ultrasonic cleaning.

3.4 Detector robustness

To test the robustness of CANTOR-1 in terms of its operating lifetime, long-term use during a carbon ENP exposition of ~ 1–7×10^4 pt cm^{-3} was performed with a recycled silicon piezoresistive cantilever integrated into CANTOR-1 within 7 workdays. It was found that the signal of CANTOR-1 started to get unstable only after 43.5 h of ENP sampling as compared to the normal device performance (Fig. 11a). Thus, ~ 40 h of continuous operations can be considered the lifetime of CANTOR-1 under typical mass concentrations in the ambience at workplaces of 20–120 μg m^{-3}. For the corresponding frequency shift and collected ENPs, we found -79.15 Hz and 0.54 μg, respectively. Using the cantilever mass of $m_{cant} \approx 32\,\mu$g, we conclude that the operating life of CANTOR-1 ends when the mass of deposited ENPs reaches a value of $\sim 2\,\%$ of the cantilever weight. This fact was supported by SEM, showing the surface of the cantilever heavily loaded by carbon ENPs (Fig. 11b). Almost all of the deposited ENPs exhibit an agglomerate shape. Thus, the unstable signal of CANTOR-1 might be explained by a loss of mass by detaching of large ENP agglomerates of high-enough inertia.

To regenerate the fully ENP-loaded cantilever, a relatively simple but efficient wet cleaning method was demonstrated using an ultrasonic cleaner with acetone solution or deionized (DI) water inside. This ultrasonic cleaning technique had been implemented in several heavily used cantilever sensors with varied geometries (i.e., down to nanoscaled devices) and exposure levels, which had always obtained high cleaning efficiencies up to $\sim 99\,\%$ (Wasisto et al., 2013c, d).

During the high pressure stage of the cleaning process using a Bandelin Sonorex Ultrasonic bath TK 52 ($f = 35$ kHz), the formed bubbles imploded, releasing enormous amounts of energy attacking every surface; hence, after 0.5–2 min, the deposited ENPs were almost completely detached from the cantilever surface. Figure 12a and b depict the surface conditions of the heavily ENP-loaded cantilever of CANTOR-1 before and after the ultrasonic wet cleaning process, respectively. Nevertheless, considering their potentially low fabrication cost (i.e., fabricated using bulk silicon instead of SOI wafers), the cantilevers may also be considered a disposable component. Since spring-loaded contact pins are used, wire-bonding is not required for electrical contact to the cantilever die. Thus, it can be replaced easily with either a new or a regenerated one after some period of the ENP sampling.

3.5 Detector comparison

To compare the developed CANTOR-1 with the other up-to-date MEMS-/NEMS-based particle detectors worldwide, Table 1 has summarized their important data (i.e., measurement principle, LOD, particle sampling time, and notes). From this comparison, it is clearly shown that only CANTOR-1 uses the electrophoretic sampling method. The other prototypes from the University of Denver (DU, USA; Hajjam et al., 2011), the Berkeley Sensor and Actuator Center (BSAC, USA; Paprotny et al., 2013), the Technical University of Denmark (DTU, Denmark; Schmid et al., 2013), and McGill University (Canada; Morris et al., 2014), employ impaction and thermophoretic methods. The slowest response is exhibited by quartz crystal microbalance (QCM) from McGill University, with a 30 min sampling time. Moreover, the detected targets of the devices from McGill University, DU, and BSAC were large particles (i.e., microparticles). They

Table 1. Comparison between the developed partially integrated CANTOR-1 and the other currently researched micro/nanomechanical particle detectors.

Developer, reference	Measurement principle	LOD and particle sampling time	Notes
DU, USA (Hajjam et al., 2011)	Direct deposition in a partial vacuum and thermally actuated silicon resonator	$3\,\mu g\,m^{-3}$, 10 s	Partial vacuum, microscope and large pump needed, microparticle detection, no sensor recycling process
BSAC, LBNL, and EPA, USA (Paprotny et al., 2013)	Thermophorectic precipitator and film bulk acoustic resonator (FBAR)	$2\,\mu g\,m^{-3}$, 10 min $10\,\mu g\,m^{-3}$, 4 min	High particle loss, fan stack needed, non-integrated electronic circuitries, no sensor recycling process
DTU, Denmark (Schmid et al., 2013)	Inertial impactor and nanomechanical resonant filter fiber	$< 1\,\mu g\,m^{-3}$, 1 s	High particle loss, external pump needed, bulky measurement tools required (i.e., lock-in amplifier and laser Doppler vibrometer), no sensor recycling process
McGill University, Canada (Morris et al., 2014)	Impactor and quartz crystal microbalance (QCM)	$5.2\,\mu g\,m^{-3}$, 30 min	Bulky measurement tools and vacuum pump needed, microparticle detection, no sensor recycling process
IHT TU Braunschweig and Fraunhofer WKI, Germany (this work)	Electrophoretic precipitator and self-sensing silicon cantilever resonator actuated with a piezoelectric actuator (CANTOR-1)	$25\,\mu g\,m^{-3}$, 1–5 min	Partially integrated, miniaturized sampler, two-stage microparticle filtrations, sensor recycling process demonstrated, sensor calibrated, digital multimeter and power supply required

do not have any filtering systems for smaller particles (i.e., NPs). Some of them even had to be operated with a large vacuum pump and the optical method to yield the aerosol flows and read out the signal, which are not applicable for development of low-cost wearable sensor systems. Although most of them provide slightly better LODs than CANTOR-1, they have not demonstrated a method to remove the deposited particles and recycle their sensors. Furthermore, for the nanowire device from DTU, it is expected that the sensor will be fully loaded by NPs within only a few minutes. As an improvement to CANTOR-1, its next generation (i.e., CANTOR-2) will be developed as a fully integrated system, where all sensing components and their supporting electronics will be miniaturized and packed into handy-format housing. Hence, it can be worn easily by workers in nanotechnology industries during their working shifts. According to the paradigm shift reported for air pollution monitoring towards lower-cost, easy-to-use, portable, and direct-reading sensors, there is further commercial potential for ENP monitoring to be expected in the near future (Kumar et al., 2015; Snyder et al., 2013).

4   Conclusions

The first generation of a low-cost portable cantilever-based airborne nanoparticle detector (i.e., CANTOR-1) has been developed as a partially integrated direct-reading device, assessed with carbon aerosol engineered nanoparticles (ENPs) under typical workplace conditions, and calibrated towards a standard ENP monitoring instrument of a fast mobility particle sizer (FMPS, TSI 3091) with a measurement precision of $< 55\,\%$ and a limit of detection (LOD) of $< 25\,\mu g\,m^{-3}$. The two-stage filtrations have also been tested, leading to a rejection of micro-sized particles ($d_P > 1\,\mu m$). The cantilever module of CANTOR-1 could also be regenerated using the ultrasonic cleaning method after being fully loaded by ENPs in a long-term ENP sampling. Further work is needed to improve the sensing performance and to pack all components into a fully wearable handy NP detector.

Acknowledgements. The authors would like to thank I. Kirsch, Q. Zhang, K. Huang, Z. Wang, W. Wu, J. Arens, D. Rümmler, and K.-H. Lachmund for their valuable technical support. This work was performed in the NanoExpo collaborative project funded by the German Federal Ministry of Education and Research (BMBF) within the NanoCare cluster under number 03X0098A/B.

References

Balbus, J. M., Florini, K., Denison, R. A., and Walsh, S. A.: Protecting workers and the environment: An environmental NGO's perspective on nanotechnology, J. Nanopart. Res., 9, 11–22, doi:10.1007/s11051-006-9173-7, 2007.

Bekker, C., Brouwer, D. H., Tielemans, E., and Pronk, A.: Industrial production and professional application of manufactured nanomaterials-enabled end products in Dutch industries: Potential for exposure, Ann. Occup. Hyg., 57, 314–327, doi:10.1093/annhyg/mes072, 2013.

Brouwer, D.: Exposure to manufactured nanoparticles in different workplaces, Toxicology, 269, 120–127, doi:10.1016/j.tox.2009.11.017, 2010.

Buonanno, G., Jayaratne, R. E., Morawska, L., and Stabile, L.: Metrological performances of a diffusion charger particle counter for personal monitoring, Aerosol Air Qual. Res., 14, 156–167, doi:10.4209/aaqr.2013.05.0152, 2014.

Cho, C.-H., Jaeger, R. C., and Suhling, J. C.: Characterization of the temperature dependence of the piezoresistive coefficients of silicon from −150 °C to +125 °C, IEEE Sens. J., 8, 1455–1468, doi:10.1109/JSEN.2008.923575, 2008.

Duarte, K., Justino, C. I. L., Freitas, A. C., Duarte, A. C., and Rocha-Santos, T. A. P.: Direct-reading methods for analysis of volatile organic compounds and nanoparticles in workplace air, Trend Anal. Chem., 53, 21–32, doi:10.1016/j.trac.2013.08.008, 2014.

Fierz, M., Houle, C., Steigmeier, P., and Burtscher, H.: Design, calibration, and field performance of a miniature diffusion size classifier, Aerosol Sci. Tech., 45, 1–10, doi:10.1080/02786826.2010.516283, 2011.

Hajjam, A., Wilson, J. C., and Pourkamali, S.: Individual airborne particle mass measurement using high-frequency micromechanical resonators, IEEE Sens. J., 11, 2883–2890, doi:10.1109/JSEN.2011.2147301, 2011.

Hullman, A.: Measuring and assessing the development of nanotechnology, Scientometrics, 70, 739–758, doi:10.1007/s11192-007-0310-6, 2007.

Krinke, T. J., Deppert, K., Magnusson, M. H., Schmidt, F., and Fissan, H.: Microscopic aspects of the deposition of nanoparticles from the gas phase, Aerosol Sci., 33, 1341–1359, doi:10.1016/S0021-8502(02)00074-5, 2002.

Kumar, P., Morawska, L., Martani, C., Biskos, G., Neophytou, M., Di Sabatino, S., Bell, M., Norford, L., and Britter, R.: The rise of low-cost sensing for managing air pollution in cities, Environ. Int., 75, 199–205, doi:10.1016/j.envint.2014.11.019, 2015.

Maynard, A. D.: Nanotechnology the next big thing, or much ado about nothing?, Ann. Occup. Hyg., 51, 1–12, doi:10.1093/annhyg/mel071, 2007.

Mehdizadeh, E., Kumar, V., Pourkamali, S., Gonzales, J., and Abdolvand, R.: A two-stage aerosol impactor with embedded MEMS resonant mass balances for particulate size segregation and mass concentration monitoring, Proceedings of IEEE SENSORS 2013, Baltimore, USA, 1–4, doi:10.1109/ICSENS.2013.6688317, 2013.

Morris, D. R. P., Fatisson, J., Olsson, A. L. J., Tufenkji, N., and Ferro, A. R.: Real-time monitoring of airborne cat allergen using a QCM-based immunosensor, Sensor. Actuat. B-Chem., 190, 851–857, doi:10.1016/j.snb.2013.09.061, 2014.

Paprotny, I., Doering, F., Solomon, P. A., White, R. M., and Gundel, L. A.: Microfabricated air-microfluidic sensor for personal monitoring of airborne particulate matter: Design, fabrication, and experimental results, Sensor. Actuat. A-Phys., 201, 506–516, doi:10.1016/j.sna.2012.12.026, 2013.

Peiner, E., Balke, M., and Doering, L.: Slender tactile sensor for contour and roughness measurements within deep and narrow holes, IEEE Sens. J., 8, 1960–1967, doi:10.1109/JSEN.2008.2006701, 2008.

Schmid, S., Kurek, M., Adolphsen, J. Q., and Boisen, A.: Real-time single airborne nanoparticle detection with nanomechanical resonant filter-fiber, Scientific Reports, 3, 1288, doi:10.1038/srep01288, 2013.

Snyder, E. G., Watkins, T. H., Solomon, P. A., Thoma, E. D., Williams, R. W., Hagler, G. S. W., Shelow, D., Hindin, D. A., Kilaru, V. J., and Preuss, P. W.: The changing paradigm of air pollution monitoring, Environ. Sci. Technol., 47, 11369–11377, doi:10.1021/es4022602, 2013.

Sökmen, Ü., Stranz, A., Waag, A., Ababneh, A., Seidel, H., Schmid, U., and Peiner, E.: Evaluation of resonating Si cantilevers sputter-deposited with AlN piezoelectric thin films for mass sensing applications, J. Micromech. Microeng., 20, 064007, doi:10.1088/0960-1317/20/6/064007, 2010.

Tammet, H., Mirme, A., and Tamm, E.: Electrical aerosol spectrometer of Tartu University, Atmos. Res., 62, 315–324, doi:10.1016/S0169-8095(02)00017-0, 2002.

Wasisto, H. S., Merzsch, S., Waag, A., Uhde, E., Salthammer, T., and Peiner, E.: Portable cantilever-based airborne nanoparticle detector, Sensor. Actuat. B-Chem., 187, 118–127, doi:10.1016/j.snb.2012.09.074, 2013a.

Wasisto, H. S., Merzsch, S., Waag, A., Uhde, E., Salthammer, T., and Peiner, E.: Evaluation of photoresist-based nanoparticle removal method for recycling silicon cantilever mass sensors, Sensor. Actuat. A-Phys., 202, 90–99, doi:10.1016/j.sna.2012.12.016, 2013b.

Wasisto, H. S., Merzsch, S., Stranz, A., Waag, A., Uhde, E., Salthammer, T., and Peiner, E.: Silicon resonant nanopillar sensors for airborne titanium dioxide engineered nanoparticle mass detection, Sensor. Actuat. B-Chem., 189, 146–156, doi:10.1016/j.snb.2013.02.053, 2013c.

Wasisto, H. S., Merzsch, S., Stranz, A., Waag, A., Uhde, E., Salthammer, T., and Peiner, E.: Femtogram aerosol nanoparticle mass sensing utilising vertical silicon nanowire resonators, IET Micro & Nano Letters, 8, 554–558, doi:10.1049/mnl.2013.0208, 2013d.

Wasisto, H. S., Zhang, Q., Merzsch, S., Waag, A., and Peiner, E.: A phase-locked loop frequency tracking system for portable microelectromechanical piezoresistive cantilever mass sensors, Microsyst. Technol., 20 559–569, doi:10.1007/s00542-013-1991-9, 2014.

Alternative strategy for manufacturing of all-solid-state reference electrodes for potentiometry

J. C. B. Fernandes and E. V. Heinke

The Municipal University of São Caetano do Sul, USCS, Centre Campus, 50, Santo Antônio Street, 09521-160, São Caetano do Sul – SP, Brazil

Correspondence to: J. C. B. Fernandes (jcbastos@uscs.edu.br, jcnandes@gmail.com)

Abstract. This paper presents an alternative strategy for manufacturing solid-state reference electrodes based on particles of graphite/silver/silver chloride synthesized by electroless deposition of metallic silver and silver chloride on graphite powder. Two kinds of reference electrodes were manufactured by mixing these particles with epoxy resin and hardener: quasi-reference and all-solid-state containing salts of alkaline or alkaline earth metals. All-solid-state reference electrodes can be sterilized with high-pressure saturated steam at 394.15 K (121 °C) using an autoclave. These electrodes presented a stable potential between pH 2 and 11. The electrode surface was characterized by scanning electron microscopy and showed the presence of silver and salt particles. The size of the silver particles was less than 2.5 μm. We successfully applied the all-solid state reference electrodes in potentiometric cells to measure pH and potassium ions in complex matrix by direct potentiometry and L-ascorbic acid by potentiometric titration.

1 Introduction

Reference electrodes have been the Achilles heel for many electrochemical systems, especially for potentiometric techniques. A good reference electrode for potentiometric measurements must maintain a constant comparative potential value and cannot respond to the presence of chloride ions in the sample solution. The most common reference electrodes have a plastic or glass body for storing the saturated solution of potassium chloride called electrolyte. A silver wire coated with silver chloride, a paste of mercury and mercury chloride (calomel), or a paste of thallium, mercury and thallium chloride (Thalamid) is dipped into this solution (Guth et al., 2009).

The calomel electrode has better technical characteristics than reference electrodes based on silver/silver chloride (Ag/AgCl, Cl$^-$). The potential drift of the calomel electrode is low and not even light influences this electrode. However, the calomel electrode cannot be used in temperatures above 353.15 K (80 °C) because it presents hysteresis at these temperatures (Janz and Taniguchi, 1953). Like Thalamid electrodes, calomel electrodes will probably be banned in the next years, because mercury is very hazardous to the envi-

ronment. Therefore, reference electrodes based on Ag/AgCl, Cl$^-$ may be the most popular in the future.

Conventionally, reference electrodes based on Ag/AgCl, Cl$^-$ are produced by electroplating of silver on the surface of a platinum wire (5–6 h) and 15–25 w/w % of the silver is converted to silver chloride by anodic treatment with hydrochloric acid for 30 min (Janz and Taniguchi, 1953).

Many patents (Jayaweera et al., 1995; Shin et al., 2003; Sorensen and Zachau-Christiansen, 2004; Cha et al., 2005; Rodes, 2008) and articles (Suzuki et al., 1998, 1999; Tymecki et al., 2004; Guth et al., 2009; Rius-Ruiz et al., 2011; Valdés-Ramírez et al., 2011) reported innumerable strategies for construction of solid-state reference electrodes. Shin et al. (2003) developed a reference electrode by depositing metal (Ag, Pt, Au or Cu) and insoluble salt (silver halide) in layers on a substrate using thick film. The inner reference solution was in the state of hydrogel with an insulating film separating it from the aqueous solution. Timecki et al. (2004) produced a miniaturized reference electrode employing a sandwich of three pastes: Ag (725A-6S-54 from Acheson) as transducer, Ag/AgCl as the sensing layer and potassium chloride (KCl) as the protective layer. Carbon nanotubes were also used as a transducer from a polyacrylate membrane containing Ag/AgCl, Cl$^-$ system (Rius-Ruiz et al., 2011). Thin film technology has also been used for microfabrication of reference electrodes (Suzuki et al., 1999). Valdés-Ramírez et al. (2011) electrodeposited silver and silver chloride on a graphite-epoxy electrode surface under adequate conditions of electrical potential and electrical current.

On the one hand, the large number of layers may complicate the miniaturization by techniques of thin film or of thick film. On the other hand, reference electrodes manufactured by electroplating using a potentiostat have an Ag/AgCl layer thickness that may be destructed by friction.

Other problems of reference electrodes include the following: classical reference electrodes can develop a potential due to outflow of the inner electrolyte through the liquid connection. One strategy to overcome this problem is to use gel electrolytes. However, gels of agar-agar or gelatin are unstable with temperature variation and reference electrodes with these gels cannot be sterilized into autoclave. Since, advanced gels such as polyvinyl acrylamide, polyvinyl alcohol, hydroxymethyl cellulose and hydroxyethyl cellulose have higher temperature resistance, these gels can be used alternatively. However, advanced gels have lower solubility for salts, therefore, a potential drift can be developed if chloride ions are in the sample solution (Guth et al., 2009).

The aim of this work is to propose an alternative strategy for manufacturing of all-solid state reference electrodes by developing silver and silver chloride particles that are deposited on graphite powder by electroless deposition without need of electroplating equipment.

2 Experimental

2.1 Materials

Trizma base buffer (minimum 99.9 %), silica gel (60 μm size), valinomycin and bis-2-ethylhexyl adipate were purchased from Sigma-Aldrich. Potassium tetraoxalate dehydrate (pH = 1.679), potassium hydrogen phthalate (pH = 4.008), potassium dihydrogen phosphate, sodium hydrogen phosphate (pH = 6.865), sodium tetraborate (pH = 9.180) and calcium hydroxide (pH = 12.454) were used for preparing primary standard pH solutions at 298.15 K (25 °C). These reagents were also purchased from Sigma-Aldrich. High molecular weight poly(vinyl)chloride (PVC) was donated by Solvay-Brazil. BQ164 ink based on Ag/AgCl particles was acquired from DuPont Microcircuit Materials. Other reagents, such as L-ascorbic acid, potassium chloride, anhydrous sodium sulfate, potassium iodide, barium chloride, potassium hydroxide, ammonium hydroxide, sodium hydroxide, silver nitrate, glucose, sodium fluoride, hydrochloric acid, sulfuric acid, propanone, tetrahydrofuran and graphite powder (50–300 μm size) were analytical grade reagents. We also acquired Brascola's epoxy resin and poly(imide) of the trademark Araldite.

Noncarbonated sports drinks and syrups were obtained from local markets or pharmacies. All solutions were prepared using distilled water.

2.2 Instrumentation and protocols

All potentiometric measurements were carried out using an ion analyser HI 223 model from Hanna instruments. All reference electrodes were characterized analysing their potential response against a commercial reference electrode (Ag/AgCl), R684-105 model from Analion. In our experiments, the reference electrolyte was potassium chloride solution at 3 mol L^{-1}. Potentiometric cells with a pH single-rod measuring cell (HI-1131P model) or a homemade potassium ion selective electrode (ISE-K) were used for testing the all-solid-state reference electrodes in real samples for direct potentiometry. The ISE-K was prepared according to previous work (Fernandes et al., 2000).

Potentiometric titrations of L-ascorbic acid were made with a mercury indicator electrode. The mercury electrode was prepared according to literature (Riyazuddin and Nazer, 1997), with some adaptations. We used a copper rod of 2 mm diameter and 5 cm length. The copper rod was immersed in a solution of epoxy : polyimide (weight proportion, 3 : 1) in propanone to form a film that electrically isolates the copper from the environment. The bottom of the copper rod was trimmed to remove the cured resin. Afterwards, this rod was polished with abrasive papers. The copper electrode was immersed in acid solution of mercury (II) nitrate at 0.02 mol L^{-1} for 10 min for deposition of the metallic mercury on the copper surface. The electrode was washed with distilled water and polished using a filter paper with a pore size of 8 μm.

Morphological studies of the surface were performed using FEI Quanta FEG 250 scanning electron microscopy. The samples were coated with a thin layer of gold (Au) by a sputter coater to reduce charging.

2.2.1 Synthesis of metallic silver and silver chloride particles on graphite powder

Metallic silver particles were obtained by electroless deposition using silver nitrate, ammonia, potassium hydroxide and glucose solutions (Koura, 1990). For electroless deposition of silver and silver chloride particles, silver nitrate solutions concentrated between 0.04 and 8.00 mol L^{-1} were used. The concentrations of the other solutions were estimated by the stoichiometry of the reactions according to the following chemical equations, from (1) to (3).

$$Ag(H_2O)_n \underset{}{\overset{NH_3}{\rightleftharpoons}} Ag(H_2O)_{n-1}NH_3 + H_2O \quad (1)$$

$$Ag(H_2O)_{n-1}NH_3 \underset{}{\overset{NH_3}{\rightleftharpoons}} Ag(H_2O)_{n-2}(NH_3)_2 + H_2O \quad (2)$$

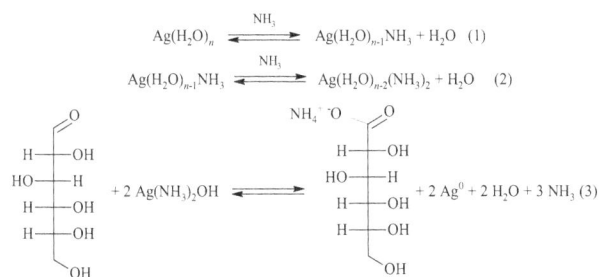

Silver chloride particles were obtained using only silver nitrate and potassium chloride solutions. Electroless silver deposition was carried out at 283.15 K (10 °C) and silver chloride precipitation at room temperature.

For preparation of the silver particles on graphite, about 8.00 g of graphite powder were weighed. The graphite powder was put uniformly into a glass container of 2 L to form a layer with thickness < 1 mm. We carefully added 10 mL of silver nitrate solution at 8.0 mol L^{-1}, 10 mL of ammonium hydroxide at 16.0 mol L^{-1}, 32 mL of glucose solution at 2.1 mol L^{-1} and 10 mL of potassium hydroxide solution at 4.0 mol L^{-1} to the graphite powder without stirring.

A similar procedure was used for deposition of silver chloride particles on the graphite/silver powder. In this case, we weighed 8.00 g of graphite/silver powder and put it into a glass container of 2 L. Afterwards, we added 100 mL of silver nitrate solution and 100 mL of potassium chloride solution (both 0.4 mol L^{-1}) into the glass container.

The graphite/silver/silver chloride particles were washed with cold distilled water and filtered under vacuum. The powder was dried in an oven at 313.15 K (40 °C) for 1 h. The graphite/silver/silver chloride particles were stocked in a dark glass flask to avoid decomposition by action of light.

2.2.2 Preparation of the quasi-reference electrodes

We manufactured quasi-reference electrodes mixing the particles of graphite/silver/silver chloride with epoxy resin and hardener (3 : 1). The weight of these particles was maintained constant at 100 mg while the weight of epoxy resin plus hardener (3 : 1) was changed to obtain a percentage of graphite/silver/silver chloride particles of between 15 and 75 % w/w. We added 2 mL of propanone to the paste in order to dissolve the adhesive. The paste was homogenized in a glass mortar with a glass pestle. Afterwards, propanone was left to evaporate until the mixture reached the proper viscosity to be transferred to a tube with a 3 mm inner diameter and 5 cm of length.

Finally, a copper wire was dipped into a mercury (II) nitrate solution (0.02 mol L^{-1}) to produce a thin layer of metallic mercury on its surface. Mercury layer avoids the copper oxidation by the silver chloride. The wire was inserted into the paste to obtain electrical contact.

All electrodes were left to dry for 8 h at 363.15 K (90 °C). The surface of the reference electrodes was polished with abrasive papers of 150–2 000 meshes before use.

We also verified the use a commercial ink (BQ164, see Sect. 2.1) for manufacturing potentiometric reference electrodes. These electrodes were fabricated in the same manner described before.

2.2.3 Preparation of the all-solid-state reference electrodes

The protocol for manufacturing all-solid-state reference electrodes was similar to the one described previously. The amount of epoxy resin plus hardener was maintained constant in a proportion of 3 : 1 in terms of weights. The concentration of graphite/silver/silver chloride particles was about 45 % w/w. Salts of halogens (F^{-}, Cl^{-} and I^{-}) with cations of alkaline metals (Na^{+} and K^{+}) or alkaline earth metals (Ba^{2+}) were added to the paste at concentrations of between 1.5 and 3.0 mol kg^{-1}. Sulfate anion (SO$_4^{2-}$) was used in some formulations as well. Silica (SiO$_2$) was studied for reducing the surface tension of the reference electrode. Table 1 shows the amount used in the preparation of all-solid-state reference electrodes with additives.

2.2.4 Characterization of the reference electrodes

Standard solutions of chloride ions (KCl) from 1×10^{-6} to 1×10^{-1} mol L^{-1} were prepared in 0.2 mol L^{-1} Trizma sulfate buffer at pH 7.0 to maintain a constant ionic strength. Manufactured reference electrodes were characterized against a conventional reference electrode based on Ag/AgCl, 3 mol L^{-1} KCl.

The effect of the pH on all-solid-state reference electrodes was studied in the pH range from 1 to 13. Primary and secondary standard solutions of pH were prepared according to the literature (Weast, 1998) and employed in these studies.

The leaching out effect of salts from an all-solid-state reference electrode was studied maintaining the reference electrode with 3.0 mol kg^{-1} of KCl continuously immersed in 1 mL of deionized water during 20 days. Afterwards, the potassium chloride content of storage water was analysed by gravimetric method using a silver nitrate solution 0.5 mol L^{-1} as a precipitating agent.

A potentiometric cell for potassium ions (ISE-K) based on valinomycin as an ionophore against an all-solid-state reference electrode was used in real samples. The potentiometric cell was calibrated with standard solutions of potassium ions (KCl) from 1×10^{-4} to 1×10^{-1} mol L^{-1}. These standard solutions and the samples of the noncarbonated sports drinks and syrups were also diluted in 0.2 mol L^{-1} Trizma sulfate buffer at pH 7.0.

Potentiometric titrations of L-ascorbic acid were studied to verify the sensitivity of a redox substance on all-solid-state reference electrodes using an indicator electrode of mer-

Table 1. Formulations of the all-solid-state reference electrodes (ASSREs).

ASSREs	$C_{graphite}$/Ag/AgCl particles (mg)	Epoxy resin (mg)	Hardener (mg)	Salt (mg)
$C_{graphite}$/Ag/AgCl, $3.0 \, mol \, kg^{-1}$ KCl	150.0	91.6	30.5	60.7 (KCl)
$C_{graphite}$/Ag/AgCl, $1.5 \, mol \, kg^{-1}$ KCl : SiO_2	150.0	91.6	30.5	30.3 (KCl) 30.3 (SiO_2)
$C_{graphite}$/Ag/AgCl, $1.5 \, mol \, kg^{-1}$ $BaCl_2$	228.4	91.6	30.5	128.4 ($BaCl_2$)
$C_{graphite}$/Ag/AgCl, $3.0 \, mol \, kg^{-1}$ Na_2SO_4	264.9	91.6	30.5	164.9 (Na_2SO_4)
$C_{graphite}$/Ag/AgCl, $3.0 \, mol \, kg^{-1}$ KI	320.3	91.6	30.5	220.3 (KI)
$C_{graphite}$/Ag/AgCl, $3.0 \, mol \, kg^{-1}$ NaF	132.0	91.6	30.5	32.0 (NaF)

cury (Hg). A weight of L-ascorbic acid of between 7.5 and 15.0 mg was weighed in a beaker (50 mL) with an analytical balance. We added 30 mL of distilled water, $500 \, \mu L$ of ammonium thiocyanate at $0.1 \, mol \, L^{-1}$ and $500 \, \mu L$ of sodium acetate/acetic acid buffer solution (pH $4.7/1 \, mol \, L^{-1}$) to the beaker. We titrated the L-ascorbic acid with a copper sulfate solution at $0.0125 \, mol \, L^{-1}$ as titrant using a burette of $10\,000 \, mL$.

All-solid-state reference electrodes were sterilized by high-pressure saturated steam at 394.15 K (121 °C) using an autoclave during 15 min in order to verify its resistance to temperature and pressure.

3   Results and discussion

Among the strategies for manufacturing reference electrodes for potentiometry, the use of silver and silver chloride seems to be considered the most suitable.

We employed graphite powder as a support of the Ag/AgCl particles. Silver particles were obtained by reducing the silver ammonia complex with glucose as shown in chemical Eq. (3), while silver chloride particles were obtained by means of a precipitating agent.

The production of small particles (less than 500 nm in diameter) by electroless deposition depends on the concentration of reagents. On the one hand, a low concentration generates small particles, because there is not a sufficient number of particles to form a crystal. On the other hand, a high concentration yields small particles since many nucleation points are formed due to the fact that a very viscose medium has a higher friction coefficient. Thus, migration of particles is difficult and, consequently, particles do not agglomerate (Har-

ris, 2007). The concentration of silver nitrate solutions used for the production of silver and silver chloride particles were respectively 8.0 and $0.4 \, mol \, L^{-1}$.

The size of a particle formed by electroless deposition can also be influenced by temperature since high temperatures provoke an agglomeration of the particles. We avoided the agglomeration of silver particles by electroless deposition by generating these particles at 283.15 K (10 °C). Silver chloride particles were produced at room temperature, because no agglomeration was observed at this temperature.

The electroless deposition of silver yielded a metalized graphite powder. Silver particles produced by chemical synthesis are charged positively due to an excess of silver ions that adsorb on the silver surface. Since graphite particles have a negative charge (Sengupta et al., 2011), silver particles are attracted to their surface.

Silver chloride particles were deposited on graphite/silver powder using a lower concentration of silver nitrate since potassium chloride has a limited solubility in water $(3 \, mol \, L^{-1})$. For silver nitrate concentrations higher than $0.4 \, mol \, L^{-1}$, the excess of silver ions in the solution adsorb onto the surface of the silver chloride particles provoking their agglomeration (Harris, 2007).

In our syntheses, the yield of silver particles and silver chloride particles varied between 60–84 and 90–97 %, respectively. The particles of graphite/silver/silver chloride exhibited a grey colour with a content of silver chloride of about 40 %.

We manufactured quasi-reference electrodes using epoxy resin and hardener as the binder of the graphite particles/silver/silver chloride. Figure 1a shows the electrical behaviour of the quasi-reference electrodes against a conven-

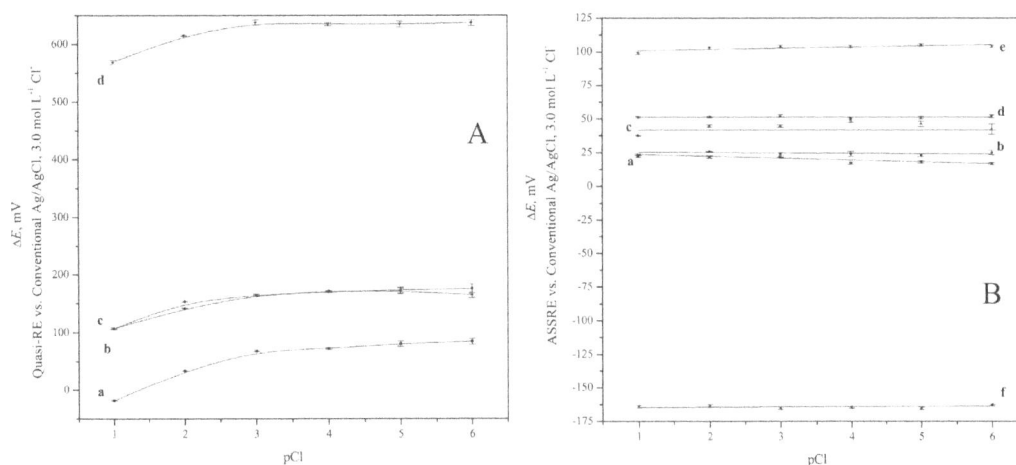

Figure 1. Potential measured with the quasi-reference electrodes (Quasi-RE) or all-solid-state reference electrodes (ASSREs) against a conventional reference electrode at room temperature. Curves obtained with **(A)** quasi-RE with an amount of $C_{graphite}$/Ag/AgCl particles in the paste equal to **(a)** 15 % w/w, **(b)** 30 % w/w, **(c)** 45 % w/w, and **(d)** BQ164 ink from Dupont; and **(B)** ASSREs containing the salts **(a)** KCl, **(b)** KCl : SiO$_2$, **(c)** BaCl$_2$, **(d)** NaF, **(e)** Na$_2$SO$_4$, and **(f)** KI.

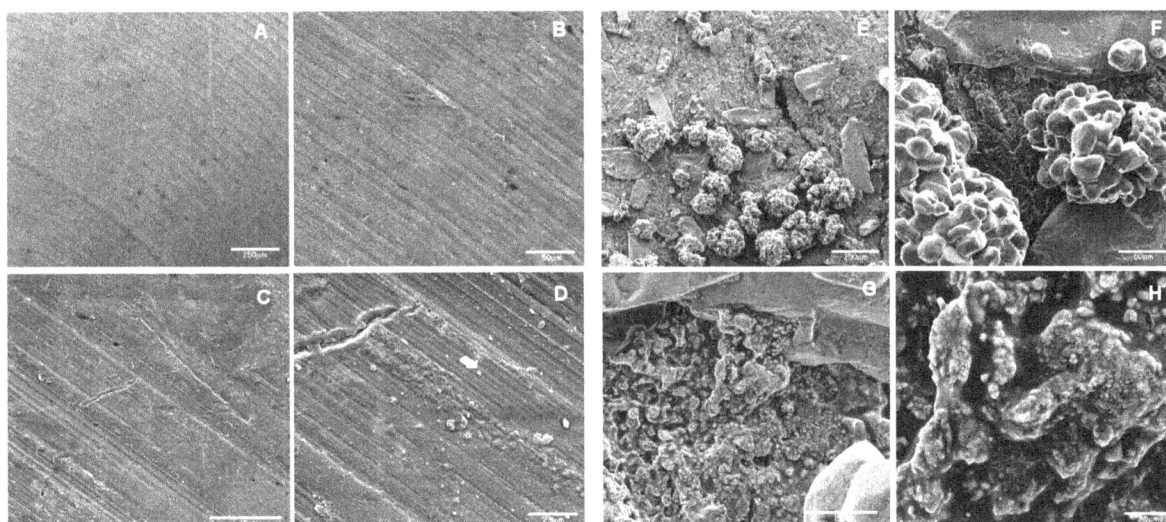

Figure 2. Surface photomicrographs of quasi- and all-solid-state reference electrodes containing 45 % w/w C$_{graphite}$/Ag/AgCl (from **a** to **d**) and 3.0 mol kg^{-1} KCl (from **e** to **h**), respectively. Magnifications are for **(a)** and **(e)** 200 ×, **(b)** and **(f)** 1000×, **(c)** and **(g)** 5000×, and **(d)** and **(h)** 20 000×.

tional reference electrode in solutions with different chloride concentrations. The quasi-reference electrodes developed a potential change to concentration of chloride ions above 1×10^{-3} mol L^{-1}. Similar behaviour was obtained with reference electrodes made from BQ 164 ink.

Although quasi-reference electrodes have been employed in strips to measure glucose based on amperometric measurements (Matsumoto et al., 2002; Mamińska et al., 2006), these electrodes are not suited for use in potentiometric cells. The stability of the potential of a potentiometric cell must be independent of the concentration of chloride ions in the sample solution. Hence, the response only depends on the ion se-

lective electrode. We added a high concentration of chloride ions, to overcome the problem of the quasi-reference electrodes, to the paste of graphite/silver/silver chloride mixed with epoxy resin and hardener. Any value between 1 and 3 mol kg^{-1} in the paste is acceptable because this concentration is about 10 times higher than the one found in real samples (whole blood, for example). Other anions such as sulfate, iodide, and fluoride were also studied. Figure 1b shows the behaviour of these all-solid-state reference electrodes containing salts in their formulation.

The potential of the reference electrodes containing salts was maintained stable and constant to any concentration of

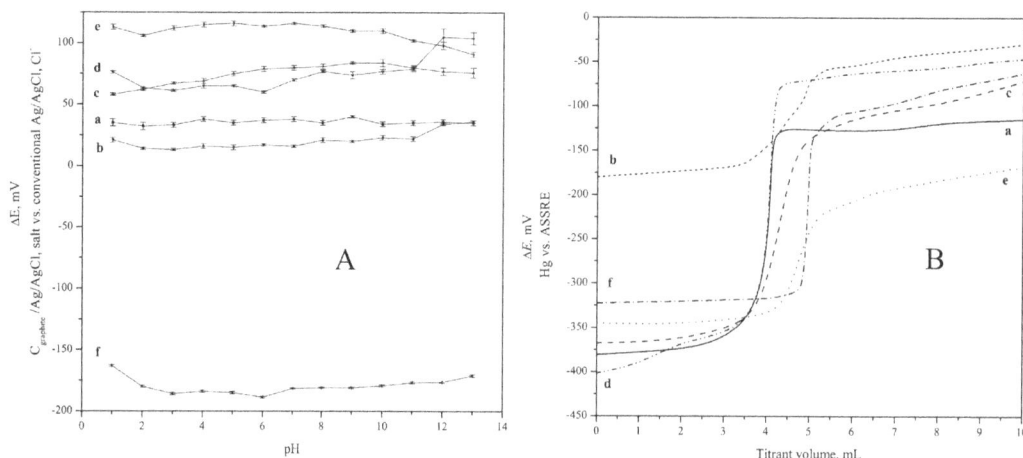

Figure 3. Influence of the pH and reducing agent on ASSREs. (**A**) Curves of pH obtained against ASSREs containing the salts (**a**) KCl, (**b**) KCl : SiO$_2$, (**c**) BaCl$_2$, (**d**) NaF, (**e**) Na$_2$SO$_4$, and (**f**) KI. (**B**) Curves of potentiometric titration of L-ascorbic acid using indicator electrode of mercury against ASSREs. Titrant: 0.0125 mol L^{-1} copper (II) sulfate solution. Conditions and results: salts in ASSRE, mass of L-ascorbic acid titrated (in parenthesis), volume of titrant in the end-point and L-ascorbic acid content (% w/w) found (in brackets). (**a**) KCl (9.1 mg) 4.102 mL [99 %], (**b**) KCl : SiO$_2$ (10.6 mg) 4.911 mL [102 %], (**c**) BaCl$_2$ (10.1 mg) 4.484 mL [98 %], (**d**) NaF (9.2 mg) 4.136 mL [99 %], (**e**) Na$_2$SO$_4$ (11.0 mg) 4.800 mL [96 %], (**f**) KI (10.3 mg) 4.901 mL [105 %]. The measurements were taken at room temperature.

Table 2. pH measured using a glass electrode against an all-solid-state reference electrode containing 1.5 mol kg^{-1} BaCl$_2$ at room temperature.

	pH		
Nominal value	Combined electrode	Glass (pH) vs. C$_{graphite}$/Ag/AgCl, 1.5 mol kg^{-1} BaCl$_2$	Relative error* (%)
2.0	1.88 ± 0.03	1.85 ± 0.08	−1.6 %
3.0	2.97 ± 0.03	3.00 ± 0.07	+1.0 %
4.0	4.03 ± 0.03	4.11 ± 0.06	+2.0 %
5.0	5.08 ± 0.02	5.12 ± 0.04	+0.8 %
6.0	6.00 ± 0.01	6.11 ± 0.03	+1.8 %
7.0	7.12 ± 0.01	7.19 ± 0.02	+1.0 %
8.0	8.15 ± 0.01	8.22 ± 0.01	+0.9 %
9.0	8.89 ± 0.02	9.09 ± 0.01	+2.3 %
10.0	10.01 ± 0.02	10.19 ± 0.01	+1.8 %
11.0	11.25 ± 0.03	11.35 ± 0.03	+0.9 %

* Average of three determinations and standard deviation estimate.

chloride ions. The presence of silica in the paste did not influence the behaviour of the reference electrode containing potassium chloride. Figure 1b reveals that all-solid-state reference electrodes containing other anions also kept a constant potential on the potentiometric cell. The potential change was lower than 5 mV at chloride concentration changes from 1×10^{-6} to 1×10^{-1} mol L^{-1}.

Figure 2 shows the surface photomicrographs of the reference electrodes obtained by means of scanning electronic microscopy. The white arrow in Fig. 2d illustrates the brilliant silver sub-particles on the surface of quasi-reference electrodes that presented diameters smaller than 2.5 μm. Figure 2

(from e to h) also shows the presence of potassium chloride particles on the surface of the all-solid-state reference electrode.

Reference electrodes have to satisfy many requirements. These electrodes have to exhibit stable potential at changing pH, redox species, complexing agents, salts and dissolved oxygen. Due to its importance for sterilization purposes, a high resistance to temperature and pressure is sometimes desirable too (Guth et al., 2009).

The effect of the pH on all-solid-state reference electrodes is shown in Fig. 3a. Silver ions may form a complex with hydroxyl ions poisoning the reference electrodes. The potential of the electrochemical cell was slightly affected in extreme regions of pH, but the potential variation within the investigated pH range was lower than 15 mV. In these regions, factors such as liquid junction potentials and ionic strength are dependent on the pH and may have contributed to this problem (Feldman, 1956).

We applied the all-solid-state reference electrode containing 1.5 mol kg^{-1} of BaCl$_2$ to measure pH using a glass electrode. The analytical performance is described by the following equation:

$$\Delta E = (841 \pm 6) - (56.0 \pm 0.7) \text{ pH}. \tag{1}$$

The potentiometric cell showed a linear response in the range of pH between 1.68 and 12.45 at 298.15 K (25 °C). The correlation coefficient (r) was 0.9997 for $n = 5$. Table 2 shows the results obtained for the determination of pH using this potententiometric cell. The results were not significantly different when compared to a conventional combined glass electrode. The relative error was lower than 2.5 %.

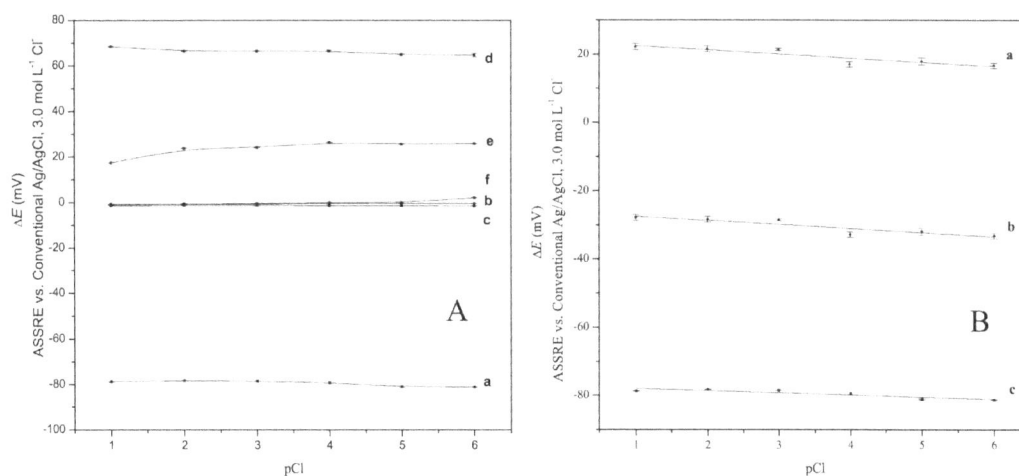

Figure 4. Behaviour of all-solid-state reference electrodes after the sterilization and diffusion process in deionized water during 20 days. Curves obtained in the conditions (**A**) after sterilization in autoclave with ASSREs containing the salts (**a**) KCl, (**b**) KCl : SiO$_2$, (**c**) BaCl$_2$, (**d**) NaF, (**e**) Na$_2$SO$_4$, and (**f**) KI. (**B**) ASSREs containing KCl at 3.0 mol kg^{-1} (**a**) after sterilization, (**b**) after diffusion process, and (**c**) before everything.

Table 3. Results of the determination of potassium ions in real samples using an all-solid-state reference electrode containing 3.0 mol kg^{-1} NaF against ion-selective electrodes for potassium (ISE-K).

Samples	Nominal value	Flame photometry (mg)	Direct potentiometry ISE-K vs. C$_{graphite}$/Ag/AgCl, 3.0 mol kg^{-1} NaF (mg)	Relative error (%)
#1	18 mg/200 mL	16 ± 4[a]	16 ± 4[b]	(−11) 0[c]
#2	50 mg/200 mL	43 ± 1	48 ± 2	(−4) +12
#3	24 mg/200 mL	22.9 ± 0.3	20 ± 1	(−16) −13
#4	44 mg/200 mL	40 ± 1	46 ± 5	(+4) +15
#5	60 mg/mL	51 ± 1	56 ± 8	(−7) +10
#6	100 mg/5 mL	90 ± 3	99 ± 13	(−1) +10

Sample composition: noncarbonated sports drinks – #1: water, sucrose, potassium dihydrogen phosphate, sodium chloride, citric acid, sodium citrate, natural tangerine flavour, dimethyl carbonate (preservative), calcium disodium EDTA, sunset yellow dye and tartrazine dye. #2: water, sucrose, potassium dihydrogen phosphate, sodium chloride, magnesium sulfate, calcium phosphate, citric acid, sodium benzoate, potassium sorbate, sodium citrate, acacia gum, calcium disodium EDTA and ester gum. #3: water, sucrose, glucose, potassium dihydrogen phosphate, sodium chloride, citric acid, flavouring, sodium citrate, 40 red dye and sunset yellow dye. #4: water, sucrose, potassium dihydrogen phosphate, sodium chloride, citric acid, sodium benzoate, potassium sorbate, arabic gum, natural carambole flavour, sodium citrate, sunset yellow dye, ascorbic acid and calcium disodium EDTA. Syrups; #5: water, potassium chloride, sodium cyclamate, sodium saccharin dihydrate, methylparaben, propylparaben, propylene glycol, hydroxyethyl cellulose, sorbitol, cherry essence and riboflavin phosphate sodium. #6: water, potassium iodide, sodium cyclamate, sodium saccharin, parabens, ethyl alcohol, sorbitol, mint essence, red dye and lobelia extract. [a] Average of three determinations and standard deviation estimate. [b] Calculated by error propagation. [c] Relative error between direct potentiometry and nominal value (parenthesis) or direct potentiometry and flame photometry.

Another all-solid-state reference electrode was applied to measure potassium ions using an ion selective electrode based on valinomycin as ionophore. Potassium ions were determined by direct potentiometry in real samples of noncarbonated sports drinks and syrups using an all-solid-state reference electrode containing fluoride ions since this anion was not present in the samples. Table 3 shows these results compared to flame photometry. The samples presented a complex matrix containing salts, EDTA, dyes and other substances that could affect the all-solid state reference electrode. However, this effect was not observed. The obtained results by

direct potentiometry agree with those determined by flame photometry within an acceptable band of deviation.

We also applied the all-solid-state reference electrodes to the potentiometric titration of L-ascorbic acid using an indicator electrode of mercury (Fig. 3b). L-ascorbic acid is a reducing agent, unstable in solutions with pH higher than 5.0 and easily oxidized by metallic ions (Fernandes et al., 1999). Hence, L-ascorbic acid could affect the all-solid-state reference electrodes. We determined L-ascorbic content in pure reagent equal to 100 ± 3 % w/w. Therefore, the influence of the redox process between L-ascorbic acid and copper (II) ions was not observed on the all-solid state reference electrodes.

The all-solid-state reference electrodes were tested for sterilization in autoclave (Fig. 4a). The reference electrodes were fully functional after sterilization. However, we observed that the potential of these electrodes differed from the one measured before sterilization. We believed that this effect occurred due to a change in the structure of the composite, with migration of salt particles to the surface. This hypothesis was based on the observation of a white powder on the surface of the electrodes after sterilization, but the potential was independent of the chloride ions concentration.

The leaching out effect of the salt was studied with all-solid-state reference electrodes containing $3.0 \, \text{mol kg}^{-1}$ KCl submitted to a diffusion process in deionized water for 20 days. We examined this water by means of a silver nitrate solution and we observed the formation of silver chloride precipitate (4.1 mg), which corresponds to 2.1 mg of potassium chloride. Every day, about 0.18 % of potassium chloride was leached out of the reference electrode. This reference electrode was re-examined in a new potentiometric cell with a commercial reference electrode to measure chloride ions (from 10^{-6} to $10^{-1} \, \text{mol L}^{-1}$). Comparing the measures, before and after of the diffusion process, we observed a potential-drift of about 50 mV (Fig. 4). Despite the loss of the chloride ions, this all-solid state reference electrode maintained a constant potential of the potentiometric cell to all range of chloride ions concentration. Therefore, this kind of electrode must be stored in dry conditions.

4 Conclusions

All-solid-state reference electrodes based on particles of graphite/silver/silver chloride, salts and epoxy resin showed good operational characteristics for potentiometry. The potential was stable and independent of any chloride ions concentration. These electrodes were not sensitive to abrasion, redox species, pH and high-pressure saturated steam at 394.15 K (121 °C). Moreover, they can be miniaturized easily by thick film technology since the components can be transformed into an ink.

The strategy of manufacturing by electroless deposition of silver and silver chloride particles on graphite powder is a simple and efficient methodology and does not require the use of a potentiostat.

The results for measuring potassium ions and pH by direct potentiometry or L-ascorbic acid by potentiometric titration indicate that the proposed methodology for manufacturing all-solid-state reference electrodes is very promising.

Author contributions. J. C. B. Fernandes designed the experiments and E. V. Heinke carried them out. J. C. B. Fernandes prepared the manuscript with contributions from all co-authors.

Acknowledgements. This research was supported in part by FAPESP (2007/556277 process) and by USCS in terms of a scholarship to E. V. Heinke. We thank the Centre of Electronic Microscopy from UNIFESP (hhttp://www.unifesp.br/centros/ceme/links.htm) for obtaining photomicrographs. We also thank Denise O. Alonso, USCS, and Peter Sussner, UNICAMP, for the English revision.

References

Cha, G. S., Cui, G., Yoo, J., Lee, J. S., and Nam, H.: Planar reference electrode. U.S. Patent 6,964,734 B2, 2005.

Feldman, I.: Use and Abuse of pH Measurements, Anal. Chem., 28, 1859–1866, 1956.

Fernandes, J. C. B., Kubota, L. T., and Neto, G. O.: Potentiometric sensor for L-ascorbic acid based on EVA membrane doped with copper (II), Electroanal., 11, 475–480, 1999.

Fernandes, J. C. B., Neto, G. O., Rohwedder, J. J. R., and Kubota, L. T.: Simultaneous determination of chloride and potassium in carbohydrate electrolyte beverages using an array of ion-selective electrodes controlled by a microcomputer, J. Braz. Chem. Soc., 11, 349–354, 2000.

Guth, U., Gerlach, F., Decker, M., Oelßner, W., and Vonau, W.: Solid-state reference electrodes for potentiometric sensors, J. Solid State Electrochem., 13, 27–39, 2009.

Harris, D. C.: Quantitative Chemical Analysis, 7th Edn., WH Freeman and Company, New York, 2007.

Janz, G. J. and Taniguchi, H.: The silver-silver halide electrodes, Chem. Rev., 3, 397–437, 1953.

Jayaweera, P., Passel, T. O., and Millett, P. J.: Solid state reference electrode for high temperature electrochemical measurements, U.S. Patent 5,425,871, 1995.

Koura, N.: Electroless plating of silver, in: Electroless plating: Fundamentals & Applications, edited by: Mallory, G. O. and Hadju, J. B., 1st Edn., American Electroplaters and Surface Finishers Society, Florida, 441–462, 1990.

Mamińska, R., Dybko, A., and Wróblewski, W.: All-solid-state miniaturised planar reference electrodes based on ionic liquids, Sens. Actuators B, 115, 552–557, 2006.

Matsumoto, T., Ohashi. A., and Ito, N.: Development of a microplanar Ag / AgCl quasi-reference electrode with long-term stability for an amperometric glucose sensor, Anal. Chim. Acta, 462, 253–259, 2002.

Rius-Ruiz, F. X., Kisiel, A., Michalska, A., Maksymiuk, K., Riu, J., and Rius, F. X.: Solid-state reference electrodes based on carbon nanotubes and polyacrylate membranes, Anal. Bioanal. Chem., 399, 3613–3622, 2011.

Riyazuddin, P. and Nazer, M. M. A. K.: Potentiometric determination of ascorbic acid in pharmaceutical preparations using a copper based mercury film electrode, J. Pharm. Biomed. Anal., 16, 545–551, 1997.

Rodes, M. L.: Solid state reference electrode, U.S. Patent 7,318,887 B2, 2008.

Sengupta, R., Bhattacharya, M., Bandyopadhyay, S., and Bhowmicka, A. K.: A review on the mechanical and electrical properties of graphite and modified graphite reinforced polymer composites, Prog. Polym. Sci., 36, 638–670, 2011.

Shin, J. H., Lee, S. D., Nam, H., Cha, G. S., and Bae, B. W.: Miniaturized solid-state reference electrode with self-diagnostic function, US Patent 6,554,982B1, 2003.

Sorensen, P. R. and Zachau-Christiansen, B.: Electrode device with a solid state reference system, U.S. Patent 2004/0163949 A1, 2004.

Suzuki, H., Hirakawa, T., Sasaki, S., and Karube, I.: Micromachined liquid-junction Ag / AgCl reference electrode, Sens. Actuators B, 46, 146–154, 1998.

Suzuki, H., Shiroishi, H., Sasaki, S., and Karube, I.: Microfabricated liquid junction Ag / AgCl reference electrode and its application to a one-chip potentiometric sensor, Anal. Chem., 71, 5069–5075, 1999.

Tymecki, L., Zwierkowska, E., and Koncki, R.: Screen-printed reference electrodes for potentiometric measurements, Anal. Chim. Acta, 526, 3–11, 2004.

Valdés-Ramírez, G., Ramírez-Silva, M. T., Palomar-Pardavé, M., Romero-Romo, M., Álvarez-Romero, G. A., Hernández-Rodríguez, P. R., Marty, J. L., and Juárez-García, J. M.: Design and construction of solid state Ag / AgCl reference electrodes through electrochemical deposition of Ag and AgCl onto a graphite/epoxy resin-based composite. Part 1: Electrochemical deposition of Ag onto a graphite/epoxy resin-based composite, Int. J. Electrochem. Sci., 6, 971–987, 2011.

Weast, R. C.: Handbook of Chemistry and Physics, 78th edition, CRC Press, Boca Raton, 1998.

17

Use of VOC sensors for air quality control of building ventilation systems

M. Großklos

Institut Wohnen und Umwelt, Darmstadt, Germany

Correspondence to: M. Großklos (m.grossklos@iwu.de)

Abstract. Air quality control with VOC (volatile organic compound) sensors in residential buildings could increase user comfort by adapting to the actual contaminant level. Preliminary tests assessed the dynamics of VOC levels in single-family passive houses with a ventilation system. At normal and exceptional usages, sufficient signal variations were measured for air quality control. An air quality control was developed and tested in four single-family passive house dwellings to control the building ventilation system via VOC sensors and a special adaptation algorithm to handle variable contaminant loads and sensor drift. Results showed good operation of the air quality control for the ventilation system, detecting changing contaminants within a few minutes and changing the air flow rate in the building immediately. The 43 VOC sensors used during the monitoring had more than 1.2 million working hours in total without any electrical failure, but with a loss of sensitivity for the calibration gas CO. The air quality control could manage that loss of sensitivity and worked well till the end of the field test. A comparison between VOC and CO_2 sensors in one building resulted in more detailed information about emission rates in the room with the VOC sensor, allowing one to get a better reaction of the ventilation system.

1 Introduction

Mechanical ventilation systems with heat recovery could reduce energy consumption and increase residents' comfort level at the same time. Ventilation systems in highly efficient residential passive houses are normally controlled by hand or by a time program. But, the individual habits of people could rarely be defined in fixed programs because there are times with low contaminants in the room air when nobody is there and other times with peak contaminants because of additional people or odors from the kitchen or the bathroom. The ventilation system normally could not manage those variable loads automatically. The residents have to interact with the control system to optimize the air quality and the energy consumption permanently. Especially at night, when residents are sleeping, manual interaction with the ventilation system is not feasible. The advantages of sensor-based demand-controlled ventilation systems to increase user comfort have been known for a long time, but only a few studies were published about the way to implement an air quality ventilation control and the usage of VOC (volatile organic compound) sensors (Fisk et al., 1998).

During a research project, an automatic air quality control was developed for ventilation systems to deal with these dynamic changes in air quality in residential buildings using VOC sensors. In a second project phase, the air quality control was tested in four single-family passive houses over a minimum time of 2 years (one building for nearly 4 years) to observe how the system works in practice. Results are presented in this paper.

1.1 Project aims and activities

The project contained a theoretical survey of the possibilities using VOC sensors in residential buildings, including first measurements in passive houses to distinguish the signal dynamic in residential applications. The second phase was a field test over a minimum period of 2 years to prove the function of the new air quality ventilation control.

Additional topics like detecting the opening stage of the windows using the ventilation system or a dynamic user feedback of their window opening behavior will not be discussed in this paper. The project started in 2006 and was finished in

Figure 1. The VOC sensor manufactured by the former ETR-Electronic.

2012. The results were published in three reports (Großklos et al., 2011, 2012; Knissel et al., 2011).

1.2 The VOC sensors

For the field test, 43 VOC sensors in total (Fig. 1), manufactured by former company ETR-Electronic in Dortmund, Germany, were placed in the buildings. The sensors are tin oxide sensors on a ceramic substrate using LuQaS technology (see FIA e.V., 2005). This means that the sensor temperature is varied between 300 and 400 °C dynamically to detect broadband oxidable gases and to reduce the influence of air humidity. The measuring range is between < 10 and 15 000 ppm as a sum of all different VOC components. For carbon monoxide, which is used for calibration, the maximum measurable concentration is about 34 ppm. The sensor signal is transmitted as an integer value between 80 and 254 digits (measured values in normal unpolluted room air start at 100–120 digits), so there is no conversion to known references or thresholds. Because the sensor output is nearly logarithmic, Eq. (1) is used for linearization:

$$C_{\mathrm{Lin}} = \left(\frac{C_{\mathrm{Digits}}}{41.4} - 1.57 \right)^{2.34}. \tag{1}$$

To reduce the influence of the casing, the sensors were mounted in small boxes made out of stainless steel.

2 Preliminary tests

Before using the sensors for building ventilation control, it was not clear whether the dynamics of the measured concentration in the exhaust air is sufficient for the control system. Numerous tests were done to characterize the behavior and signal amplitudes in ventilated single-family homes with lower emission rates than in office buildings or in schools,

which were tested in other publications (e.g., Kopiske et al., 2004). Additionally, the repeatability and the time-dependent drift between the sensors in room air was tested. Finally, the measurement of outside air VOC signals at different places within a particular region was tested.

2.1 VOC sensor in a passive house sleeping room compared to exhaust air of the ventilation system

Figure 2 shows the measurement of the VOC and CO_2 values in a passive house bedroom over 2 days. At the beginning of the period shown, the flow rate of the ventilation system was slightly reduced to 86 % compared with the design flow rate (to simulate pollution of the filters). During the first day, the VOC levels in the bedroom and the exhaust air were quite constant, but on a different level. The CO_2 level in the bedroom began to rise in the evening because of open doors in the building. When the bedroom was occupied at about 11 p.m. in the evening, VOC and CO_2 levels increased significantly. The dynamics of the VOC signal was up to 50 % (based on a signal range of the VOC sensor of between 80 and 254 digits). Relative humidity (rh) varied only little around a mean value of 43.7 % rh (standard deviation 1.58 % rh), so an influence on the sensor signal is not likely. Additional chemical analyses were not executed. During the night, the door of the bedroom was then opened, and an additional passive air exchange to the stairway occurred. The VOC level as well as the CO_2 concentration lowered directly. In the morning, when the occupants left the bedroom, there was an exponential decline in the concentration of both, VOC and CO_2, in the room.

Interesting for the ventilation control are the VOC levels of the exhaust air. At the beginning, it stayed on a quite constant level and then increased clearly in the evening when two persons were in the building. The moving of the users from the living room to the bedroom has limited influence on the VOC level in the exhaust air. The VOC peaks in the bedroom could also be found in the exhaust air. Additionally, there are peaks, for example, when the residents were showering. The signal dynamics between daytime and night is about 20 %. That means that there is sufficient variation for an air quality ventilation control.

2.2 Vertical sensor position and relative sensor drift

The influence of the vertical sensor position was tested in an office room with three sensors: directly at the floor, at 1 m height at the desk and at 2.20 m above the floor. Figure 3 shows that the measured VOC values at 0 and 1 m are very similar and have a comparable dynamic. The sensor at 2.20 m height has a slightly higher variation and a signal between 2 and 3 digits higher than at floor level. Before that test, all three sensors had been in permanent operation for 3 weeks at the same place, and the results at 0 and 2.20 m were calibrated onto the difference in the level of the sensor

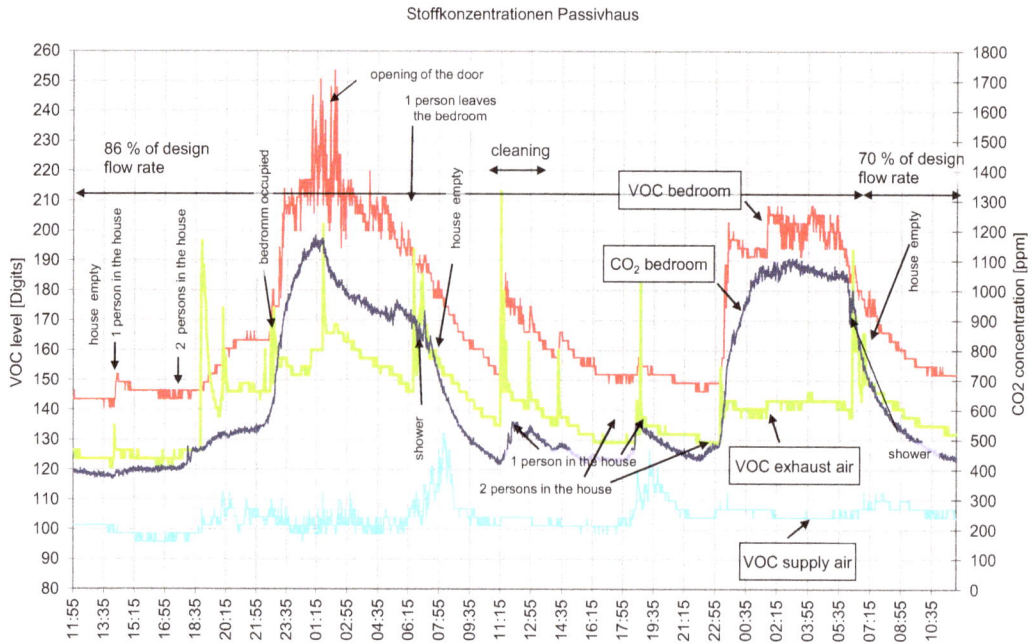

Figure 2. Measurements of VOC and CO_2 in room air and exhaust air in a passive house bedroom over 2 days.

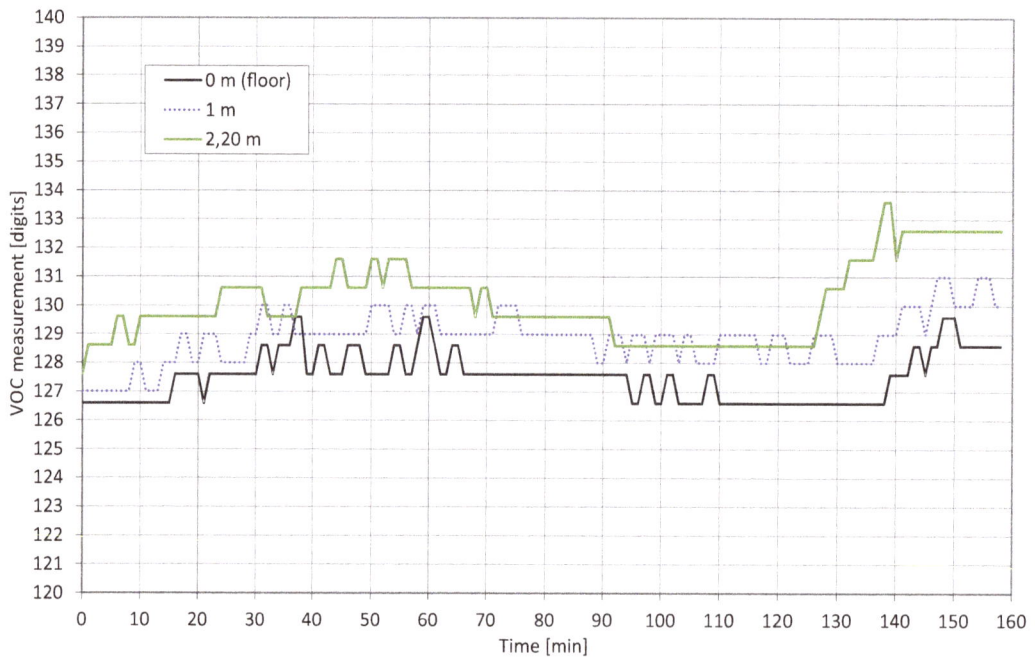

Figure 3. VOC signal at different heights in an office room.

at 1 m height directly before the measurement of height dependency. For signal dynamics, a higher position of the sensor could be better than a lower one. To detect those VOCs at the level of the head of a person, a medium height seems to be better.

The three sensors had different operation times at the beginning of the 5 week comparison phase. Figure 4 shows the difference in time compared to the sensor at 1 m height,

which had probably the longest operation time until then. After a 5 digit higher signal level at the beginning for sensors 2 and 3, the mean difference decreased in time, but sensor 2 showed a bigger variation. At the end of the 5 weeks, the differences of sensors 2 and 3 remained at 1.7 or 1.1 digits, respectively. That shows that absolute measurement values were very difficult to evaluate in a ventilation control with these sensors.

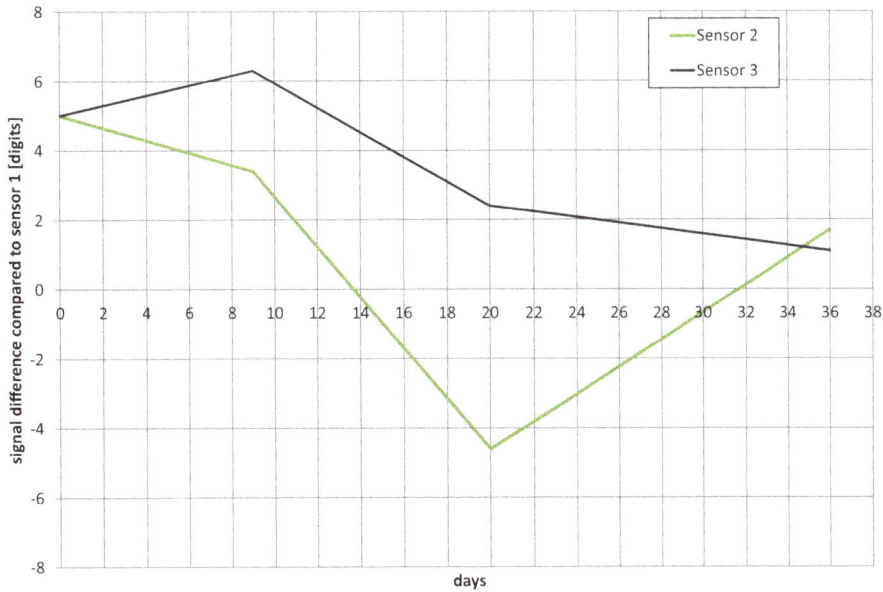

Figure 4. Relative differences between three VOC sensors over a period of 5 weeks.

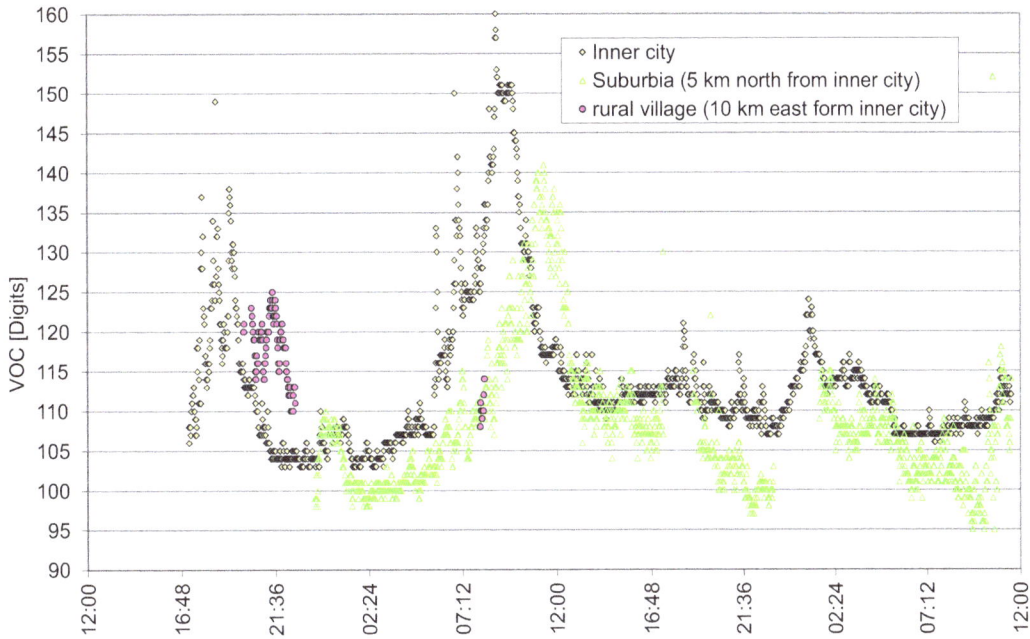

Figure 5. Outside VOC measurements over 2 days at three different locations nearby.

2.3 Outside air VOC measurement in different regional positions

To distinguish the variation of the outside air VOC level, the three sensors measured over several days in outside air simultaneously in different places. One sensor was placed at an inner city location, another 5 km north of the inner city sensor in suburbia, and the third one in a rural area 10 km east of the inner city. During the time period shown in Fig. 5, the measured VOC level has a very similar trend between the inner city and suburbia, with slightly different signal levels. The values for rural areas are less complete, but have also comparable dynamics. It is important to mention that an intensive slurry odor at the rural measurement could hardly be seen in the measured values. That demonstrates the limitations of the VOC sensors used for the test, because the sensitivity of the human nose is partly higher than those of the electronic sensors.

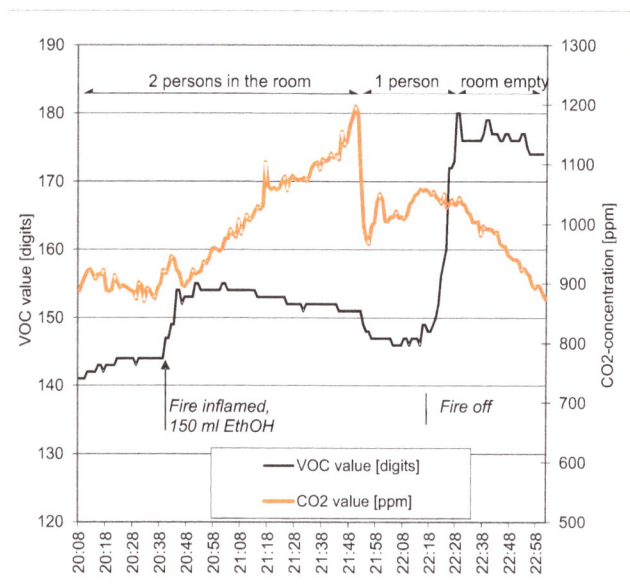

Figure 6. VOC and CO_2 measurements with an ethanol oven in a passive house living room with $159\,m^3$ volume.

2.4 Influence of an ethanol oven on room air quality

To simulate a high source strength for room air, contamination tests with an ethanol oven in the living room of a passive house with a mechanical ventilation system were done. The room has a living area of $60\,m^2$, 2.65 m height, and the ethanol oven a mean heating power of 600 W. Figure 6 shows the measured VOC and CO_2 values in the room. Before inflaming the oven, the CO_2 value was around 900 ppm with two people in the room. The VOC level rose slightly. When filling and inflaming the oven, there is a leap in VOC measurement, while CO_2 starts to rise constantly. After half of the burning time, one person leaves the room. That could be seen in both values simultaneously, probably because of air mixing in the room. It is interesting that the VOC level rises drastically when the oven expired combustion at the end. The reasons are probably increased evaporation of the remaining ethanol from the hot combustion chamber when the flame expires, or bad burning. In parallel, there is no information about that event in the CO_2 measurements. This survey gives hints for the additional advantages using VOC sensors compared to CO_2 sensors.

3 Air quality control

The results from the preliminary tests influenced the development of the new air quality control (see Fig. 7) that controls the building ventilation system in three steps. Two VOC sensors were used during the field test. The function is as follows.

The VOC values measured in the exhaust air are linearized to correct the logarithmic sensitivity of the sensor. Then, the emission rate of the contaminants is calculated using addi-

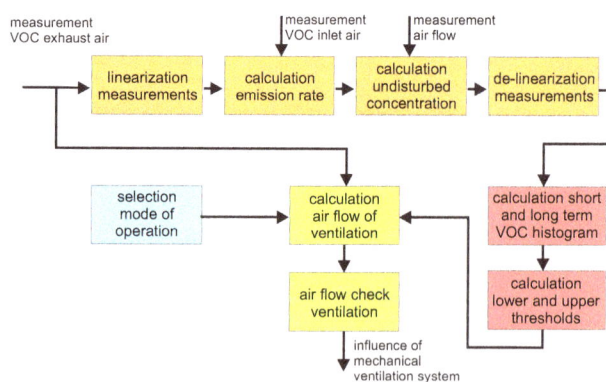

Figure 7. Schematic drawing of the air quality control.

tionally the linearized VOC concentration in the inlet air. With this emission rate and the measured air flow, an undisturbed VOC concentration is calculated for the building without the influence of the changing air flow rates from the air quality control. These concentrations are de-linearized and afterwards two histograms are calculated continuously. One short-term histogram collects the VOC distribution in the building over a period of 14 days and one long-term histogram collects data over 41 days (originally 3 months, which turned out to be too long). The short-term histogram is then used to calculate the threshold for the air quality control, and the long-term histogram for limiting the thresholds against drift because of times of unusual use in the building (e.g., residents are on holiday). The lower and upper thresholds are then calculated as, e.g., the 20 % and 80 % quantiles of the histograms.

For air quality control, the measured VOC value in the exhaust air is compared with the calculated thresholds. If the current VOC value is over the upper threshold, the air flow rate in the building is increased to reduce the contaminants. A VOC value under the lower threshold leads to a reduction in the air flow rate, because the VOC concentration is lower than during normal utilization. Between the lower and upper thresholds, the ventilation system is in normal operation.

This air quality control was implemented with the dasy-lab data acquisition and control software, collecting all data, calculating the necessary ventilation level and handling the result of the standard ventilation control.

3.1 Test buildings

For the field test, four passive houses were chosen, occupied by one family each (normally four persons), situated in three different cities in Germany. All passive houses were built by Schwörer Haus KG, a leading German manufacturer of prefab houses.

The sensors used for the air quality control were placed at the inlet and exhaust air ducts of the ventilation system (Fig. 8). Additionally, in all rooms of the building, VOC sen-

Figure 8. VOC sensors in the rooms were mounted in a box in the wall and are covered by a metal plate fixed 10 mm in front of the sensor.

Figure 9. Position of the sensors in the ventilation system.

sors were mounted to compare the concentration measured in the individual room with the concentration in the exhaust air, where air from all rooms is mixed. The room sensors were placed in the walls with a metal plate as coverage 10 mm in front of the sensor (Fig. 9). Because of the high temperature of the sensor, there is buoyancy-driven permanent thermic ventilation of room air to the sensor. Preliminary tests resulted in no sensitivity of the VOC sensors to the material of the boxes and the metal plate.

3.2 Results from the field monitoring of the air quality control

Detailed analyses of the behavior of the air quality control in the four field test passive houses showed a quick reaction to emission peaks in the buildings. The time delay between an increase in VOC concentration and the reaction of the air quality control was about 2–3 min normally (the VOC sensors measured the gas concentration every minute). Also, the short-term (during the day) or long-term (holidays) absence of the residents was detected very well. Additionally, the calculation of the thresholds was not affected by a longer absence of the residents.

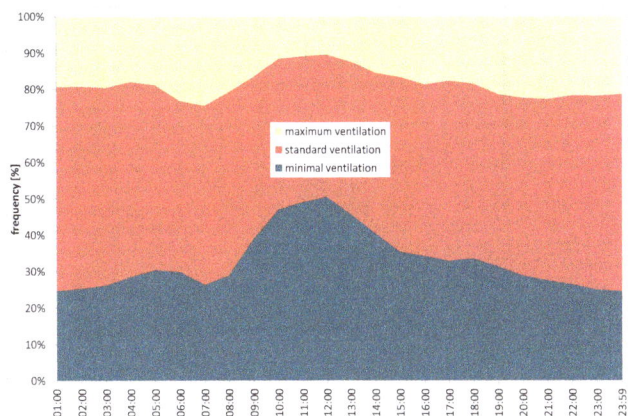

Figure 10. Mean distribution of the three air flow levels in the ventilation system during the day for one building.

Figure 10 shows the mean distribution of the three air flow levels in the ventilation system during the day in one of the test buildings. Standard ventilation is the most frequent mode of operation during the night. Sometimes the ventilation is reduced, for example, when there are no residents in the building, and sometimes the ventilation system is run with the maximum flow rate to reduce VOC levels during the night. During the day, when the residents are absent, the ventilation is reduced to the minimum level, which occurs mainly on working days. This distribution is different in each of the field test buildings because the indoor and outdoor air quality is different, the furnishing is different and, in particular, the residents' daily routines are different.

In Fig. 11, an exemplary day profile of building 4 on 12 September 2011 is shown. Illustrated are the measured VOC concentration in exhaust air, ventilation mode, the lower and upper short-term thresholds and the opening of the front door. Beginning at midnight, the ventilation runs in standard mode interjected by short times of minimal ventilation. Before breakfast, the emissions rise, and so the ventilation has increased to the maximum level. Vice versa, the ventilation is reduced to minimum level after the residents have left the building (which could be seen in the status of the front door). After 1.00 p.m., more people enter the dwelling, and so the ventilation flow rate increased to standard level subsequently.

The users had the ability to increase the air flow rate of the ventilation system by manual input ("party button") if needed. The assessment of the situations, when the residents pressed the button for maximum air flow, showed that in most situations the air quality control had already increased the air flow to the maximum value. That means that air quality control and user sense worked very similarly.

The air quality control resulted in a slightly lower ventilation flow rate in mean than in standard operation mode (1–6 % less). But, the reduction leads only to a low decrease in heat and auxiliary energy consumption because of the heat

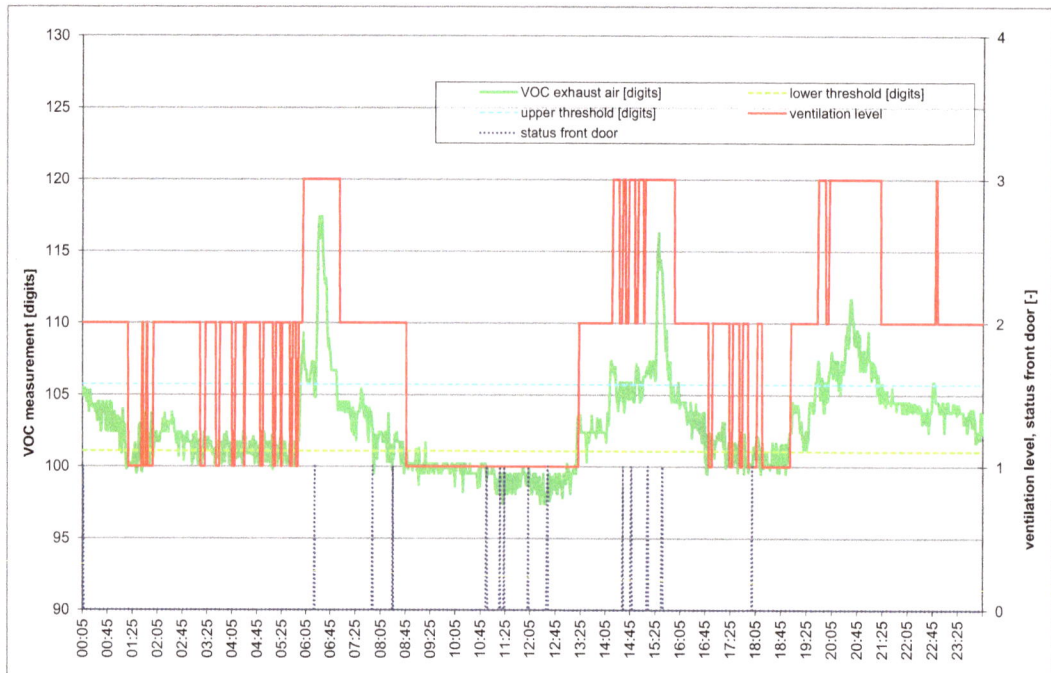

Figure 11. Example daily profile of VOC concentration in exhaust air, ventilation mode.

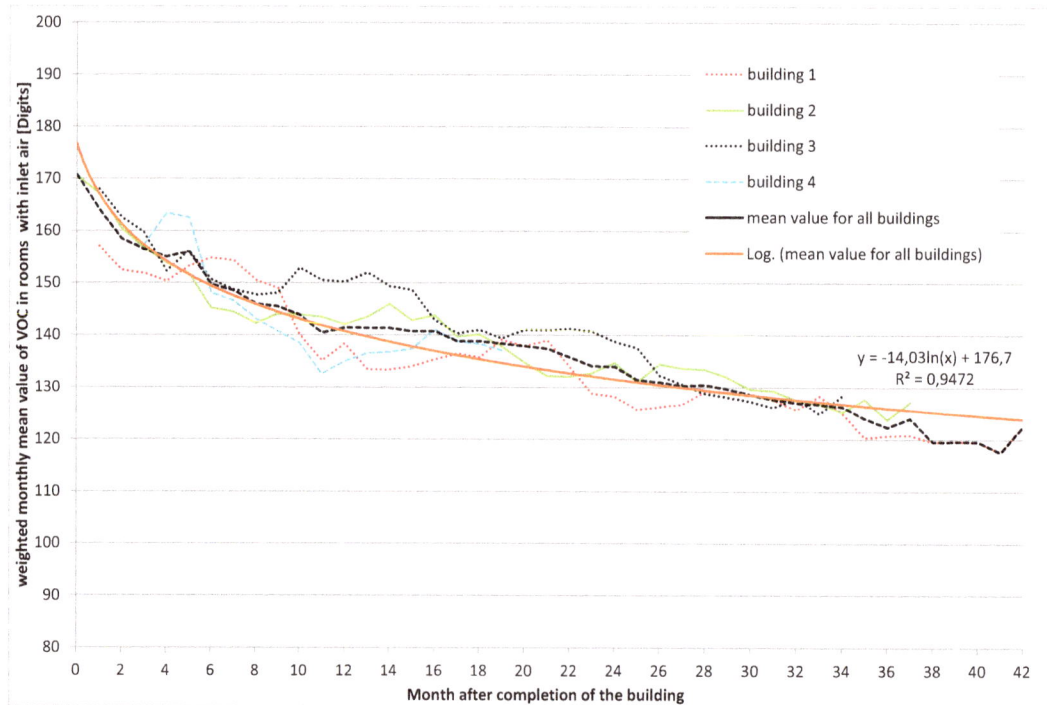

Figure 12. Weighted mean VOC concentration in the four test buildings in time elapsed after move-in.

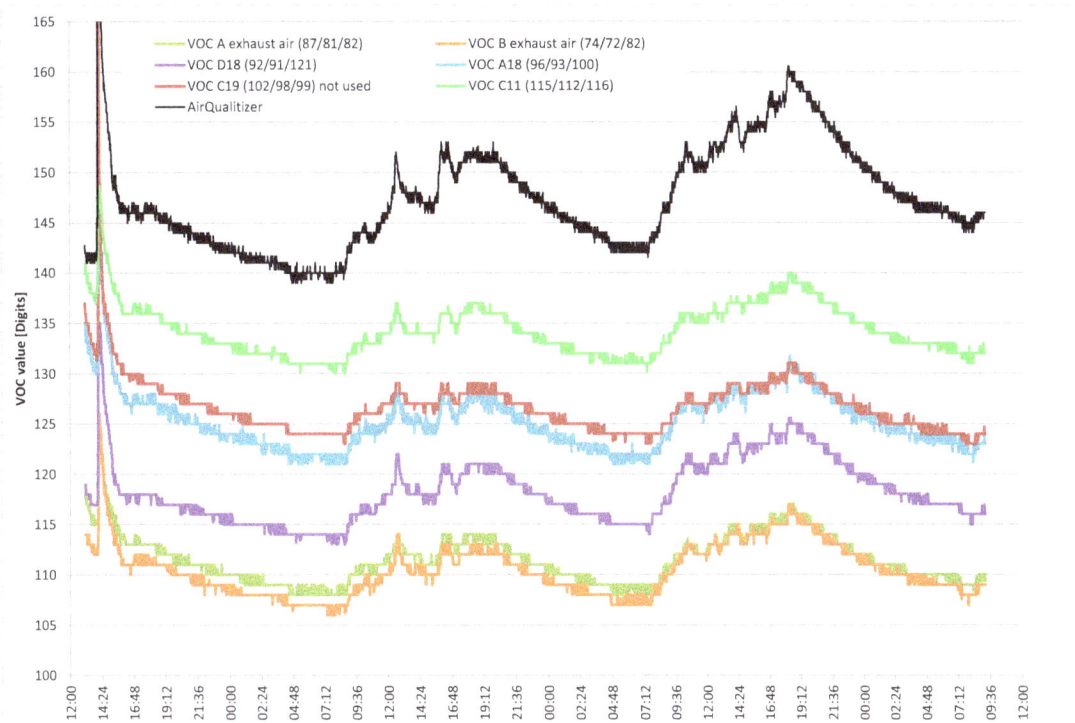

Figure 13. Signal dynamics of six used VOC sensors from the field test and one AirQualitizer (AirQualitizer, 2012) in an office room.

recovery and the highly efficient fans. In buildings like passive houses, the advantage of air quality control focuses on an increase in comfort for the residents. For the use in other buildings, see Sect. 5.

4 Experiences with the VOC sensors in long-term use and comparison with CO_2 sensors

4.1 Long-term development of VOC level

Measuring the VOC concentration in the exhaust air yielded a good mean VOC concentration in the building, but with an emphasis on the exhaust air rooms (kitchen, bathroom, WC). The measured VOC concentration decreased after the mounting of the building during the first months. It is not known till when the changing of the calibration of the sensors (see below) had a significant influence on the measurements (Fig. 12).

The measured VOC concentrations in the inlet air showed a significant dependency on the air humidity, rapidly changing in the test buildings because of the inlet air heating with a heat pump. During the year, a fluctuation of the fresh air VOC concentration was detected, probably resulting from the changing humidity of the fresh air. The conclusion drawn from these results is that there was a non-negligible influence of air humidity on the measured VOC level.

4.2 Recalibration results

All VOC sensors were recalibrated after the end of the field test to determine the drift between delivery and end of the monitoring phase (span of time about 4 years). The calibration gas was carbon monoxide (CO). The VOC sensors mounted in the inlet air had a deviation from the original calibration, but showed similar characteristics. The sensors mounted in the exhaust air and in all rooms had hardly any sensitivity to the gas CO left. Anyway, the sensors had a nameable signal dynamics in room air, so the air quality control could still influence the flow rate of the ventilation system. Figure 13 shows a comparison of six VOC sensors from the field test and one new AirQualitizer (a very similar sensor by AL-KO Therm GmbH, Jettingen-Scheppach, Germany) during 3 days in a normal office room. The signal level is different, but the characteristic signal dynamics is still there, even if the individual sensor has no notable reaction to the calibration gas CO (values in brackets for 0, 5 and 20 ppm CO during the recalibration).

The air quality control appeared to be robust against sensor drift because of the continuous recalculation of the thresholds.

The operation time of the 43 VOC sensors added up to 1 200 000 h of operation during the field test without an electrical failure of a sensor. The electrical stability seems to be very good. In inlet and exhaust air, additional VOC sensors packed in an M18 screw thread were mounted. They were used parallel to the ones described above, but did not have

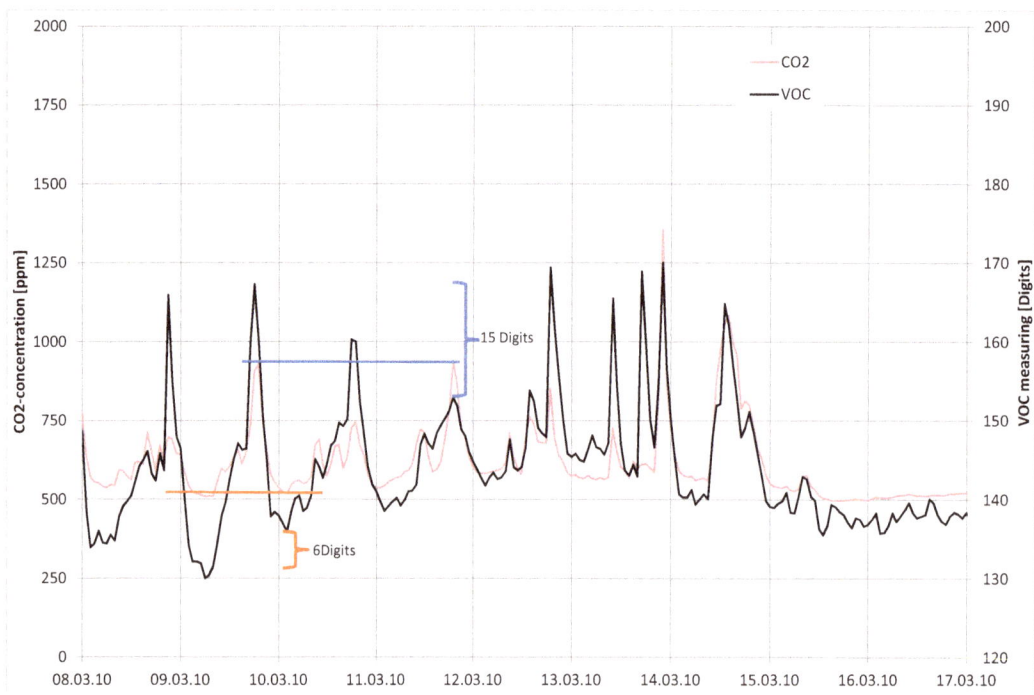

Figure 14. VOC and CO_2 measurements in the living room of one test building over 1 week.

their stability. Of those mounted in the inlet air, all broke after a while. Maybe, the temperature leap while heating the inlet air led to electrical problems.

In principle, an air quality control could also be constructed with CO_2 sensors. A comparison between the measured dynamics of the VOC sensors and CO_2 sensors mounted parallel in the rooms of one building showed some similar signals resulting from peoples' activities in the building (e.g., during the night time). Other VOC sources like emissions from the building materials or the furniture could not be detected with CO_2 sensors. This is the reason why VOC sensors have some advantages for use in air quality control.

Figure 14 shows the comparison of measured CO_2 and VOC values during 1 week. Some peaks are very similar at the CO_2 and VOC sensors, but others with the same CO_2 measurement show wide differences in VOC value. Odors especially could not be detected with CO_2 sensors.

5   User feedback

Introducing new techniques in buildings could lead to user discontentment when their needs are not considered. For that reason, user interviews had been held during the monitoring phase with the residents of the buildings. In addition to the users' global contentment with their passive houses, questions about their contentment with indoor climate, ventilation systems and air quality resulting from the new air quality control had been asked.

The residents said that the air quality was good even after a longer absence of people in the building and that they could not distinguish between indoor and outdoor air quality. Odors, for example in the kitchen, could be removed quickly. Smaller problems with air quality in individual rooms were solved by correcting the air flow rates, which did not match the planned values. The noise from the changing ventilation levels was low and not disturbing.

6   Conclusion and outlook

In highly efficient buildings, the air quality control leads to improvements in buildings' air quality, but there are only a few possibilities to reduce energy consumption because of the existing heat recovery. In contrast, energy efficiency could be increased significantly when implementing the air quality control in an exhaust air ventilation system (with no inlet air ducts and no heat recovery), e.g., for retrofitted buildings, to reduce costs and work in rented flats. By means of the automatic detection of air quality, the air flow rate in the flat or building could be adapted to the actual demand, which reduces ventilation heat energy losses and saves nameable energy. With this combination, low refurbishing costs could be combined with low heating costs. This application could support the energy transition in the building sector.

Nevertheless, long-term stability of the VOC sensors is an important issue for industrial use in buildings, where usage times of 10–20 years are expected. Newer generations of VOC sensors could possibly match these requirements and be used for building air quality control for ventilation systems.

Acknowledgements. The author thanks the German ministry for economics and technology for funding the research project (funding code 0327 398A). Additional thanks go to the project partners: the former ETR-Electronic in Dortmund, now AL-KO Therm GmbH, who delivered the sensors, developed an interface for a bus system and did the recalibration, as well as Schwörer Haus KG, who provided the integration of the control into the ventilation and heating system of the field test buildings, and engineering consultant ebök GbR in Tübingen, who assisted in topics of changing the pressure balance in the building and did measurements of noise and air tightness in the test buildings. Special thanks go to Jens Knissel, University of Kassel, Germany, who initiated and led the project at the Institute for Housing and the Environment (IWU) in Darmstadt for a long time.

References

FIA e.V.: LUQUASII Triplesensor, Abschlussbericht, ISBN 3-938210-12-5, FIA e.V., Bietigheim-Bissingen, Germany, 2005.

Fisk, W. J. and de Almeida, A. T.: Sensor-based demand-controlled ventilation: a review, Energ. Buildings, 29, 35–45, 1998.

Großklos, M., Ebel, W., and Knissel, J.: Entwicklung energieeffizienter Komfortlüftungsanlagen, Technical Report 1: Einsatz des LuQaS-Triple-Sensors zur luftqualitätsgeführten Volumenstromregelung von mechanischen Lüftungsanlagen in Wohngebäuden, Institut Wohnen und Umwelt (IWU), Darmstadt, Germany, 2011.

Großklos, M. and Hacke, U.: Entwicklung energieeffizienter Komfortlüftungsanlagen, Technical Report 3: Endbericht und Dokumentation der Feldphase, Institut Wohnen und Umwelt (IWU), Darmstadt, Germany, 2012.

Knissel, J., Großklos, M., and Werner, J.: Entwicklung energieeffizienter Komfortlüftungsanlagen, Technical Report 2: Theoretische Untersuchungen zur Druckdifferenz-Methode, Institut Wohnen und Umwelt (IWU), Darmstadt, Germany, 2011.

Kopiske, G., Imann, C., Clausnitzer, K.-H., Simers, U., and Llamas, M.: Wissenschaftliche Begleitung sowie Verifizierung einer Lüftungsampel für den Einsatz im Mietwohnungsbau und in Schulen, UTEC, Bremer Energieinstitut und Bremer Umweltinstitut, Bremen, Germany, 2004.

18

Combination of clustering algorithms to maximize the lifespan of distributed wireless sensors

Derssie D. Mebratu and Charles Kim

Electrical and Computer Engineering, Howard University, Washington DC, 20059, USA

Correspondence to: Derssie D. Mebratu (mebratu@scs.howard.edu)

Abstract. Increasing the lifespan of a group of distributed wireless sensors is one of the major challenges in research. This is especially important for distributed wireless sensor nodes used in harsh environments since it is not feasible to replace or recharge their batteries. Thus, the popular low-energy adaptive clustering hierarchy (LEACH) algorithm uses the "computation and communication energy model" to increase the lifespan of distributed wireless sensor nodes. As an improved method, we present here that a combination of three clustering algorithms performs better than the LEACH algorithm. The clustering algorithms included in the combination are the k-means^{++}, k-means, and gap statistics algorithms. These three algorithms are used selectively in the following manner: the k-means^{++} algorithm initializes the center for the k-means algorithm, the k-means algorithm computes the optimal center of the clusters, and the gap statistics algorithm selects the optimal number of clusters in a distributed wireless sensor network. Our simulation shows that the approach of using a combination of clustering algorithms increases the lifespan of the wireless sensor nodes by 15 % compared with the LEACH algorithm. This paper reports the details of the clustering algorithms selected for use in the combination approach and, based on the simulation results, compares the performance of the combination approach with that of the LEACH algorithm.

1 Introduction

Wireless sensor networks are being used for many different applications, such as monitoring chemical spills, detecting and assessing the extent of environmental contamination, and monitoring the movement of soldiers and weapons on the battlefield. However, their limited lifespan is a great concern when they are used in remote locations or in harsh environments.

Many different techniques have been introduced in an effort to maximize their lifespan, but these techniques have focused on having the nodes in a cluster send their data to a selected cluster head node that, in turn, reports the data to the base station. Therefore, the choice of the number of clusters and the way the cluster head node is selected are the main focuses of these techniques. Clustering and the use of cluster heads in wireless sensor networks have the potential to enhance the lifespans of a group of sensor nodes and to minimize the generation of noise in the signals exchanged between the sensor nodes and the base station (sink)

(Heinzelman et al., 2000). In this approach, the cluster head organizes a reservation scheme to improve communication with the sensor nodes in the cluster, and the cluster head uses this scheme to aggregate, compress, and transmit the cluster's sensing data to the base station. Several technologies have been designed to improve the lifespan of the sensors. For example, algorithms were developed for this purpose by the energy efficient heterogeneous clustered scheme (EEHC) (Kumar et al., 2009) by the design of a distributed energy efficient clustering (DEEC)(Qing et al., 2006), and by the low-energy adaptive clustering hierarchy (LEACH) (Heinzelman et al., 2000). These goals of these algorithms were to determine the optimal number of clusters in a given number of sensor nodes and to selecting a head in a cluster of sensors. The low energy consumption clustering routing protocol (Kumar et al., 2009) improved the LEACH algorithm by utilizing the k-means algorithm that divides the sensor nodes into k clusters in the setup and steady-state phases. A major problem of the k-means algorithm was that it could not ac-

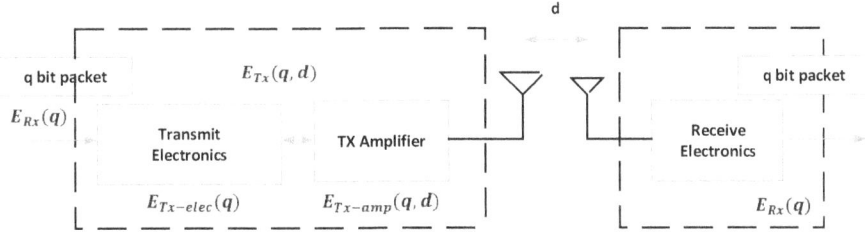

Figure 1. Radio energy model.

commodate the inevitable situation that the number of clusters gradually changed as the energy levels of the nodes decreased. Also, the method did not solve the sticky issue of initialization of the k-means process (Zhong et al., 2012). However, k-harmonic means (KHM) clustering solved the initialization problem by providing "soft membership", which assumes that a data element belongs to more than one cluster; also, if a data point is not close to any center or cluster, a "dynamic weighting function" provides a higher weight to the data element in the next iteration so that it becomes a candidate for all of the clusters. However, the KHM algorithm cannot provide an optimal number of clusters.

In this paper, we have provided detailed discussions of clustering algorithms; the combination of k-means^{++}, k-means, and gap statistics algorithms; the selective ways in which each is used and combined; and how, using the combination, the optimal number of clusters is generated, which leads to the maximum lifespan of a group of distributed wireless sensors. Before discussing the clustering algorithms and their combination, in the next section, we discuss a popular clustering algorithm, known as the low-energy adaptive clustering hierarchy algorithm (LEACH), for extending the lifespan of wireless sensors. Section 3 describes the selected clustering algorithms and their combination for determining the optimal number of clusters. Last, Sect. 4 provides the simulation results and compares the results provided by a combined clustering algorithm and the LEACH algorithm. Section 5 presents the conclusion.

2 Clustering algorithms

2.1 Low-energy adaptive clustering algorithm (LEACH)

The LEACH algorithm was developed to minimize the power consumption of wireless sensor nodes by determining the optimal number of clusters, k, in a group of distributed homogeneous wireless sensors based on the "computation and communication energy model" (Heinzelman et al., 2000). In order to determine the optimal number of clusters, k, first, the algorithm considers how much energy the head of a cluster consumes using the radio energy model depicted in Fig. 1. In the radio energy model, for a single bit transmission over a unit distance, E_{Tx} is the transmission energy dissipated, which is composed of two components, i.e., $E_{Tx-elec}(q)$, the

electrical energy consumed for digital coding, modulation, and filtering a signal and $E_{Tx-amp}(q,d)$, the energy required for amplification.

Then, the total energy used to transmit a q bit message over a distance d is expressed by

$$E_{Tx}(q,d) = E_{Tx-elec}(q) + E_{Tx-amp}(q,d). \tag{1}$$

The energy for k bit amplification is expressed by $E_{Tx-amp}(q,d) = q\varepsilon_{fs}d^2$, in a free space path ($\varepsilon_{fs}$) with distance squared (d^2). When a multi-path is considered, the amplification energy is defined as $E_{Tx-amp}(q,d) = q\varepsilon_{mp}d^4$, for q bits with distance to the fourth power (d^4). The LEACH algorithm proposed that the free space (fs) model be used when the distance between the transmitter and the receiver is less than the threshold distance d_o (base station distance); otherwise, the multipath (mp) model is used, as summarized below:

$$E_{Tx}(q,d) = E_{Tx-elec} + q\varepsilon_{fs}d^2, \quad d < d_o, \tag{2}$$

$$E_{Tx}(q,d) = E_{Tx-elec} + q\varepsilon_{mp}d^4, \quad d \geq d_o. \tag{3}$$

The receiver's energy for a q bit receipt is calculated by

$$E_{Rx}(k) = q E_{elec}. \tag{4}$$

Let us now consider energy consumption by the sensor nodes in a cluster of a multi-cluster sensor network. Assuming that there are N wireless sensor nodes uniformly distributed in a square region of $M \times M$ geographical units that have k clusters, there are N/k nodes per cluster, and, in each cluster, there is one cluster head node and $(N/k) - 1$ non-cluster-head nodes (or "cluster member nodes"). In a cluster, during the steady-state phase, data transfer from the nodes to a cluster head as well as from the cluster head to the sink, which is located a long distance away, so the energy of the cluster head's battery is being depleted faster that of any of the member nodes, because the cluster head receives data from the member nodes, aggregates and compresses them, and transmits the compressed data to the sink. The energy consumption of a cluster head is calculated by

$$E_{CH} = q E_{elec}\left(\frac{N}{k} - 1\right) + q E_{DA}\left(\frac{N}{k}\right) \tag{5}$$

$$+ q\left(E_{Tx-elec} + \varepsilon_{mp}d^4_{to\,BS}\right),$$

where $d_{\text{to BS}}$ is the distance between the cluster head and the base station, and (E_{DA}) is the energy dissipation per bit for data aggregation and compression.

The energy consumption by a member node for transmitting a q bit message to the cluster head is defined as

$$E_{\text{Non-CH}} = q\left(E_{\text{Tx-elec}} + q \epsilon_{\text{fs}} d_{\text{CH}}^2\right), \tag{6}$$

where d_{CH}^2 is the distance between the member nodes and the cluster head.

Now, let us calculate the energy consumption in a cluster in the aforementioned sensor network, i.e., N sensors distributed uniformly in an $M \times M$ geographical unit square area that is divided into k clusters. First, we can say that each cluster in the area takes up approximately (M^2/k) of the geographical region. Second, the location of a sensor node can be described by a Cartesian coordinate $\rho(x, y)$ (Heinzelman et al., 2000). If the area is a circle, the sensor's location can be described by a polar coordinate $\rho(r, \theta)$, where r is the radius and θ is an angle, with the radius defined by $r = M/\sqrt{\pi k}$. Third, the expected square distance in a circular area between the cluster head and the member sensor nodes is calculated by

$$E\left[d_{\text{to CH}}^2\right] = \rho \int\limits_{\theta=0}^{2\pi} \int\limits_{r=0}^{M} /\sqrt{\pi k} r^3 \, dr \, d\theta = \frac{\rho M^4}{2\pi k^2}, \tag{7}$$

where due to the uniform region of a node,

$$\rho = \frac{1}{(M^2/k)}, \quad E\left[d_{\text{to CH}}^2\right] = \frac{M^2}{2\pi k}, \tag{8}$$

$$E_{\text{Non-CH}} = q E_{\text{Tx-elec}} + \frac{q\varepsilon_{\text{fs}} M^2}{2\pi p}. \tag{9}$$

Fourth, the total energy consumption for a cluster is the sum of that for the cluster head and for the non-cluster head member nodes:

$$E_{\text{total}} = E_{\text{CH}} + E_{\text{Non-CH}}, \tag{10}$$

$$E_{\text{total}} = q\left(E_{\text{elec}}\left(N/k - 1\right) + E_{\text{DA}}\left(N/k\right) + 2E_{\text{Tx-elec}}\right. \tag{11}$$

$$\left. + \varepsilon_{\text{mp}} d_{\text{to BS}}^4 + q\varepsilon_{\text{fs}} M^2/2\pi k\right).$$

Finally, the optimal number of clusters, k, can be determined by setting the derivative of E_{total} with respect k to zero, resulting in

$$k = \frac{\sqrt{N}\sqrt{\epsilon_{\text{fs}}} M}{\sqrt{2\pi}\sqrt{\epsilon_{\text{mp}}} d_{\text{to BS}}^2} \tag{12}$$

$$\epsilon_{\text{fs}} = 10\,\text{pJ bit}^{-1}\,\text{m}^{-2}, \quad \epsilon_{\text{mp}} = 0.0013\,\text{pJ bit}^{-1}\,\text{m}^{-4}.$$

Based on Eq. (12), let us assume that the number of sensor nodes (N) and the network region (M) are constant, but the base station distance (d) increases; subsequently, the optimal

number of clusters (k) decreases. Ultimately, some clusters have many sensor nodes when the number of clusters decreases due to k is the inverse squared distance. As Haibo et al. (2010) described, a cluster head with many sensor nodes consumes more energy than a cluster head with a few sensor nodes, because it aggregates, receives, and compresses more sensing information than a cluster head with few sensor nodes. In addition, if there is a large distance between a cluster head and the base station, the cluster head node consumes more energy than it would if the distance were shorter. If the current cluster head runs out of energy, the entire wireless sensor network is no longer operational. The main challenge is to minimize the power consumption of the cluster head, especially when many sensor nodes are allocated to a single cluster.

2.2 k-means^{++} algorithm

The k-means^{++} algorithm is used to assign the initial center of the k-means algorithm. Since the k-means algorithm randomly chooses the initial centroid, it is not guaranteed that clustering by the k-means algorithm is optimal. For example, if the initial random centroid is far away from the cluster's true center, the number of iterations required to optimize the centroid takes longer, and an incorrect clustering result may be obtained (Arthur et al., 2007; Avros et al., 2012). To remedy these problems, the k-means^{++} algorithm randomly selects the initial center from the sensor nodes' locations, but their location depends on their squared distances from the closest center that already has been selected.

For example, the first single initial center (c_1) is selected randomly; however, the remaining centers, such as those in the range from (c_2) to (c_l), are calculated based on the steps described below.

First, let us assume that the sensor nodes are represented by $X = (x_1, \ldots, x_n)$ and that l centers are represented as C, where $C = (c_1, \ldots, c_l)$. The distance between each sensor node and (c_1) is calculated by

$$D_1 = \|x_1 - c_1\|^2, \quad D_2 = \|x_2 - c_1\|^2 \tag{13}$$

and $D_n = \|x_n - c_1\|^2$.

The distance of each sensor nodes and over the average distance is calculated by

$$p(x_1) = \frac{D_1^2}{D_1^2}, \quad p(x_2) = \frac{D_2^2}{D_1^2 + D_2^2}, \tag{14}$$

$$p(x_n) = \frac{D_n^2}{D_1^2 + D_2^2 + \ldots + D_n^2}.$$

Second, the algorithm generates a random number. Then, one of the values of $p(x_1), p(x_2), \ldots, p(x_n)$ close to a random number (i.e., x_i) becomes the second center. For example, for the random number of $R \approx p(x_4)$, the sensor node x_4 becomes (c_2); otherwise, the algorithm generates another value. The third step is to choose the third center (c_3). The distance is calculated as

$$D_1^2 = \min(\|x_1 - c_1\|^2 \|x_1 - c_2\|^2),$$
$$D_2^2 = \min(\|x_2 - c_1\|^2 \|x_2 - c_2\|^2), \qquad (15)$$
$$D_3^2 = \min(\|x_3 - c_1\|^2 \|x_3 - c_2\|^2).$$

The distance of each sensor node and over the average distance of sensor nodes is also calculated as

$$p(x_1) = \frac{D_1^2}{D_1^2}, \quad p(x_2) = \frac{D_2^2}{D_1^2 + D_2^2}, \qquad (16)$$
$$p(x_n) = \frac{D_n^2}{D_1^2 + D_2^2 + \ldots + D_n^2}.$$

Again, the algorithm generates a random number to choose one of the values of $p(x_1), p(x_2), \ldots, p(x_n)$. The process of selecting the initial centers using the above steps continues until l centers are selected.

Moreover, Arthur et al. (2007) chose the initialization center of a data set one by one in a controlled fashion using the k-means^{++} algorithm. For example, the first initial center was selected randomly in a sensing region, but the subsequent centers depended on the value of the previous center. For example, c_2 depends on c_1, and c_3 depends on c_2 and c_1. If we expand the illustrative example, the k-means^{++} algorithm can be conveniently generalized for any number of nodes and clusters.

The first step is to choose the first single initial center (c_1) randomly. The second step is to compute the distance between all sensor nodes and (c_1) and choose c_2 by the following:

$$D_i = \|x_i - c_1\|^2, \qquad (17)$$
$$p(x_1) = \frac{D_1^2}{D_1^2}, \quad p(x_2) = \frac{D_2^2}{D_1^2 + D_2^2}, \qquad (18)$$
$$p(x_n) = \frac{D_n^2}{D_1^2 + D_2^2 + \ldots + D_n^2}.$$

The algorithm generates a random number. Then, one of the values of $p(x_1), p(x_2), \ldots, p(x_n)$ close to the random number, x_i, becomes a second center, c_2. Third, recompute the distance vector to choose the third center as

$$D_i^2 = \min(\|x_i - c_1\|^2 \|x_i - c_2\|^2, \ldots, \|x_i - c_l\|^2). \qquad (19)$$

Calculate $p(x_1), \ldots, p(x_n)$ as Eq. (16) and generate a random number close to $p(x_1), \ldots, p(x_n)$ to choose the third center (c_3). The difference between Eqs. (17) and (19) is that Eq. (17) is used to calculate the distance between the initial center (c_1) and the sensor nodes, whereas Eq. (19) is used to calculate the distance based on (c_1) and (c_2). In general, all remaining centers, such as c_l, are calculated as

$$D_n^2 = \min(\|x_i - c_1\|^2, \ldots, \|x_i - c_l\|^2), \qquad (20)$$

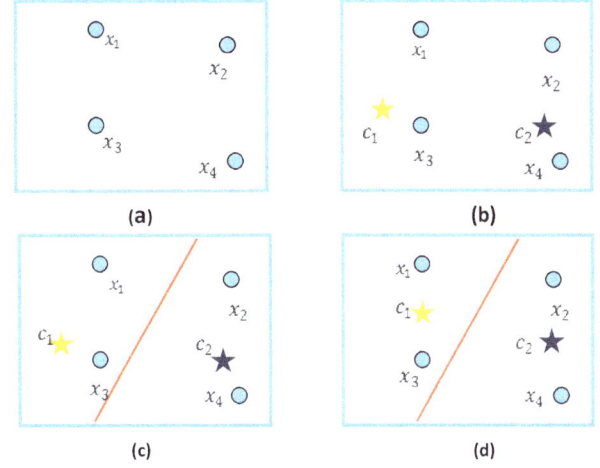

Figure 2. k-means algorithm diagram: **(a)** location of the sensor nodes; **(b)** initial centers; **(c)** new center after multiple iterations; **(d)** optimal centers.

$$p(x_1) = \frac{D_1^2}{D_1^2}, \quad p(x_2) = \frac{D_2^2}{D_1^2 D_2^2}, \ldots, \qquad (21)$$
$$p(x_n) = \frac{D_n^2}{D_1^2 + D_2^2 + \ldots + D_n^2}.$$

2.3　k-means algorithm

The k-means algorithm is a method of grouping or classifying sensor nodes into k numbers of groups/clusters (Zhong et al., 2012). This technique selects an optimal center location of a cluster from which the sum of the squared distances to the locations of the sensor nodes is minimized.

Figure 2 illustrates how the k-mean algorithm is used to select an optimal center. First, sensor nodes are represented as (x_1, x_2, x_3, x_4) in Fig. 2a, and let us randomly choose two centers, called c_1 and c_2 (Fig. 2b). Next, calculate the distance between each sensor node to the two centers, $\|x_1 - c_1\|^2, \ldots, \|x_4 - c_1\|^2$ and $\|x_1 - c_2\|^2, \ldots, \|x_4 - c_2\|^2$. Third, group sensor nodes are based on sensor nodes' minimum distance to the centers. For example, if x_1 and x_2 are closest to c_1, then x_1 and x_2 will be in the same group. Similarly, if x_3 and x_4 are closest to c_2, then x_3 and x_4 will be in the same group. In addition, Fig. 2c shows that sensor nodes are grouped based on the closest distance to the centers. Four, calculate a new center for sensor nodes, which are in the same group. For example, $c_{1_{\text{new}}} = \frac{1}{2}\{(x_1 - c_1)^2 + (x_2 - c_1)^2\}$ and $c_{2_{\text{new}}} = \frac{1}{2}\{(x_1 - c_2)^2 + (x_2 - c_2)^2\}$. Last, we continue to calculate the center based on the previous equation until the new center is the same as the previous center location. When the previous and the new center location are the same, the centers are optimal, shown in Fig. 2d.

If we expand the illustrative example, the k-means algorithm can be generalized conveniently for any number of nodes and clusters. In general, the locations of n sensor nodes are represented by X, where $X = (x_1, \ldots, x_n)$, and l centers are represented by C, where $C = (c_1, \ldots, c_l)$. The k-means objective function, which minimizes the distance between sensor node (x_i) and the cluster center (c_j), is defined as

$$KM(X, C) = \sum_{i=1}^{n} \| x_i - c_j \|^2 \qquad (22)$$

$i = 1, \ldots, n$ and $j = 1, \ldots, l$,

where

$$c_j = \frac{1}{u_j} \sum_{x_i \epsilon u_j} x_i. \qquad (23)$$

The cluster center c_j represents the current estimation of the location of the center of cluster j, and u_j is the number of sensor nodes in cluster j.

2.4 Gap statistics

"Gap statistics" is a standard technique for determining the optimal number of clusters for a data set (or a group of sensor nodes) by comparing the observed weight curve to the expectation of a referenced weight curve (Tibshirani et al., 2001).

The observed weight is the sum of the distance between all observed sensor nodes (actual data) and the center of the cluster; the referenced weight is the sum of the distance between all referenced sensor nodes (ideal) and the center of the cluster (Yan, 2005; Zhang, 2001). The observed weight and the expectation of the referenced weight can be derived mathematically as shown below.

First, let us assume that the sensor nodes are represented by $X = (x_1, \ldots, x_n)$. Also, if there are sensor nodes in a cluster, the distance between each of them is defined by

$$D_k = \sum_{ii}^{l} d'_{ii'} \qquad i = (1, \ldots, n) \qquad (24)$$

$$= \sum_{i=1}^{n} \sum_{i'=1}^{n} \| x_i - x_{i'} \|^2,$$

$$= (x_1 - x_1)^2 + (x_1 - x_2)^2 + (x_1 - x_3)^2 + (x_2 - x_1)^2$$

$$+ (x_2 - x_2)^2 + \ldots + (x_n - x_n)^2$$

$x_1 - x_1 = 0$, $x_2 - x_2 = 0, \ldots, x_n - x_n = 0$.

Therefore, $D_k = 2n_k \sum_{i=1}^{n} \| x_i - \overline{x} \|^2$,

where $\overline{x} = \frac{x_{1'} + x_{2'} + \ldots + x_{n'}}{n}$, and $\overline{x}$ is the center of the cluster, n is the number of sensor nodes, and $d_{i,i'}$ is the distance between two nodes (i and i'), k is the number of clusters,

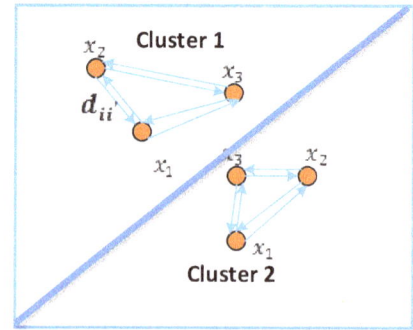

Figure 3. Sensor nodes in a cluster.

$(k = 1, \ldots, g)$, and g is the maximum number of clusters. The weight in the k cluster is defined by

$$W_k = \sum_{k=1}^{g} \frac{1}{2n_k} D_k. \qquad (25)$$

Figure 3 also illustrates the distance between each of the sensor nodes, the number of clusters, and the number of sensor nodes in a cluster. For example, $k = 2$, $n_1 = 3$, and $n_2 = 3$, D_1 is the total distance between sensor nodes to the center at cluster 1, and D_2 is the total distance between sensor nodes to the center at cluster 2.

Second, the algorithm generates the referenced weight by adding a small noise into the original sensor nodes or the observed sensor nodes. The referenced weight is W_k, and the referenced weight dispersion is W_{kb}^*; k is the number of clusters, $k = 1, \ldots, g$, and b refers to the reference data sets, $b = (1, 2, \ldots, B)$, where B is the maximum number of data sets. For example, when $k = 3$ and $b = 5$, the algorithm generates five different locations for sensor nodes which are distributed across three clusters.[8]

Third, the algorithm calculates the expected value of the referenced weight, $E_n^*(W_{kb})$, and n is the number of sensor nodes. In order to analyze the difference between observed weight and the expected value of referenced weight, the algorithm uses the logarithmic scale graph since it shows a visual differentiation between observed and referenced weight. Therefore, the observed weight is represented as log (W_k), and the expected referenced weight is represented as $E_n^*(\log(W_{kb}))$.

As expressed above, the main goal of the gap statistics method is to compare the curve of the observed weight $(\log(W_k))$ to the curve that represents the expectation of a referenced weight $(E_n^*\{\log(W_{kb})\})$ to determine the optimal number of clusters based on the maximum gap between the two curves. As Yan (2005) and Zhang (2001) describe, the number of optimal clusters can be found when $(\log(W_k))$ falls the farthest below the expected referenced weight dispersion curve.

However, when there is a small gap between the $\log(W_k)$ curve and the expected referenced weight curve

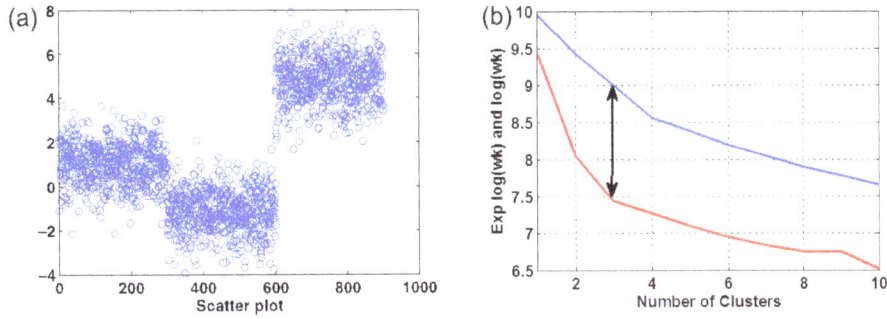

Figure 4. Results of the example with three clusters: **(a)** sensor nodes; **(b)** weight dispersion, W_k, as a function of k number of clusters.

$(E_n^*\{\log(W_{kb})\})$, the cluster is not optimal because the observed sensor nodes have noise that is the same as that of the referenced weight sensor nodes. Conversely, when there is a maximum gap between the $\log(W_k)$ curves and the expected referenced weight curve $(E_n^*\{\log(W_{kb})\})$, the cluster is optimal. In other words, the observed sensor nodes have very small noise at the maximum gap compared to that of the referenced sensor nodes, which are generated with noise. In this discussion, the term "noise" indicates that the sensor nodes are not close to each other and that they do not form the optimal number of clusters.

For example, Fig. 4a shows a scatter graph in which the sensor nodes are distributed across three clusters; one cluster is well separated from the other two clusters, which are connected. Figure 4b shows that using the gap statistics algorithm determines the optimal number of clusters in Fig. 4a. As Fig. 4b shows, the increased number of clusters results in decreased weight. The red line indicates the location of the original sensor nodes within the cluster and has observed weight $(\log(W_k))$; the graph shows a rapid decrease up to cluster number 2, and, then, it decreases slowly from cluster numbers 3–10. In addition, the blue line is the referenced weight, $(E_n^*\{\log(W_{kb})\})$. The optimal number of clusters is determined to be three, because, at that point, the gap between the two lines is at its maximum.

3   Combination of the clustering algorithms

As summarized above, the LEACH (Heinzelman et al., 2000) algorithm uses a computation and communication energy model to increase the lifespan of the sensor nodes. But the method is still far from being a complete and optimal solution to the problem. For example, the LEACH algorithm selects a fixed number of clusters, but it ignores the fact that some of the sensor nodes in a cluster can be reallocated to another cluster. It also ignores the fact that the cluster head's energy will be depleted quickly when too many sensor nodes remain in a single cluster, because more energy is required for aggregating, compressing, and transmitting more information. With this background of partial solutions to the problem, our intention was to attain a complete solution by using other

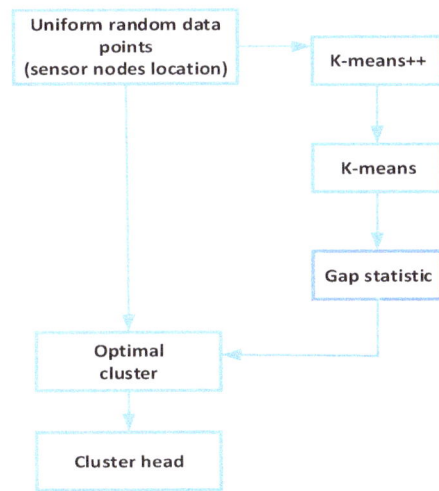

Figure 5. Combination of the three clustering algorithms.

clustering algorithms that were developed for other purposes. This section provides details concerning how they were used. The operation of wireless sensor nodes is divided into three phases, i.e., setup, advertisement, and steady state. In this research, we focused only on the setup phase. During the setup phase, first, the sensor nodes identify their locations and positions and then transmit the information to a base station. At the base station, where this combined algorithm is located and runs, the k-means^{++} algorithm generates the initial center for the sensor nodes' location. Second, the k-means algorithm chooses the optimal centers of the clusters. Finally, the gap statistics algorithm is used to select the optimal number of clusters for the nodes.

Figure 5 shows the steps that are used to choose the optimal number of clusters based on the three clustering algorithms (k-means^{++}, k-means, and gap statistics).

In the first step, we represent the location of the sensor node. In the second step, we initialize the cluster's center based on the k-means^{++} algorithm. In the third step, we choose the optimal center for the cluster based on the k-means algorithm. In the fourth step, we used the gap statistic algorithm to calculate the optimal number of clusters.

The first step starts with a number of sensor nodes represented by $X = (x_1, \ldots, x_n)$ In the second step, we calculate the initial centers for the sensor nodes based on Eqs. (17)–(19).

Third, we calculate the optimal centers of the distributed sensor network based on the k-means algorithm using Eqs. (22) and (23). The k number of clusters is defined as $(k = 1, \ldots, g)$. Using Eqs. (24)–(25), the sum of the clusters' weight (W_k) is calculated, and the mean of a reference weight (W_{kb}^*) is generated. b refers to the reference data sets, $b = (1, 2, \ldots, B)$, where B is the maximum number of data sets. In order to analyze the difference between the observed weight and the expected value of the referenced weight, the algorithm uses a logarithmic scale graph since it shows a visual differentiation between the observed weight and the referenced weight. Therefore, the observed weight is represented as $\log(W_k)$, and the expected referenced weight is represented as $E_n^*(\log(W_{kb}))$. The gap statistics is defined by

$$\text{Gap}_n(k) = E_n^*\{\log(W_{kb})\} - \log(W_k). \quad (26)$$

As expressed above, the main goal of the gap statistics method is to compare the curve of the observed weight ($\log(W_k)$) to the curve that represents the expected reference weight ($E_n^*\{\log(W_{kb})\}$) to determine the optimal number of clusters based on the maximum gap between the two curves.

$$\max(\text{Gap}_n(k)) \approx \hat{k}_{\text{opt}} \quad (27)$$

4 Simulation and discussion

4.1 Test sensor network and scope of simulation

The test sensor network is of the sensor nodes randomly distributed between $u(0, 0)$ and $u(100\,\text{m}, 100\,\text{m})$ as illustrated in Fig. 6, with their location expressed as $X = [(x_{ij})]$, where $(i = 1, 2, \ldots, n)$ and $(j = 1, 2, \ldots, k)$. In addition, the base station (sink) is assumed to be at $(50\,\text{m}, 175\,\text{m})$.

For the simulation of the test sensor network, we used the LEACH algorithm's simulation parameters, as indicated in Table 1. For example, the initial energy for each of the sensor nodes was set to 0.5 J. Each of the data messages were 525 bytes long, and the broadcast packet size header was 25 bytes long.

The radio electronics energy was 50 nJ bit^{-1}, and the radio transmitter energy was set to 10 pJ bit^{-1} m^{-2} or 0.0013 pJ bit^{-1} m^4. The cluster head collects data from the sensor nodes and aggregates those data prior to sending them to the base station. The energy used to aggregate the data (E_{DA}) was 5 nJ bit^{-1} signal^{-1}.

4.2 Code structures for the clustering algorithms

The simulation steps of the three combined algorithms are described in Table 2. First, the k-means^{++} algorithm simulates choosing the initial center of the sensor nodes; second,

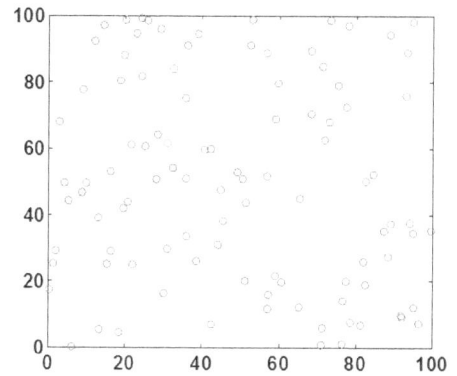

Figure 6. 100 wireless sensor nodes in the area of the sensing network.

Table 1. Simulation parameters.

Parameter	Value
Network field	From (0,0) to (100,100)
Number of nodes	100
Base station	At (50,175)
Initial energy	0.5 J
Data packet size	525 bytes
Broadcast packet size	25 bytes
E_{elec}	50 nJ bit^{-1}
ϵ_{fs}	10 pJ bit^{-1} m^2
ϵ_{emp}	0.0013 pJ bit^{-1} m^{-4}
E_{DA}	5 nJ bit^{-1} signal^{-1}
Threshold distance (d_o)	75 m

the k-means algorithm simulates the calculation of the optimal center. Third, the gap statistics algorithm simulates the calculation of the optimal number of clusters.

4.3 Simulation

Step 1: determination of the optimal center of the cluster using the k-means algorithm

The k-means^{++} algorithm and the k-means algorithm were used to generate the optimal location of the center of the sensor nodes. For example, in Fig. 7, the optimal center is marked by "X", and the sensor nodes are marked by gray, blue, green, cyan, dark blue, black, and Red.

Step 2: determination of the optimal number of clusters using gap statistics

After the optimal location of the center of the sensor nodes was calculated, the gap statistics algorithm determined the optimal number of sensor nodes by comparing the observed weight curve ($\log(W_k)$) to the expected reference weight curve $E_n^*\{\log(W_{kb})\}$.

Table 2. Combination of clustering algorithms.

Algorithm 1: k-means^{++}

Require: generate a uniform random number sensor nodes' location
1: $c_1 \leftarrow$ select a single center from uniformly distributed sensor node location X
2: while $c_i < k\,d_o$ $> k$ is the number of cluster
3: sample $x \in X$ with probability $\dfrac{D_i^2}{\sum\limits_{i=1} D_i^2}$
4: $c_i \leftarrow c_i \cup \{x\}$ end while $>$ select a new center
5: end while

Algorithm 2: k-means

6: use Initial center from k-means^{++} $C \subset X$ $> C = c_1 \ldots c_l$
7: repeat
8: for all $x \in X$ find $\mathrm{KM}(X,C)$ (closet center $c \in C$ to x)
9: for all $i \in k$ let $c_j =$ average $\{x \in X | \mathrm{KM}(X,C) = c_j\}$ $> j = 1,\ldots,l$
10: until The set C is unchanged

Algorithm 3: gap statistics

Require: cluster the observed data, with the number of clusters fixed at $k = 1, 2, \ldots, g$
11: for $k = 1 \rightarrow g\,d_o$
12: $D_k \leftarrow \sum\limits_{i,i'} d_{ii'}$
13: $W_k \leftarrow \sum_{k=1}^{g} \frac{1}{2n_k} D_k$ $>$ total distance within clusters
14: end for

Require: generate reference data W_{kb}^*, $> b = 1, 2 \ldots B,\ k = 1, 2, \ldots, g$
15: for $k = 1 \leftarrow g\,d_o$
16: for $b = 1 \rightarrow B\,d_o$
17: $D_k \leftarrow \sum\limits_{i,i'} d_{ii'}$
18: $W_{kb}^* \leftarrow \sum_{k=1}^{g} \frac{1}{2n_k} D_k$
19: end for
20: end for
21: $\mathrm{Gap}_n(k) = E_n^* \{\log(W_{kb})\} - \log(W_k)$
22: $\max(\mathrm{Gap}_n(k)) \approx \hat{k}_{\mathrm{opt}}$

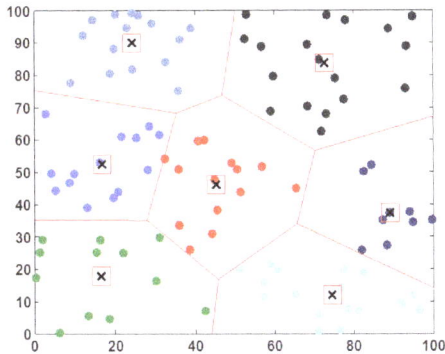

Figure 7. Sensor nodes grouped in seven clusters.

Figure 8 shows the observed and reference weight functions versus the number of clusters. In addition, the red dots on the red curve are the observed weight curve ($\log(W_k)$). The blue curve is the reference weight curve $E_n^* \{\log(W_{kb})\}$

for different numbers of clusters, which was used to calculate the gap statistics. The optimal number of clusters was estimated to be seven because the maximum gap between the reference (blue) and the observed (red) curves reached its maximum at the seven-cluster point.

Step 3: Comparison of the LEACH algorithm and the combination of clustering algorithms

We compare our approaches with the LEACH algorithm's approaches to determine which method provided a longer lifespan for the wireless sensor nodes.

As discussed in Sect. 2, the LEACH algorithm determines the optimal number of clusters, k, in a group of distributed homogeneous wireless sensors based on the "computation and communication energy model".

To assess the two methods, we used the LEACH algorithm to choose a cluster head within sensor nodes in a cluster. For example, sensor nodes randomly chosen from 0 to 1. When

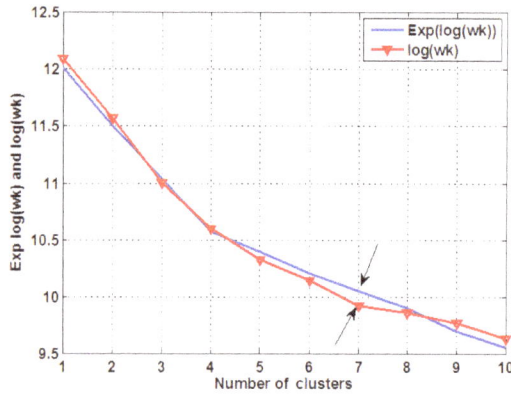

Figure 8. log(mean) dispersion of reference and log dispersion original data sets.

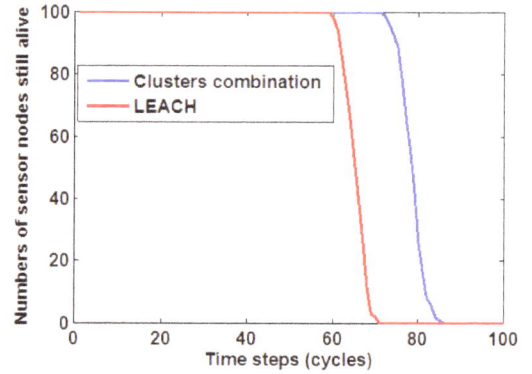

Figure 9. Lifespans of homogenous wireless sensor nodes: (red) LEACH algorithm; (blue) combination of clustering algorithms.

the randomly chosen value is less than the $T(n)$, the sensor node becomes a cluster head; otherwise, a different sensor node chooses another random number to become a cluster head.

The value of $T(n)$ is calculated based on the probability of a sensor node becoming a cluster head and the number of rounds. For example, if there are 20 sensor nodes in a cluster, the probability of becoming a cluster head for each sensor node is $p = 1/20 = 0.05$. After the first cluster head is chosen, the probability of 1 of the remaining 19 sensor nodes becoming a cluster head in the next round is $1/19$. Thus, the number of rounds required for every sensor node to become a cluster head is $r = 1/p$.

$$T(n) = \begin{cases} \dfrac{p}{1 - p \cdot (r \cdot \bmod 1/p)} & \text{if } n \epsilon G \\ 0 & \text{otherwise} \end{cases}, \qquad (28)$$

where r is the number of rounds remaining, G is a group of sensor nodes that have not yet become cluster heads in the previous rounds, p is the expected probability to become a cluster head, and n is a sensor node.

The operation of the LEACH algorithm depends on the rounds. Each round has two phases, i.e., a setup phase and a steady-state phase. During the setup phase, the number of clusters and the cluster head are selected. In the steady-state phase, data are transferred from the sensor nodes to cluster head, which sends them to the base station.

4.4   Comparison of performance

Figure 9 shows the number of sensors still alive over time and shows the advantage of using the combination of the clustering algorithms (blue curve) over the LEACH algorithm (red curve). The energy of the sensor nodes begins to diminish at $t = 62$ cycles using the LEACH algorithm, while it begins to diminish at $t = 75$ cycles using the combination of clustering algorithms. In the LEACH algorithm, all of sensor nodes became inactive at $t = 73$ cycles, whereas they lasted up to

87 cycles in the combination of clustering algorithms. Overall, the combination of clustering algorithms provided 15 % greater lifespan for the sensor nodes than the LEACH algorithm.

5   Conclusions

To improve the lifespan of sensor networks, we proposed using a combination of clustering algorithms, i.e., the k-means algorithm, the k-means^{++} algorithm, and gap statistics, and we compared that approach with the use of the popular LEACH algorithm. In applying the clustering algorithms, the k-means algorithm was used to classify or group sensor nodes into k clusters based on their locations. Also, the k-means^{++} algorithm obtained more appropriate initial center locations for the k-means algorithm, which allowed the optimization of the cluster's center, and gap statistics was used to select the optimal number of clusters for a wireless sensor network.

Our simulation demonstrated the advantage of using the combination of clustering algorithms over using the LEACH algorithm in that the lifespan of the wireless sensor nodes was increased by 15 %.

References

Arthur, D. and Vassilvitskii, S.: k-means^{++}: the advantage of careful seeding, 18th Symposium on Discrete Algorithms, New Orleans, Louisiana, 7–9 January 2007, 1027–1035, 2007.

Avros, R., Granichin, O., Shalymov, D. Volkovich, Z., and Weber, G.: Randomized Algorithm of Finding the True Number of Clusters Based on Chebychev Polynomial Approximation, in: Data Mining: Foundation and Intelligent Paradigms, Springer, 23, doi:10.1007/978-3-642-23166-7, 2012.

Haibo, Z., Wu, Y., Hu, Y., and Xie, G.: A novel stable selection and reliable transmission protocol for clustered heterogeneous wireless sensor networks, Comput. Commun., 33, 1843–1849, 2010.

Heinzelman, W., Chandrakasan, A., and Balakrishnan, H.: Energy-efficient communication protocol for wireless microsensor networks, The 33rd Hawaii International Conference on System Science, Maui, Hawaii, 4–7 January 2000, p. 8020, doi:10.1109/HICSS.2000.926982, 2000.

Kumar, D., Aseri, T., and Patel, R. B.: EEHC: Energy efficient heterogeneous clustered scheme for wireless sensor networks, Comput. Commun., 32, 662–667, 2009.

Qing, L., Zhu, Q., and Wang, M.: DEEC: Design of a distributed energy-efficient clustering algorithm for Heterogeneous wireless sensor networks, Comput. Commun., 29, 2230–2237, 2006.

Tibshirani, R., Walther, G., and Haste, T.: Estimating the number of clusters in a data set via the gap statistic, J. Roy. Stat. Soc. B, 63, 411–423, 2001.

Yan, M.: Method of Determining the Number of Clusters in a Data Set and a New Clustering Criterion, PhD dissertation, Virginia Polytechnic Institute and State University, 23–73, 2005.

Zhang, B.: Generalized K-Harmonic Means – Dynamic weighting of Data in Unsupervised Learning, 1st SIAM international Conference on Data Mining (SDM'2001), Chicago, USA, 5–7 April, 1–13, 2001.

Zhong, S., Wang, G., Leng, X., Wang, X., Xue, L., and Gu, Y.: A Low Energy Consumption Clustering Routing Protocol Based on K-Means, Software Engineering and Applications, 5, 1013–1015, 2012.

19

Atmospheric transmission coefficient modelling in the infrared for thermovision measurements

W. Minkina and D. Klecha

The Faculty of Electrical Engineering, Częstochowa University of Technology, Częstochowa, 42–200, Poland

Correspondence to: W. Minkina (minkina@el.pcz.czest.pl)

Abstract. The aim of this paper is to discuss different models that describe atmospheric transmission in the infrared. They were compared in order to choose the most appropriate one for certain atmospheric conditions. Universal models and different inaccuracies connected with them were analysed in this paper. It is well known that all these models are different, but the aim of this paper is calculate how big the differences are between the characteristics of atmospheric transmission as a function of the distance. There have been models analysed from the literature, and these are used in infrared cameras. Correctly measured atmospheric transmission allows the correct temperature of an object to be determined, which is very vast problem that is discussed in paper.

1 Introduction

The atmospheric transmission in the infrared (IR) is an important parameter in thermovision measurements. This is due to the fact that, when the temperature of an object is measured, the atmosphere which is between the thermal imaging camera and the object attenuates infrared radiation emitted by the object. Additionally, it has been observed, even in laboratory conditions, that at distance of 1–10 m, the atmospheric absorption, caused by water vapour and carbon dioxide, is noticeable. The most important role in absorption of the infrared radiation for the wavelength $\lambda = 4.3\,\mu m$ is played by carbon dioxide, present in the exhaled air (Rudowski, 1978). It was stated, for example, that after 3 h of two persons being in a closed room, about $40\,cm^3$ in volume, the concentration of CO_2 was such that, at distance $d = 0.8\,m$, 70 % of radiation of the wavelength $\lambda = 4.3\,\mu m$ was absorbed by the air (Rudowski, 1978).

Correctly measured atmospheric transmission allows the correct temperature of an object to be determined. In the case of there being no precise model describing the atmospheric transmission in the thermal imager microcontroller (Minkina and Dudzik, 2006, 2009; Minkina et al., 2010), the obtained temperature of the object would be wrong, lower or higher. The atmospheric attenuation depends strongly on the wavelength. For some wavelengths there is very low attenuation over distance of several kilometres, whereas for other wavelengths the radiation is attenuated to close to nothing over a few metres. The attenuation in the atmosphere does not allow the total original radiation from the object to reach the camera. If no correction for the attenuation is applied, then the measured apparent temperature will be lower and lower with increased distance. The influence of distance on the temperature measurement for the short-wave (SW) and long-wave (LW) camera, without taking into account correction of the impact of the atmosphere on the measurement, can be clearly seen in Fig. 1.

The paper compares different methods of calculating the atmospheric transmission coefficient in the infrared which can be found in practice and in the literature. When a model is chosen, such factors as accuracy and the time needed to do the measurements should be taken into consideration. Greater accuracy means a longer time needed for the calculations. In fact, that subject of research about the atmospheric transmission has a wide range, but this paper has some limitations. The paper concentrates only on the atmosphere's impact on the measurement. The effect of the IR radiation emitted by the absorbing atmosphere (Kirchhoff's law) was skipped in this case and will be considered in the subsequent paper. Thermal imaging cameras operate in a particular infrared range, for which the atmospheric transmission coefficient will be different than for the whole band.

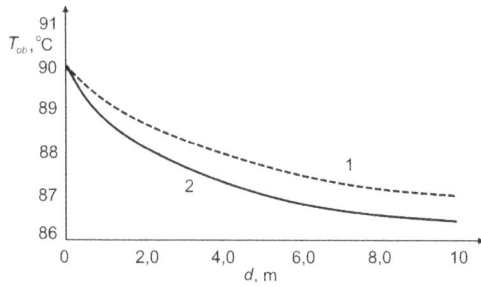

Figure 1. The influence of distance to the temperature measurement for the thermal imaging cameras without taking into account correction of the impact of the atmosphere on the measurement: (1) LW, 8–12 μm and (2) SW, 2–5 μm – an example (IR-Book, 2000).

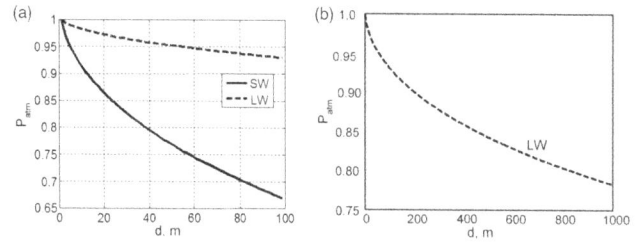

Figure 2. Characteristics of atmospheric transmission coefficient $P_{atm} = f(d)$ for AGEMA 880 as a function of the camera–object distance d for LOWTRAN model (Eq. 1) and $T_{atm} = 15\,°C$, $\omega_\% = 50\,\%$, for **(a)** $d < 100$ m (LW and SW camera) and **(b)** $d < 1000$ m (LW camera for further comparisons).

2 Thermal imaging cameras AGEMA 880 LW and AGEMA 470 Pro SW

Depending on a thermovision camera model, there are several different models of the atmosphere transmittance, such as FASCODE, MODTRAN and SENTRAN (Anderson et al., 1995; Rothman et al., 2005; Vollmer and Möllmann, 2010, Pręgowski and Świderski, 1996; Pręgowski, 2001). For example, in the AGEMA 470 Pro SW and AGEMA 880 LW systems, the manufacturer employs the following simplified equation that describes the atmospheric transmission in the infrared, using the LOWTRAN model:

$$P_{atm}(d) = \exp\left[-\alpha \cdot \left(\sqrt{d} - \sqrt{d_{cal}}\right) - \beta \cdot (d - d_{cal})\right], \quad (1)$$

where P_{atm} is atmospheric transmission; d is camera–object distance (in m); d_{cal} is camera–object distance (in m) (in calibration process – the value of 1 m); and α and β are coefficients specified for normal conditions: atmospheric temperature $T_{atm} = 15\,°C$, relative humidity $\omega_\% = 50\,\%$. For SW bands, $\alpha = 0.393$ and $\beta = 0.00049$; for LW bands, $\alpha = 0.008$ and $\beta = 0$.

The given values are determined in normal conditions of atmospheric temperature $T_{atm} = 15\,°C$ and relative humidity $\omega_\% = 50\,\%$. Under different conditions, the atmospheric transmittance model will be different. The value of coefficient P_{atm} vs. distance d between the camera and the object is shown in Fig. 2a for a LW camera (1) and a SW camera (2). These relationships were obtained from numerical computations using Eq. (1). One can see that the atmosphere has greater transmittance within the LW infrared band. Very similar results are presented in Narasimhan and Nayar (2002) and Orlove (1982).

3 ThermaCAM PM 595 LW

The transmittance model defined by FLIR for the ThermaCAM PM 595 camera is a function of three variables: atmospheric relative humidity $\omega_\%$, camera-to-object distance

d and atmospheric temperature T_{atm} (Toolkit IC2, 2001):

$$P_{atm} = f(\omega_\%, d, T_{atm}). \quad (2)$$

This model was applied to error and uncertainty analysis in the monograph (Minkina and Dudzik, 2009). It is actually very complex. It includes, among others, nine coefficients adjusted empirically. The explicit form of function (Eq. 2) is copyrighted and reserved by the camera manufacturer (Toolkit IC2, 2001). It was made available to the authors only for research purposes, so we may not publish it here. We are allowed to present the characteristics of the atmospheric transmittance P_{atm} as a function of camera-to-object distance d. The results, shown in Fig. 3, were obtained by numerical simulations using the full form of Eq. (2).

It should be emphasized that the model described with Eq. (2) concerns most of the infrared cameras produced by the AGEMA company (e.g. 900 series) and FLIR company (e.g. ThermaCAM PM 595 LW).

Finally, the thermovision camera measurement model is defined as a function of five variables (Minkina and Klecha, 2015):

$$T_{ob} = f(\varepsilon_{ob}, T_{atm}, T_o, \omega_\%, d), \quad (3)$$

where ε_{ob} is emissivity of the object, T_o is ambient temperature, and T_{atm} is atmospheric temperature, the same as in Eq. (2).

We want to emphasize that the model derived above is a simplified model. In reality, the camera detector receives radiation not only from the object but also from other sources. The simplification can be explained looking at Fig. 4.

The signal proportional to the ambient radiation intensity and dependent on ambient temperature T_o is in reality an average response to radiation coming from clouds of temperature T_{cl}, buildings of temperature T_b, ground of temperature T_{gr} and from the atmosphere of temperature T_{atm}. All these temperatures differ a little from each other (Orlove, 1982; DeWitt, 1983; Saunders, 1999).

Equations (2) and (3) for calculating atmospheric transmission coefficient P_{atm} include all parameters affecting

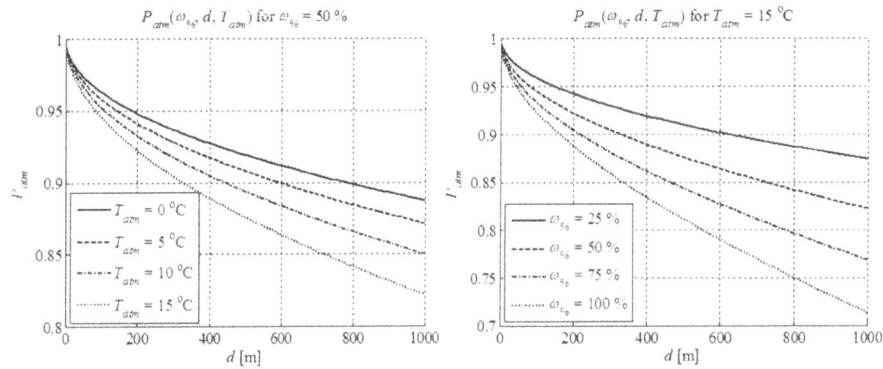

Figure 3. Simulation characteristics of the atmospheric transmittance $P_{atm} = f(d)$ vs. camera-to-object distance d for full form of model (Eq. 2) as a function of the (**a**) temperature of the atmosphere T_{atm}, $\omega_\% = 50\%$, and (**b**) relative humidity of the atmosphere $\omega_\%$, $T_{atm} = 15\,°C$.

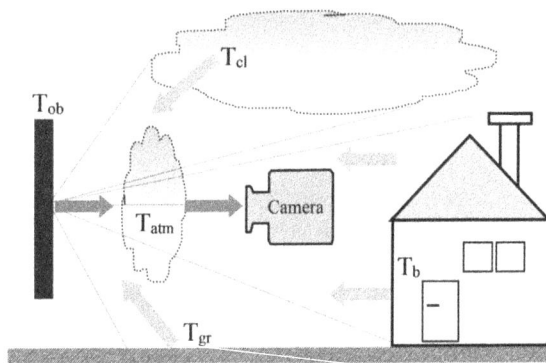

Figure 4. Explanation of simplifications assumed in the thermovision camera measurement model (2, 3); ambient temperature T_0 is an average of temperatures of clouds T_{cl}, atmosphere T_{atm}, ground T_{gr} and e.g. buildings T_b (Minkina and Dudzik, 2009).

thermal measurement; therefore they are universal equations which can be used in practice by FLIR (Toolkit IC2, 2001).

4   Passman–Larmore tables

Using experimental studies conducted by Passman and Larmore (1956), the characteristics of the transmission coefficient can be calculated precisely. Gas composition influences the results of measurements carried out using a thermovision camera. In this case, the most important are the absorbance coefficients: vapour absorbance (P_{H_2O}) and carbon dioxide absorbance (P_{CO_2}). According to Eq. (4) we have (Gaussorgues, 1994)

$$P_{atm} \cong P_{H_2O} \cdot P_{CO_2}. \tag{4}$$

The vapour absorbance depends on the number of absorbing molecules, i.e. on the partial pressure of water vapour, and the distance d travelled by radiation in the absorbing medium. It is usually defined as height h of the cylinder with diameter D. The volume of cylinder is equal to the volume of

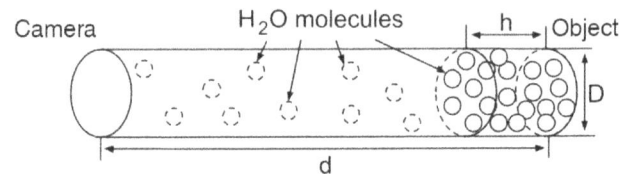

Figure 5. The figure shows the cylinder with height d and diameter D; it includes amount of water vapour, and it is at the distance of measurement using the thermovision camera (Gaussorgues, 1994).

liquid obtained by condensation of water vapour contained in the cylinder with diameter D and height d, reflecting the atmosphere which is at the distance d, where the thermovision camera measurement is taken for $d = 1$ km. It is also shown in Fig. 5. Taking into consideration the equations described above, the following equation can be used:

$$h = \frac{4 \cdot V_{H_2O}}{\pi D^2}, \tag{5}$$

where V_{H_2O} is the volume of liquid obtained by condensation of water vapour in the absorbing medium (m^3) and D is the diameter of the cylinder representing the absorbing medium (m^2).

Vapour absorbance P_{H_2O} depends on molecular processes which are responsible for a selective absorption spectrum. P_{H_2O} also depends on temperature and total pressure of the gas mixture which regulates the width of absorbing lines as a result of molecular collisions and the Doppler effect.

There is a relationship between the height of the cylinder with water h (Eq. 2), temperature T_{atm} and relative humidity $\omega_\%$. An approximation of a function was obtained using of the Newton method $h(d)$ described in Gaussorgues (1994) and its dependency on relative humidity and distance was taken into consideration according to Eq. (3). See the function in Fig. 6.

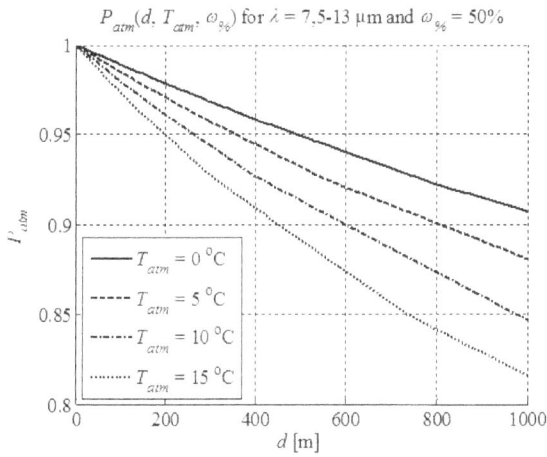

$P_{atm}(d, T_{atm}, \omega_{\%})$ for $\lambda = 7.5\text{-}13\,\mu m$ and $\omega_{\%} = 50\%$

Figure 6. Characteristics of coefficient h, $m\,km^{-1}$ for $d = 1\,km$, according to the Eq. (6).

Table 1. A part of the Passman–Larmore table for vapour absorbance P_{H_2O} (Gaussorgues, 1994; Passman and Larmore, 1956).

$\lambda, \mu m$	P_{H_2O}						
	h, $mm\,km^{-1}$						
	0.2	0.5	1	2	5	10	20
7.5	0.947	0.874	0.762	0.582	0.258	0.066	0
8.0	0.990	0.975	0.951	0.904	0.777	0.603	0.365
8.5	0.994	0.986	0.972	0.944	0.866	0.750	0.562
9.0	0.997	0.992	0.984	0.968	0.921	0.848	0.719
9.5	0.997	0.993	0.987	0.973	0.934	0.873	0.762
10.0	0.998	0.994	0.988	0.975	0.940	0.883	0.780
10.5	0.998	0.994	0.988	0.976	0.941	0.886	0.784
11.0	0.998	0.994	0.988	0.975	0.940	0.883	0.779
11.5	0.997	0.993	0.986	0.972	0.932	0.868	0.753
12.0	0.997	0.993	0.987	0.974	0.937	0.878	0.770
12.5	0.997	0.993	0.986	0.973	0.933	0.871	0.759
13.0	0.997	0.992	0.984	0.967	0.921	0.846	0.718

$$h(T_{atm}, d, \omega_{\%}) = \left(1.6667 \cdot 10^{-4} \cdot T_{atm}^3 + 10^{-2} \cdot T_{atm}^2\right.$$
$$\left. + 3.8333 \cdot 10^{-1} \cdot T_{atm} + 5\right) \omega_{\%} \cdot d \cdot 10^{-3}, \quad (6)$$

where h is height of the cylinder with water (in $m\,km^{-1}$); T_{atm} is the ambient temperature (in $°C$); $\omega_{\%}$ is relative humidity (%); and d is distance (in km).

Taking into account distance d, height of the cylinder with water h and wavelength λ, the appropriate values of P_{H_2O} and P_{CO_2} can be determined using the Passman–Larmore tables. When these values are placed into Eq. (4), it is possible to calculate the atmospheric transmission coefficient P_{atm}. The appropriate part of Passman–Larmore tables is shown in Tables 1 and 2.

Table 2. A part of the Passman–Larmore table for carbon dioxide absorbance P_{CO_2} (Gaussorgues, 1994; Passman and Larmore, 1956).

$\lambda, \mu m$	P_{CO_2}						
	d, km						
	0.2	0.5	1	2	5	10	20
7.5	1.000	1.000	1.000	1.000	1.000	1.000	1.000
8.0	1.000	1.000	1.000	1.000	1.000	1.000	1.000
8.5	1.000	1.000	1.000	1.000	1.000	1.000	1.000
9.0	1.000	1.000	1.000	1.000	1.000	1.000	1.000
9.5	0.993	0.983	0.967	0.935	0.842	0.715	0.512
10.0	1.000	1.000	0.999	0.997	0.994	0.989	0.978
10.5	1.000	1.000	0.999	0.998	0.998	0.995	0.991
11.0	1.000	0.999	0.999	0.997	0.993	0.986	0.973
11.5	0.999	0.998	0.996	0.992	0.980	0.960	0.921
12.0	1.000	1.000	0.999	0.999	0.997	0.993	0.986
12.5	0.987	0.968	0.936	0.877	0.719	0.517	0.268
13.0	0.991	0.977	0.955	0.912	0.794	0.630	0.397

Example.

Calculate the atmospheric transmission coefficient for $\lambda = 13\,\mu m$, $d = 500\,m$, $\omega_{\%} = 50\%$, ($\omega = 0.5$) and $T_{atm} = 20\,°C$.

Using Eq. (3) the result is $h(T_{atm}, d, \omega_{\%}) = (1.6667 \times 10^{-4} \times 20^3 + 10^{-2} \times 20^2 + 3.8333 \times 10^{-1} \times 20 + 5) \times 0.5 \times 0.5 \times 10^{-3} = 4.5 \times 10^{-3} \approx 5 \times 10^{-3}$. When the values are placed in Table 1 and 2, the results are 0.921 and 0.977. Next they are placed into Eq. (4); in this way the value of atmospheric transmission coefficient is calculated, and it is equal approximately to 0.900.

In order to adapt the results to models described in paragraphs 2 and 3, the average characteristics for the wavelength $\lambda = 7.5\text{-}13\,\mu m$ are shown in Fig. 7.

The tables were obtained on the basis of experimental studies, and that is why the model seems to give the most accurate value of the atmospheric transmission coefficient P_{atm}. Calculating atmospheric transmission coefficient by means of the final value of P_{atm} is time-consuming.

5 Approach adopted in the paper of Więcek (2011)

The paper of Więcek (2011) gives another atmospheric transmission model in the infrared P_{atm}. First, Eq. (7) describing saturation vapour pressure was introduced:

$$p_s(T_{atm}) = \begin{cases} 6.112 \cdot e^{21.874 \frac{T_{atm}-273.17}{T_{atm}-7.66}} \\ \qquad \text{for } T_{atm} < 273.15\,K \\ 6.112 \cdot e^{17.269 \frac{T_{atm}-273.17}{T_{atm}-35.86}} \\ \qquad \text{for } T_{atm} \geq 273.15\,K \end{cases}, \quad (7)$$

where p_s is saturation vapour pressure (in Pa) and T_{atm} is temperature of the atmosphere (in K).

Next, the value of the coefficient α_λ was calculated. It describes the cross section of the vapour molecule

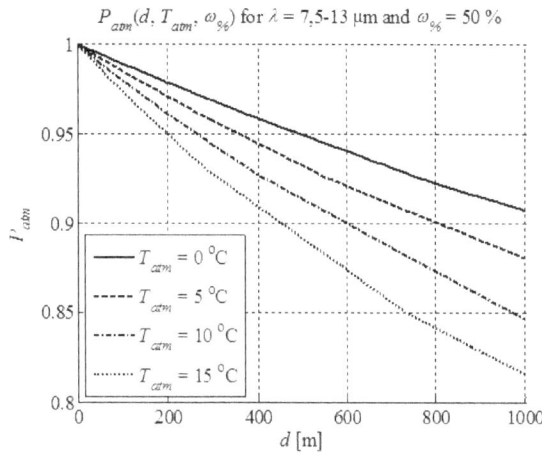

Figure 7. Characteristics of atmospheric transmission coefficient $P_{atm} = f(d)$ calculated using Passman–Larmore tables.

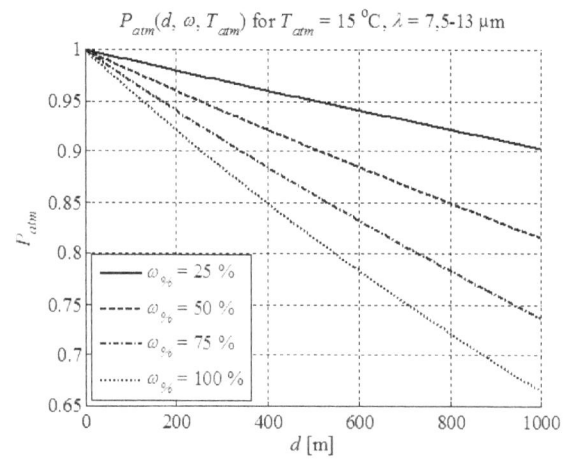

Figure 8. Characteristics of atmospheric transmission coefficient $P_{atm} = f(d)$ calculated according to the approach given in (Więcek, 2011).

which depends on the wavelength (Eq. 8). In the paper of Więcek (2011) Eq. (8) describes the correlation resulting from Beer's law and ideal gas law:

$$\alpha_\lambda = -\frac{k_B \cdot T_{atm}}{\omega_\% \cdot p_s \cdot d} \ln(P_{atm}), \tag{8}$$

where $k_B = 1.28 \times 10^{-23}$ is the Boltzmann constant (J/K); T_{atm} is temperature of the atmosphere (in K); $\omega_\%$ is relative humidity (%); p_s is saturation vapour pressure (in Pa); d is the camera–object distance (in m); and P_{atm} is the value of the atmospheric transmission coefficient resulting from calibration measurements or literature data (e.g. Passman–Larmore tables; see the example above).

Using Eqs. (4) and (5) and assuming constant wavelength λ, the atmospheric transmission coefficient P_{atm} can be calculated according to the equation (Więcek, 2011)

$$P_{atm}\left(d, T_{atm,\omega_\%}\right) = e^{-\alpha_\lambda d \frac{p_s \omega_\%}{k_B T_{atm}}}. \tag{9}$$

The calibration value is $P_{atm} = 0.8160$. For the method of calculating this value, see example. It was calculated using Passman–Larmore tables for $d = 1000$ m, $\omega_\% = 50\%$, $T_{atm} = 15\,°C$ and $\lambda = 7.5$–$13\,\mu$m. The distance selected in the calculation of the calibration value was chosen experimentally. Figure 8 presents the functions of the coefficient.

Equation (6) gives satisfactory results of P_{atm}, but it is a huge obstacle to set the calibration point properly.

6 Comparison of the functions

In order to compare all models describing atmospheric transmission P_{atm} in the infrared analysed in this paper, the same parameters were applied, e.g. $T_{atm} = 15\,°C$, $\omega_\% = 50\%$ and $\lambda = 7$–$13.5\,\mu$m. The first two models were defined for LW bands, where the average wavelength was used. To adapt the scope to the Passman–Larmore model, Tables 1 and 2 in the

range 7–$13.5\,\mu$m and the average result were used. For h values which were not in the tables we have carried out linear averaging in relation to the two nearest points. The model (Eq. 6) shown in this paper (Więcek, 2011) depends on the calibration value. The calibration value that was calculated in paragraph 5 and which amounted to 0.8160 was used.

7 Conclusions

The atmosphere attenuates infrared radiation of an object whose temperature is measured. In the case of there being no properly designed model including the atmospheric transmission coefficient in the infrared $P_{atm} = f(d)$ in the thermal imaging camera microcontroller, the results given by the camera are inappropriate – see Fig. 1. It is very important to use a good model of the atmospheric transmission coefficient $P_{atm} = f(d)$. The paper has some limitations. Thermal imaging camera models should take into account additional factors affecting on the measurement, such as the spectral response. The dependence of the atmospheric transmission for various types of thermal imaging cameras' matrices cannot be brought into one universal plot. Bearing in mind described limitations, in the paper only comparisons of the plots were made.

The function based on Passman–Larmore measurements was introduced as the initial model. All models have similar characteristics (Fig. 9, which becomes more different with the distance; Fig. 10). The model described in the paper of Więcek (2011) can be much more different from the others if the calibration point is not appropriate. Practical and experimental models give similar values of the atmospheric transmission coefficient P_{atm}. This is due to the fact that the model of the atmospheric transmission coefficient is chosen properly. As a result similar results can be obtained using simple calculations. In Figs. 9 and 10 similar conditions

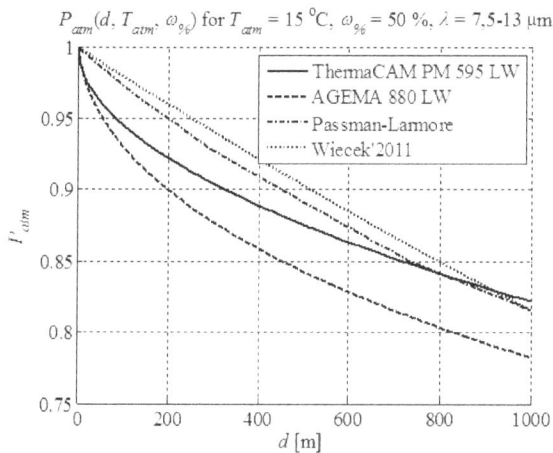

Figure 9. Characteristics of atmospheric transmission coefficient $P_{atm} = f(d)$ for a few models described in this paper for a distance up to 1000 m.

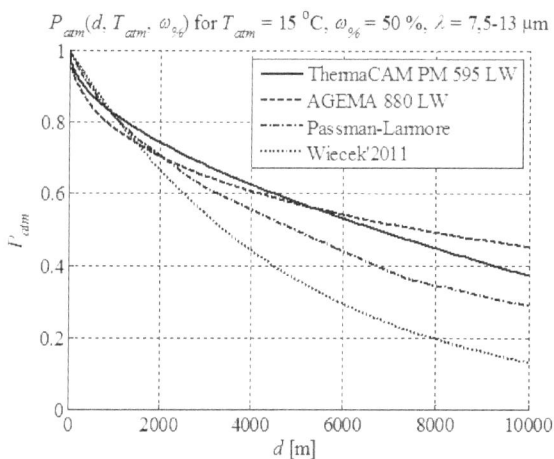

Figure 10. Characteristics of atmospheric transmission coefficient $P_{atm} = f(d)$ for a few models described in this paper for a distance up to 10 000 m.

were applied for all the models in typical measurement situations. In different circumstances comparative characteristics can be different. This results from the fact that the model described in the paper of Więcek (2011) can be used in practice only when the calibration point is chosen properly. Figure 10 shows huge disparities in characteristics of the atmospheric transmission coefficient $P_{atm} = f(d)$. This is particularly noticeable for distances over 1000 m (Fig. 10). For a distance up to 1000 m (Fig. 9) all characteristics are similar. All these models are dedicated to a certain type of cameras. The Passman–Larmore model is the most universal, because it is possible to use tables for different wavelength λ.

It is well known, that all these models are different, but the aim of this paper is to calculate how big the differences are between the characteristics of the atmospheric transmission as a function of the distance.

It should be noted that the properly calculated atmospheric transmission coefficient in the infrared $P_{atm} = f(d)$ makes it possible to read the right temperature of the object. A correct value of atmospheric transmission coefficient set in the camera microcontroller is of fundamental importance for accuracy of contact-less measurement of the object temperature (Minkina and Dudzik, 2006, 2009). Additional information about the atmospheric transmission coefficient in the infrared can be found in Anderson et al. (1995), DeWitt (1983), Narasimhan and Nayar (2002) and Rothman (2005).

References

Anderson, G. P., Kneizys, F. X., Chetwynd, J. H., Wang, J., Hoke, M. L., Rothman, L. S., Kimball L. M., McClatchey, R. A., Shettle, E. P., Clough, S. A., Gallery, W. O., Abreu, L. W., and Selby, J. E. A.: FASCODE, MODTRAN, LOWTRAN: past, present, future, Proceedings of the 18th Annual Review Conference on Atmospheric Transmission Models, Boston 6–8 June 1995, 101–120, edited by: Anderson, G. P., Picard, R. H., Chetwynd, J. H., 1995.

DeWitt, D. P.: Inferring temperature from optical radiation measurements, Proc. of SPIE, Vol. 0446, Thermosense VI (ÙOctober 1983): An Int. Conf. on Thermal Infrared Sensing for Diagnostics and Control, edited by: Burrer, G. J., 226–233, 1983.

Gaussorgues, G.: Infrared Thermography. Springer Science+Business Media, B. V., Dordrecht, 552 pp., ISBN: 978-94-010-4306-9, 1994.

IR-Book: FLIR Training Proceedings, Level II (Infrared Training Center – International, itc-i), 120 pp., 2000.

Minkina, W. and Dudzik, S.: Simulation analysis of uncertainty of infrared camera measurement and processing path, Measurement, Vol. 39, Nr. 8, 758–763, Elsevier Ltd, 2006.

Minkina, W. and Dudzik, S.: Infrared thermography – errors and uncertainties, John Wiley & Sons Ltd, Chichester, 222 pp., ISBN: 978-0-470-74718-6, 2009.

Minkina, W. and Klecha, D.: Modeling of atmospheric transmission coefficient in infrared for thermovision measurements 14th International Conference on Infrared Sensors & Systems (IRS²'2015), 19–21 May 2015, 903–907, Nürnberg, materials on CD-R, doi:10.5162/irs2015/1.4, 2015.

Minkina, W., Dudzik, S., and Gryś, S.: Errors of thermographic measurements – exercises, Proceedings of 10th International Conference on Quantitative Infrared Thermography (QIRT'2010) 27–30 July 2010, Québec, Canada, organization: Université Laval, Electrical and Computer Engineering Department, 503–509, ISBN: 978-2-9809199-1-6, 2010.

Narasimhan, S. G. and Nayar, S. K.: Vision and the atmosphere, Int. J. Comput. Vision, 48, 233–254, 2002.

Orlove, G. L.: Practical thermal measurement techniques, Proc. of SPIE, Vol. 371, Thermosense V (October 1982), Thermal Infrared Sensing Diagnostics, edited by: Courville, G. E., Detroit, Michigan, 72–81, 1982.

Passman, S. and Larmore, L.: Atmospheric Transmission, Rand Paper, Rand Corporation, Santa Monica, 897 pp., 1956.

Pręgowski, P.: Spectral analysis of radiant signals in processes of tele-thermodetection, Proc. of SPIE, Vol. 4360, Thermosense XXIII, An Int. Conf. on Thermal Sensing and Imaging Diagnostic Applications, edited by: Rozlosnik, A. E., Dinwiddie, R. B., 1–12, 2001.

Pręgowski, P. and Świderski, W.: Experimental determination of the transmission of the atmosphere – based on thermographic measurements, Proc. of 50 Eurotherm Seminar, Quantitative Infrared Thermography (QIRT'1996), Stuttgart (RFN), edited by: Balageas, D., Busse, G., Carlomagno, G. M., 363–367, 1996.

Rothman, L. S.: Jacquemart D., Barbe A., Benner D. Ch. and others: The HITRAN 2004 molecular spectroscopic database, J. Quant. Spectrosc. Ra., 96, 139–204, 2005.

Rudowski, G.: Thermovision and its application, Wydawnictwo Komunikacji i Łączności, Warszawa, 212 pp., 1978 (in Polish).

Saunders, P.: Reflection errors in industrial radiation thermometry, Proc. of 7th Int. Symposium on Temperature and Thermal Measurements in Industry and Science (TEMPMEKO'1999 TC 12), edited by: Dubbeldam, J. F., de Groot, M. J., Delft, 631–636, 1–3 June 1999.

Toolkit IC2: Dig16 Developers Guide 1.01 for Agema 550/570, ThermaCAM PM 5×5 and the ThermoVision Family, FLIR Systems 2001 – in the daytime: 20 February 2014, available at: http://u88.n24.queensu.ca/exiftool/forum/index.php?action=dlattach;topic=4898.0;attach=1035 (last access: 15 January 2016), 2001.

Vollmer, M. and Möllmann, K.-P.: Infrared Thermal Imaging – Fundamentals, Research and Applications, Wiley-VCH Verlag GmbH & Co. KGaA, Weinheim, 594 pp., ISBN: 978-3-527-40717-0, 2010.

Więcek, B.: Thermovision in infrared – basics and applications, Measurement Automation Monitoring Publishing House, Warszawa, 372 pp., ISBN: 978-83-926319-7-2 (in Polish), 2011.

20

Tracer gas experiments in subways using an integrated measuring and analysis system for sulfur hexafluoride

M. Brüne[1], J. Spiegel[1], K. Potje-Kamloth[2], C. Stein[3], and A. Pflitsch[1]

[1]Department of Geography, Ruhr-Universität Bochum, Universitätsstraße 150, 44801 Bochum, Germany
[2]Fraunhofer ICT-IMM, Carl-Zeiss-Straße 18–20, 55129 Mainz, Germany
[3]smartGAS Mikrosensorik GmbH, Kreuzenstraße 98, 74076 Heilbronn, Germany

Correspondence to: M. Brüne (markus.bruene@rub.de)

Abstract. Several sulfur hexafluoride (SF_6) tracer gas experiments were conducted in a subway system to measure the possible pathways of toxic gas for subway tunnels and stations empirically. A new mobile integrated measuring and analysis system was used to achieve high sample rates and a long measurement time. Due to the mobility of the sensors, tracer gas experiments were also carried out inside running subway coaches. All experiments showed a common pattern: the pathways of tracer gas dispersion overlapped with some escape routes, which were contaminated within a few minutes. So in case of catastrophic circumstances like terrorist attacks or subway fires, some escape routes will become deathly traps, but the results also showed free escape routes. With the new sensor technique it will be possible to conduct safety assessments for escape routes in underground transportation facilities.

1 Introduction and motivation

The results of tracer gas experiments help to understand the possible spread of toxic airborne substances for example in subway stations, which are vulnerable in the face of terrorism. In contrast to the period of 1982–1991, where deliberate acts of malice caused 1327 deaths among air travelers and none among subway commuters, the pattern reversed between 2002 and 2011 (Barnett, 2015). As preventing terrorists from entering subway stations is very difficult, preparedness and response are very important. This includes the pathways of airborne toxic substances in subway stations needing to be known and not overlap with emergency escape routes. Sulfur hexafluoride (SF_6) has become an accepted standard in underground ventilation studies (Kennedy et al., 1987). In the past, the contamination of air with SF_6 was often determined by manual air probes with a 60 mL syringe sand subsequent analysis with a gas chromatograph. This method, however, has some disadvantages: the number of syringes is limited to the laboratory capacity, which results in a short measurement time and a low sample rate. Notably, the time period between taking and analyzing the samples can pollute the probes. The involvement of a large number of people is also a possible source of error. The development of a mobile battery powered integrated measuring and analysis system for SF_6 recording with a sample rate of 2 s solves these problems.

2 Tracer gas experiments

In February 2014, several tracer gas experiments were conducted in the subway system of a major European city. The objectives of these experiments were the following.

- Determine pathways of gas dispersion inside subway stations.

- Determine the influence of train traffic.

- Determine the effect on large parts of a subway system if a tracer gas is released inside a running subway coach.

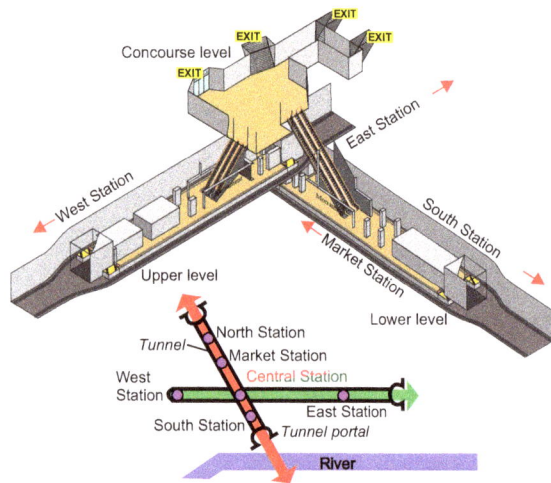

Figure 1. Three-dimensional view of Central Station and overview of underground stations of the subway system.

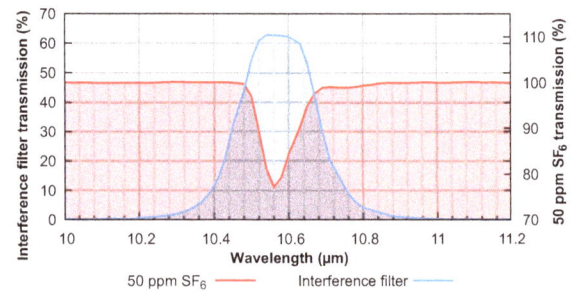

Figure 2. Transmission of the interference filter and absorption bands of SF_6.

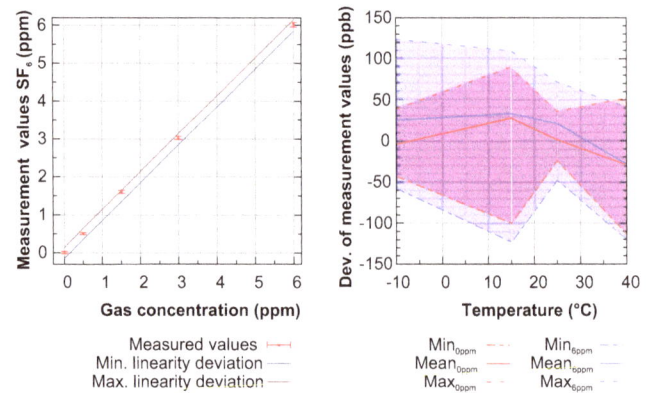

Figure 3. Designated representation of nonlinearity in the sub-ppm range over a temperature range of -10–$40\,^\circ$C.

2.1 The experimental site: the subway system

The experiments were carried out inside the underground station of the subway of a major European city. The underground network consists basically of two lines, running north to south and east to west with a total length of about 4 km. At Central Station, both lines cross. This station has a concourse level connecting all platforms and also a shopping mall (see Fig. 1). Due to orography, the tunnel of the east–west line climbs from East Station to West Station by 15 m, while the north–south line climbs approximately 30 m towards North Station.

2.2 The gas sensor

In contrast to former experiments, which only reflect rough pictures of the gas dispersion (Pflitsch et al., 2010), the new mobile integrated measuring and analysis system can provide much more detailed data (Potje-Kamloth, 2014). As the sensor platform is a battery powered hand-held device, more experiment settings are possible. The device can be connected to temperature sensors and ultra-sonic anemometers, and is thus able to record all parameters to feed a numerical simulation (CFD) with margin conditions. The tracer gas data can be used to validate dispersion forecasts of numerical simulations with empirical data. The heart of the device for measuring the smallest SF_6 concentrations consists of the integrated gas sensor. This is a robust nondispersive infrared sensor (NDIR) assembly optimized for reliable detection of concentrations ≤ 50 ppb SF_6. The performance of the gas sensor had to be improved by a factor of 4 compared to the state of the art. The method of measurement of infrared absorption is based on the Lambert–Beer law, so there are several basic approaches to improve the signal/noise performance. According to the theory for this purpose, the radiation in-

tensity $I_{(0)}$ must be increased substantially. The optical key components for this are a powerful and modular blackbody radiator, an interference filter tuned to the absorption of SF_6 in the range of $10.6\,\mu$m (see Fig. 2), a sensor cuvette with low optical attenuation and a low-noise and highly sensitive detector. Novel amplifying and signal evaluation topologies with performance electronic components as well as the emulation of a lock-in amplifier with software also provide an important contribution. Particularly noteworthy here is the specially developed map-correction algorithm (matrix calibration). The result is that the measurement setup during exposure to ambient air is ideally if possible decoupled from either static temperature or pressure fluctuations. Therefore a very low nonlinearity is achieved (the excessive amount maximum in the measuring range is 0.2 ppm, or 0.4 % of the upper full scale) even in the sub-ppm range for a temperature range of -10 to $40\,^\circ$C (see Fig. 3).

2.3 The influence of train traffic

During the night of 22 February, two tracer gas experiments were conducted to track the gas propagation for a station and its adjacent tunnels and stations. Therefore, Central Station was chosen, as it consists of two platform levels and a mezzanine. The experiment was carried out twice. The first gas

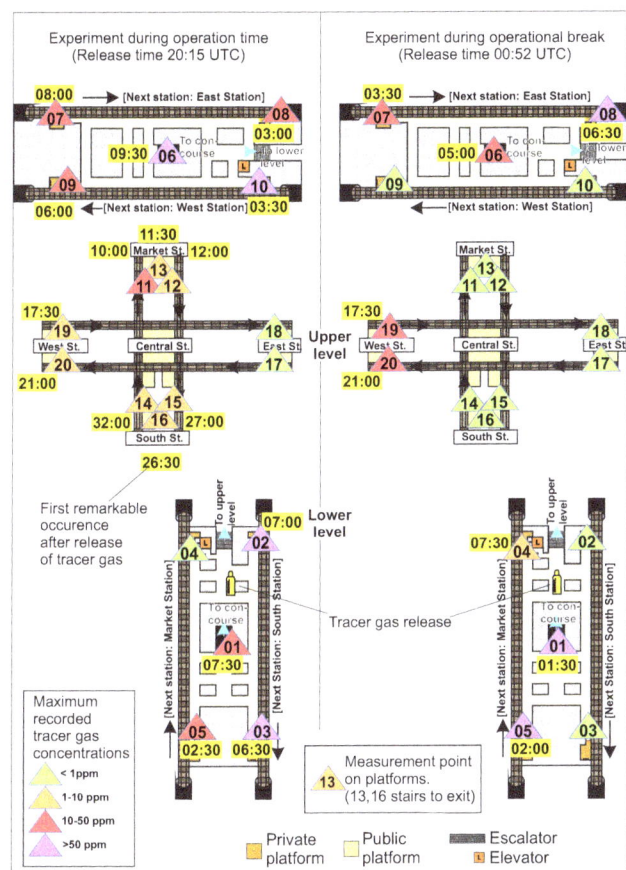

Figure 4. Comparison of tracer gas experiments during traffic times (left) and operational break (right).

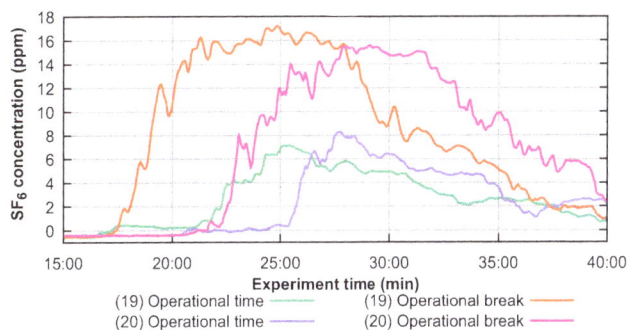

Figure 5. SF$_6$ concentrations at North Station. The numbers in parentheses contribute to the measurement points (see Fig. 4).

steeper rise of the values occurred approximately 5 min later. In contrast, during operational break this steep rise coincided with the first remarkable increase in concentration (see Fig. 5). Market Station and South Station were affected by traffic during the experiment. As the track rises from South Station to North Station, a natural background air flow from south to north is established due to the buoyancy effect. Of course, running trains disturbed the background airflow, but the gas reached the station in the direction of the natural flow 10 min after the gas release. At South Station, on the other hand, it took three times longer for gas to be transported against the direction of natural flow by the piston effect of the trains.

3 Tracer gas release inside a subway train

The hand-held integrated analysis and measuring system offered the possibility to measure tracer gas inside operating trains. An SF$_6$ bottle was boarded in the back coach of a two-wagon subway train. At 2 min before the train reached Central Station, in the back coach tracer gas was released continuously with a relatively low flow rate. The release was stopped 7 min later when the doors closed at North Station. By then, an amount of 1.49 kg SF$_6$ was released from the gas bottle. The contaminated coach reached very high values of 800 ppm for a period of 2 min. This experiment was repeated in the opposite direction about 90 min later, releasing 1.91 kg SF$_6$.

3.1 Contamination of stations

Moreover, the highest concentration inside the station was found in the staircases to the concourse levels and the exits astonishingly quickly. This is a further verification that staircases in subway stations are sucking air from the lower parts of the structure (see Fig. 6). During the repetition of this experiment, a slightly lower concentration was recorded, but the patterns of the spatial and temporal distribution were confirmed. An additional measurement sensor records the tracer gas concentration on the concourse level of Central Station,

release was done during the operational time of the subway at 20:15 GMT. For 10 min, an amount of 2.15 kg SF$_6$ was released. In order to compare traffic times with the operational break, the experiment was repeated on the same night. At 00:52 GMT 2.06 kg of SF$_6$ were released for about 8 min. The upper level (platforms 3/4) was contaminated within 3 min during the train traffic experiment and within 5 min without traffic. This confirms an experiment undertaken 1 day earlier (Spiegel et al., 2014a), which focused only on the station. The maximum value of recorded tracer gas concentration was higher and larger parts of the platform levels were affected. The propagation paths were chaotic due to several train movements (see Fig. 4). A similar picture emerged when focusing on the adjacent stations, but here the maximum recorded concentration was found at West Station during the operational break. During the operational time, sensor (19) detected the first values at 17:30 min after the gas release. Sensor (20) followed at 21:00 min. However, the first measured concentrations were on a very low level. As a subway train reached West Station from Central Station at around the same time, it can be assumed that this first detected concentration is influenced by train movements. A

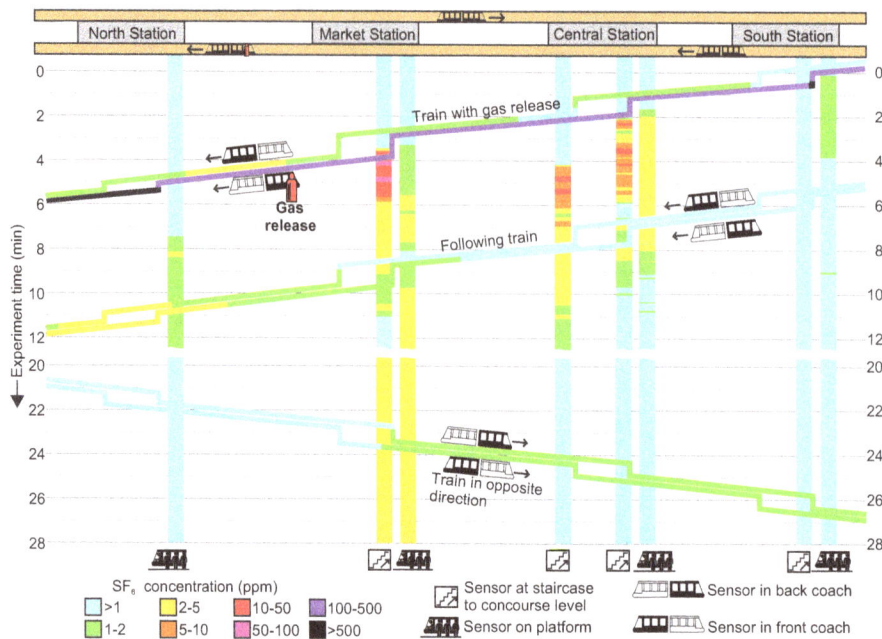

Figure 6. Tracer gas release inside a running train in the back coach. The lines represent the measured concentration while the train is running. The bars represent sensors on the platforms inside the station.

which represents the only escape route from the lower platform levels. As the contaminated train reaches Central Station, some remarkable concentrations were observed on the concourse level within 3 min (see Fig. 7).

3.2 Contamination of trains

The trains of the investigated subway system consist of two separate coaches. After the first stop, a small contamination was observed. At the third stop significantly higher values were logged. Using the mobile advantage of the new sensors, SF_6 was also measured inside the following train. At each station this train opens the doors and some contaminated air flows into the coaches. Consequently, the concentration accumulates up to 3 ppm after the passage of Market Station. The mobile sensor students were located in the trains and after riding from South Station to North Station they were replaced in the next train in the opposite direction, riding through the contaminated underground stations again. At Market Station, the concentration stabilized at 2–5 ppm. As trains called there some tracer gas was sucked inside the train and the concentration considerably increased (see Fig. 6). In order to measure the concentration over a longer time period, one mobile sensor remained in the contaminated train. After stopping at North Station, the train continues over ground. Concentrations remained at nearly 1000 ppm for three further stops. At 7 min after the end of gas release it decreased to 100 ppm, and after 11 further minutes, calling at the ninth overground station, values dropped to 10 ppm. At 50 min after the beginning of the gas release and 20 calls at overground

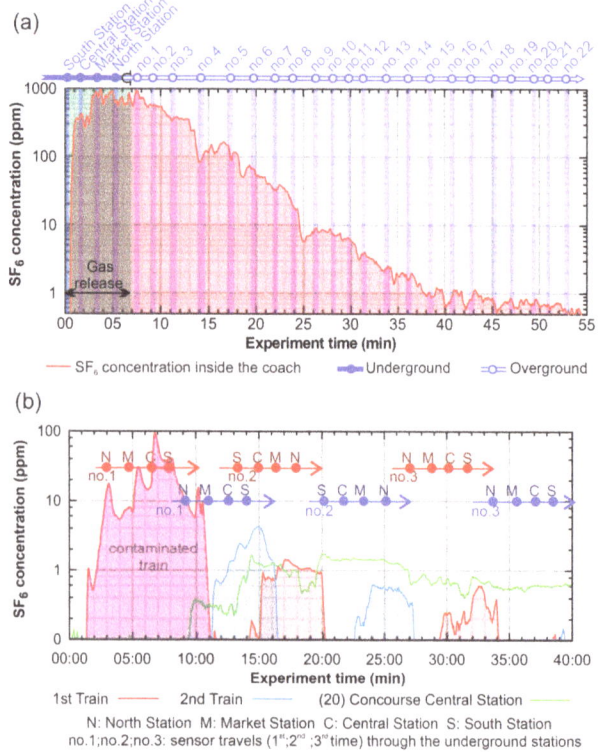

Figure 7. (a) Remaining concentration level inside a subway train after the initial gas release; **(b)** contamination of running subway trains while passing the contaminated underground track and of the concourse level at Central Station (see Fig. 1).

stations, there were still measurably elevated levels in gas concentration (see Fig. 7). The sensors traveled three times on two trains through the underground station in the repetition experiment. The sensor's positions were changed to trains in the opposite direction as they leave the experiment area. Whereas the concentrations were slightly lower in the repetition, the tracer gas remained in the system for a longer period. Therefore, even during the third transit 30 min after the gas release, a demonstrable increase in concentration levels was observed in a train. Furthermore, about 40 min after the release, some notable peaks were recorded (see Fig. 7).

4 Main results of the tracer gas experiment

Subway stations are mostly relatively overwarmed in the temperate climatic zone. With increasing depths of stations, and lengths of staircases, the buoyancy effect, which pushes air upwards, also increases. The experiments have shown that the tracer gas propagates to upper parts within a few minutes. Unfortunately, the propagation path overlaps with escape routes for passengers. The geometry of tunnels has an effect, as differences in elevation drive a natural background air flow. However, more detailed climatological data have to be considered to forecast the spreading of toxic gas in subway systems. The gas release inside a running subway coach can affect wide parts of stations and even harm passengers in following trains up to 30 min later.

5 Conclusions

Using the new integrated analysis and measuring system opens new opportunities, as more complex field tests can be conducted with less effort. During the measurement campaign five tracer gas experiments were conducted. Collecting this amount of data with previous methods like air samples using syringes, over 6000 samples were necessary to cover the same measuring time, with a sample rate of only 1 min. The new sensors could measure with a sample rate of 2 s, which is a big advantage. Experiment costs are reduced to a tenth with the method covered in this paper. The threshold limit of SF_6 is 1000 ppm for an average working day of 8 h, so it is possible to conduct such tracer gas experiments during operational times without harming the passengers. Gas chromatographs cannot be used in high numbers. Existing hand-held measuring systems are used for leak detection of high voltage switch gears, the measuring range is too big, or the high power consumption for the devices makes them utilizable for a long-term continuous mobile measurement. The new sensors can accurately detect SF_6 contamination from 0.05 to 50 ppm. A very low detection limit is warranted due to the high greenhouse potential of SF_6. Results of the field, as mentioned in this paper, can be coupled with numerical simulations (Spiegel et al., 2014b) or with pedestrian simula-

tions (Qian et al., 2014; Brüne et al., 2014) to assess subway systems in many safety questions.

Acknowledgements. Conducting research into critical infrastructures like subway systems is very difficult and only possible with the permission of the operator. Our research caused a lot of effort on the part of the operators. We are grateful that subway operators provide us with their tunnels and stations. We also focused on safety questions and we are very pleased that we find operators who have the courage to work with us on these issues. This work was part of research project "Measuring system for the determination of the dispersal of hazardous materials in critical infrastructures and complex building for the prevention of civil disasters" (MAusKat) by the German Federal Ministry of Education and Research (funding code 13N11673-678).

References

Barnett, A.: Has successful terror gone to ground?, Risk analysis: an official publication of the Society for Risk Analysis, 35, 732–740, doi:10.1111/risa.12352, 2015.

Brüne, M., Charlton, J., Pflitsch, A., and Agnew, B.: Coupling tracer gas experiments with evacuation simulation: Am empirical approach to assess the effectiveness of evacuation routes in subway stations, in: Proceedings of the 6th International Symposium on Tunnel Safetyand Security, 12–14 March 2014, 505–512, Marseille, 2014.

Kennedy, D. J., Stokes, A. W., and Klinowski, W. G.: Resolving Complex Mine Ventilation Problems With Multiple Tracer Gases, in: Proceedings of the 3rd Mine Ventilation Symposium, 12–14 October 1987, Pennsylvania State University, 213–218, 1987.

Pflitsch, A., Brüne, M., Ringeis, J., and Killing-Heinze, M.: ORGAMIR – Development of a safety system for reaction of an event with emission of hazardous airborne substances - like a terror attack or fire - based on subway climatology, in: Proceedings from the fourth International Symposium on Tunnel Safety and Security, edited by: Lönnermark, A. and Ingason, H., vol. 1, 451–462, SP Technical Research Institute of Sweden, Borås and Sweden, 2010.

Potje-Kamloth, K.: Finding the ideal escape and emergency route, available at: http://www.imm.fraunhofer.de/content/dam/imm/de/documents/pdfs/PD_MAusKat_final.pdf (last access: 2 February 2016), 2014.

Qian, Z., Agnew, B., and Thompson, E. M.: Simulation of Air flow, Smoke Dispersion and Evacuation of the Monument Metro Station based on Subway Climatology, in: Fusion, Proceedings of the 32nd International Conference on Education and research in Computer Aided Architectural Design in Europe : Fusion, Proceedings of the 32nd International Conference on Education and research in Computer Aided Architectural Design in Europe, vol. 1 of eCAADe: Conferences, 119–128, Northumbria University, Newcastle upon Tyne, UK, 2014.

Spiegel, J., Brüne, M., Dering, N., Pflitsch, A., Qian, Z., Agnew, B., Paliacin R., and Irving, M.: Propagation of tracer gas in a subway station controlled by natural ventilation, Journal of Heat Island Institute Internationl, Vol. 9, 103–107, 2014a.

Spiegel, J., Letzel, M., Flassak, T., and Pflitsch, A.: Dispersion of Airborne Toxins in a highly complex Subway Station, in: Proceedings of the 6th International Symposium on Tunnel Safetyand Security, 12–14 March 2014, 255–264, Marseille, 2014b.

21

Flexible free-standing SU-8 microfluidic impedance spectroscopy sensor for 3-D molded interconnect devices application

Marc-Peter Schmidt[1], **Aleksandr Oseev**[1], **Christian Engel**[2], **Andreas Brose**[1], **Bertram Schmidt**[1], **and Sören Hirsch**[3]

[1]Institute of Micro and Sensor Systems, Otto von Guericke University Magdeburg, Universitätsplatz 2, 39106 Magdeburg, Germany
[2]TEPROSA GmbH, Paul-Ecke-Str. 6, 39114 Magdeburg, Germany
[3]Department of Engineering, University of Applied Sciences Brandenburg, Magdeburger Str. 50, 14770 Brandenburg an der Havel, Germany

Correspondence to: M.-P. Schmidt (marc-peter.schmidt@ovgu.de)

Abstract. The current contribution reports about the fabrication technology for the development of novel microfluidic impedance spectroscopy sensors that are directly attachable on 3-D molded interconnect devices (3D-MID) that provides an opportunity to create reduced-scale sensor devices for 3-D applications. Advantages of the MID technology in particular for an automotive industry application were recently discussed (Moser and Krause, 2006). An ability to integrate electrical and fluidic parts into the 3D-MID platform brings a sensor device to a new level of the miniaturization. The demonstrated sensor is made of a flexible polymer material featuring a system of electrodes that are structured on and embedded in the SU-8 polymer. The sensor chips can be directly soldered on the MID due to the electroless plated contact pads. A flip chip process based on the opposite electrode design and the implementation of all fluidic and electrical connections at one side of the sensors can be used to assemble the sensor to a three-dimensional substrate. The developed microfluidic sensor demonstrated a predictable impedance spectrum behavior and a sufficient sensitivity to the concentration of ethanol in deionized water. To the best of our knowledge, there is no report regarding such sensor fabrication technology.

1 Introduction

Impedance spectroscopy is a well-known method for liquid analysis (Barsoukov and Macdonald, 2005). For decades it is applied in microfluidic systems for determination of liquid properties (Gawad et al., 2001; Gómez et al., 2001). Depending on the microchannel material, microfluidic sensors can be fabricated on the wafer level (Schmidt et al., 2014a, b) or they can be completed as a free-standing structure for the attachment on a 3-D substrate. There are a number of advantages when microfluidic sensors are implemented on the wafer level, but at the same time they remain non-applicable for the usage within a 3-D-shaped system.

During the last years, molded interconnection devices (MIDs) have attracted considerable attention. Several advantages applicable for the automotive industry such as a high level of integration ability, a compact packaging, etc. were recently highlighted (Moser and Krause, 2006). The possibility of applying 3D-MID-based sensor systems for in-line gasoline analysis can raise considerable interest especially regarding the control of ethanol-containing fuels (Oseev et al., 2013). Structures completed within the MID technology can be applied as a detail of a 3-D system and simultaneously have directly all required integrated electrical and fluidic interconnections. Due to the injection molding and metallization process, a wide range of 3-D-shaped substrates can be fabricated (an example is shown in Fig. 1). In order to be applied in such 3-D systems, the sensor should have a flexible origin with the further possibility of a 3-D direct MID attachment.

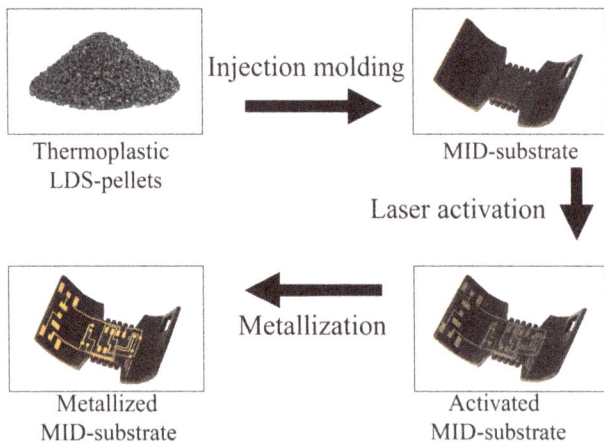

Figure 1. Manufacturing process of a 3D-MID substrate with fluidic channels and conducting paths fabricated by laser direct structuring (LDS) (Schmidt et al., 2012).

The manufacturing of free-standing microfluidic structures can be completed under different approaches (Liu, 2007). Polydimethylsiloxane (PDMS) is well known as a material for the manufacturing of microfluidic structures (Jo et al., 2000). It is a considerably low-cost material, but it cannot be directly structured in a photolithography process. In order to structure PDMS, different techniques such as micromolding need to be applied. These processes can considerably affect the resolution of the final microfluidic structures.

SU-8 is a well-known photoresist that is widely used as a construction material for microfluidic structures, sensors, and a variety of different applications. Based on EPON-Resin, SU-8 is chemically stable and an optically transparent material with controllable mechanical properties through its processing and possibility of being directly structured with a standard photolithography process (Lorenz et al., 1997; Despont et al., 1997).

A variety of wafer-based SU-8 microfluidic structures have already been published (Lin et al., 2002; Zhang et al., 2001), but completely free-standing SU-8 microfluidic structures are rarely declared due to the complex structuring process of metal layers on SU-8 and the releasing step technology for centimeter-scale structures.

In Vilares et al. (2010), the fabrication of a SU-8-based thermal flow sensor with a manual Kapton film releasing was demonstrated. The releasing process was completed only from the single side of the fluidic channel and finalized with the pure SU-8 encapsulation of microfluidic structures. After the releasing step, the entire microfluidic sensor still remained on a second PMMA handling wafer; therefore, it was not applicable for the 3-D attachment. The same Kapton film releasing steps were used in Ezkerra et al. (2007) where SU-8 free-standing cantilever structures were embedded in SU-8/Si-based structures and in Agirregabiria et al. (2005) where SU-8 multilayer microstructures were adhesive bonded and

released with the help of a Kapton film. In all referred contributions the final microfluidic structures remained on a handling wafer that makes them inapplicable for a three-dimensional attachment. Additionally to that, the Kapton foil is unsuitable to be a whole wafer releasing solution because it causes a considerable mechanical stress. In some cases, this mismatch can completely destroy the free-standing SU-8 structures.

Recently, the new nanoscale sacrificial releasing layer Omnicoat™ was introduced and applied for SU-8 structure releasing (Wang et al., 2009). It was initially aimed to be applied for stripping of SU-8 molds in which metal structures are electroplated, so it has become a very convenient and considerably clean method for the wet releasing of SU-8 microfluidic structures (Pesantez et al., 2008). In the current contribution, we demonstrate a SU-8-based technology with an Omnicoat™ releasing step that allows completely free-standing SU-8-based microfluidic sensors.

In comparison to any previously published results, we demonstrate an impedance spectroscopy sensor with integrated opposite metal electrode structures in SU-8 and released from both sides to achieve a free-standing and flexible SU-8-based sensor. The manufactured free-standing SU-8 microfluidic sensor can be used for a subsequent 3-D-MID attachment that provides an opportunity to create reduced-scale sensor devices for 3-D applications.

2   3D-MID interposer

The microfluidic interposer that has a three-dimensional shape was injection molded in an Arburg 320 S 500. As molding material, the thermoplastic liquid crystal polymer Vectra E840i LDS was used. It also can be directly applied for a laser direct structuring (LDS) (Leneke et al., 2009). The polymer material implies an additive of a metal–organic copper-based complex. This complex can be split into metal atoms (surface activation) which act as a nucleus in a chemical copper plating by a focused IR-laser beam. For this reason, the LPKF MicroLine 3-D 160i system was used for the laser structuring of the MID surface. The Nd : YAG laser with a wavelength of 1064 nm was utilized. The following laser parameters were applied for the process: power 2.5 W, speed 2000 mm s^{-1}, frequency 100 kHz, and a spot size of 60 μm with a writing overlap of 50 %. The complete LDS technological process is shown in Fig. 1.

After the laser structuring of the polymer, a mechanical precleaning step was performed to eliminate residuals from the surface. By electroless plating of copper, nickel and gold on the irradiated structures a solderable surface was achieved. This metallization process contains four main steps. At first, an intensive cleaning step removed any pollution on the surface in an alcohol-based solution with the help of an ultrasonic bath. Further, a first metallization bath of copper(II) sulfate, Cupralux INI, was performed in order

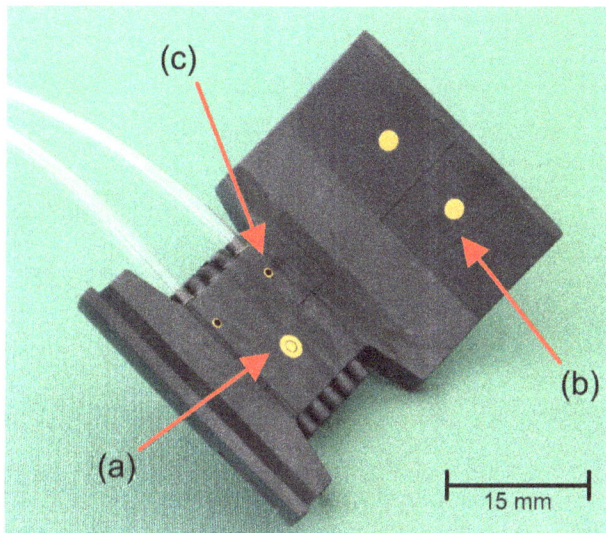

Figure 2. 3D-MID carrier for the flexible impedance spectroscopy microfluidic sensor; central sensor contact (**a**), mechanical stabilization pads (**b**), and fluidic contact (**c**).

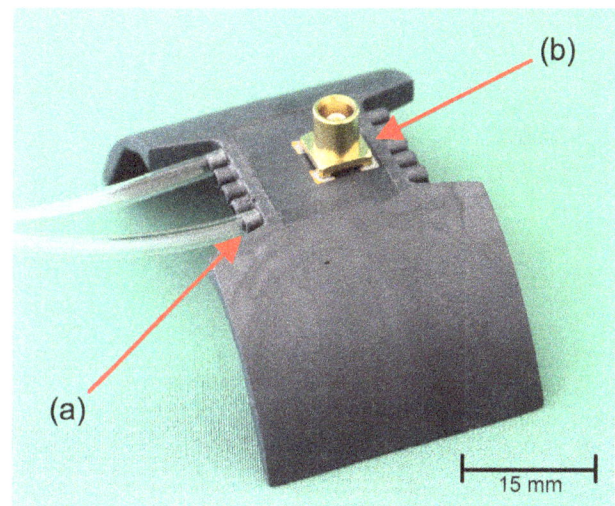

Figure 3. Back side of the 3D-MID carrier with connected fluidic tubes (**a**) and mounted MCX connector (**b**).

to deposit a 5–7 μm thick copper layer on the substrate to ensure a proper conductivity of the metallization. Then, after short treatment in a palladium chloride solution, the surface of the plated copper was catalyzed for a subsequent nickel layer deposition. The MID Nickel HP solution was used for the following nickel plating step. It was heated up to 83 °C, which enabled reaching a nickel layer with a thickness of a minimum of 5 μm atop of the previously plated copper within 40 min. The metallization finish was performed by an immersion gold bath of the MID Gold MP to grow a 0.1 μm layer atop to prevent any possible oxidation of the free nickel surface. The completed plating process of the MID guaranties the compatibility with standard packaging technologies such as flip chip and lead free soldering.

The three-dimensional injection molded device has fluidic channels with a diameter of 1.5 mm and a length of 20 mm for the fluidic supply to provide an easy attachment of the commercial tubes (Fig. 2). To establish a connection between the fluidic channels of the MID platform and the microfluidic test chip two ring-shaped solder connections define a sealed fluidic interconnection with an inner diameter of 500 μm on the MID surface. The fluidic ports were laser drilled with the same 160i IR laser system in a spiral form and a rise of 30 μm rotation^{-1} but with different laser parameters. The used laser power was increased up to 4.5 W; the speed and frequency were lowered to 500 mm s^{-1} and 50 kHz. Besides the fluidic ports, a circular solderable pad was created with the LDS process to establish the electrical connection of the MID carrier and the flexible sensor chip.

To improve the mechanical stability between the sensor chip and the MID, four metallized pads with a diameter of

2 mm were fabricated on the curved side areas of the LCP substrate, shown in Fig. 2.

In the middle of the main mounting sector, a central electric VIA was laser drilled from the top to the back side of the MID and filled during the electroless plating process to establish an electrical connection for the signal transfer to a standard MCX connector at the back side, shown in Figs. 2 and 3. A circular shielding pad around the central sensor via is connected on the front side of the MID with four additional metallized vias to the grounding pads of the MCX connector at the back side. These grounded contacts minimize the influence of external fields on the sensor signal between the sensor and the connector (Fig. 3).

3 Fabrication of the sensor device

The fabrication process of the sensor was divided into three parts (Fig. 4). The first fabrication processes were completed on a silicon wafer where the SU-8 microfluidic structures and the sensor bottom electrodes were defined, shown in Fig. 4 (Part 1). The first part starts with the cleaning of the 100 mm silicon wafer in a HF 1 % dip for 1 min. An adhesion layer of the Omnicoat™ was spun onto the clean substrate and baked for 1 min on a hotplate at 200 °C. In order to improve the releasing speed, the thickness of the Omnicoat™ layer was enlarged by subsequent multiple coating. Then, the handling SU-8-50 layer was coated on the Si wafer in a thickness of 50 μm without any structures. After the first SU-8 layer coating, the bottom electrode made of titanium (30 nm) and aluminum (300 nm) was deposited on the polymer and structured via lift-off. The surface of the SU-8 was pretreated with a plasma activation step in a mixture of O_2/SF_6 to improve the adhesion between the polymer and metal electrodes. After stripping the photoresist layer, a 50 μm thick

Part 1	Part 2	Part 3

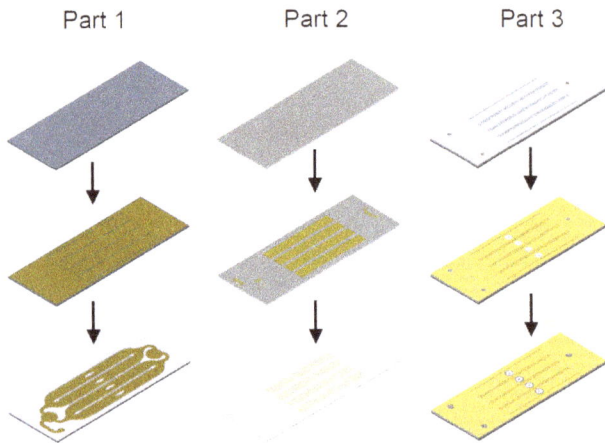

Figure 4. Fabrication of the complete sensor. Part 1: fabrication of the bottom electrode and the microfluidic channels of the sensor chip; part 2: manufacturing of the top electrode and the SU-8 top cap; part 3: establishment of the bonding, dry etching, and plating of the contact pad structures.

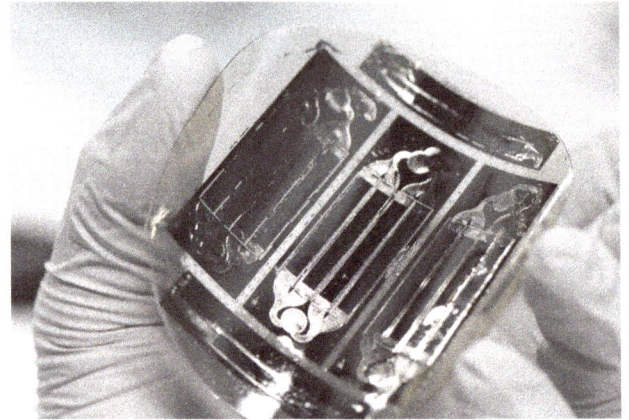

Figure 5. Completed SU-8-based free-standing impedance spectroscopy sensors with different designs on a 100 mm substrate.

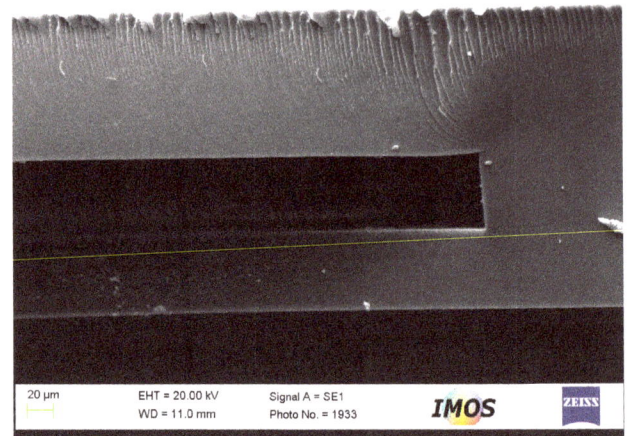

Figure 6. SEM cross-section image of a free-standing microfluidic channel.

SU-8-50 layer was spin coated and structured to implement the microfluidic channels. To avoid any thermo-mechanical stress, all baking processes are needed to be temperature ramped. The parameters of the SU-8 processing also should be adjusted in order to obtain the flexible free-standing microfluidic structures. It is known that the softbake, exposure, and post exposure bake are the most influencing factors on the SU-8 layer mechanical properties. In order to complete the described microfluidic structure, the bake temperatures were lowered to 85 °C and the exposure dose was reduced to 160 mJ cm^{-2}.

The next part of the sensor manufacturing included the fabrication of the sensor top electrode structures on the SU-8/glass wafer, shown in Fig. 4 (Part 2). A borosilicate glass was cleaned in a solution of H_2SO_4 and H_2O_2 at 135 °C for 10 min and then coated with Omnicoat™ for the further releasing of the next 50 μm thick layer of SU-8-50. An additional plasma treatment was performed to activate the surface of the polymer layer to improve the adhesion of the subsequently sputtered titanium and aluminum metal layers that were further lift-off structured. The SU-8/metal layer was further covered with a thin film of SU-8-5. After the covering of the metal structure, the silicon and the glass wafer were adhesive bonded under assistance of a SUSS MA6/BA6 mask aligner and a SUSS SB6e substrate bonder. Finally, after the bonding step the glass wafer was released in a wet etchant.

On top of the free SU-8 surface a 200 nm aluminum layer was sputtered and structured via lithography and etching. The Al mask was used in an Oxford Instruments PlasmaSystem100 dry etching process to open the polymer layers and establish the entry for the top sensor electrode contact and the sidewalls for the shielding. To avoid any thermal mechanical stress in the SU-8 layers during the plasma etching process

with a mixture of SF_6/O_2, the Si handling wafer was backside cooled with helium and the etching process was divided into steps of 30 s etching and 120 s pause. After stripping the masking material, a new layer of 500 nm Al was structured with a lift-off process to connect the top electrodes and establish the 3-D-shielding. To realize the solder connections, a final photoresist TI Spray was spray coated with a SUSS Delta Altaspray on the Al / SU-8 surface. The final layers of nickel and gold were plated on the contact pads by the use of an electroless metallization bath. After the metallization, the completed SU-8 multilayer structures were separated from the Si handling wafer in a releasing step of the Omnicoat™ layer with the help of the developer MF-319.

4 Experimental results

The fully processed free-standing microfluidic sensors intended for the further separation are shown in Fig. 5. The sensors are fabricated on a 4 in. wafer level and demonstrate

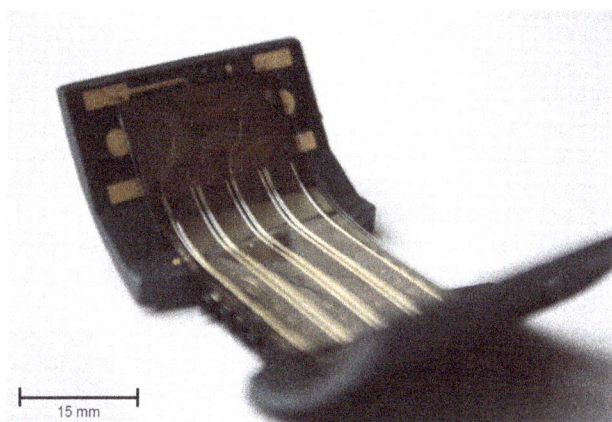

Figure 7. Microfluidic impedance spectroscopy sensor directly attached on the 3D-MID carrier.

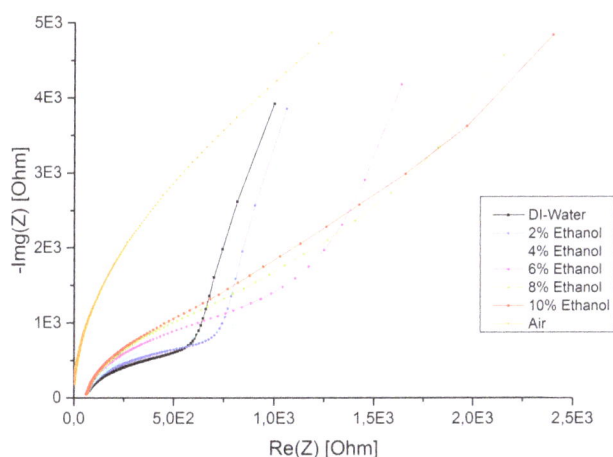

Figure 8. Impedance measurement of water, air, and a mixture of water and ethanol in a Cole–Cole plot for the frequency range of 10 kHz–10 MHz.

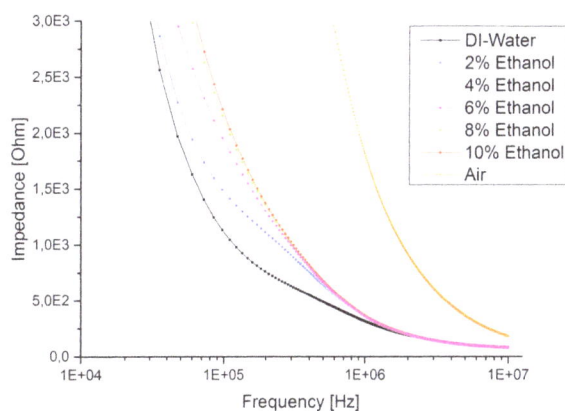

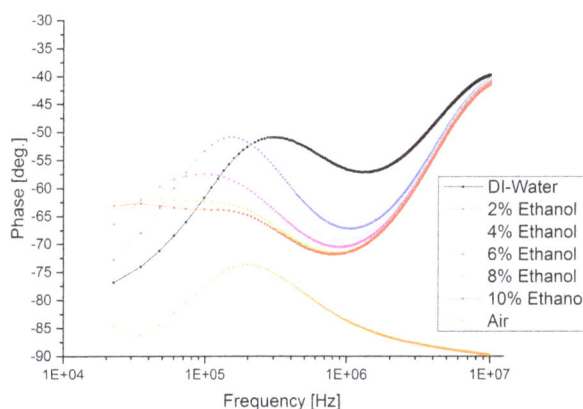

Figure 9. Impedance and phase measurement of water, air, and a mixture of water and ethanol in a series of Bode plots.

a sufficient flexibility for an attachment on a 3-D substrate. A SEM cross-section view of the free-standing microfluidic channel is shown in Fig. 6. As it can be seen, there are no boundaries between the different SU-8 layers and the sidewalls are almost vertical. The bonding area is completely closed without any defects even after a sample dicing.

The self-developed 3D-MID platform was injection molded with the help of ARBURG 320s equipment. The 3D-MID substrate, shown in Figs. 2 and 3, integrates all electrical as well as fluidic interconnections between the microfluidic sensor chip and the other devices that are necessary for the measurement.

Due to plated contact pads of the sensor, a direct attachment to the 3D-MID platform by a solder process without any additional chip wire bonding was possible. The microfluidic part of the 3D-MID was connected to a fluid system that contained a micropump, a reservoir, and a temperature control unit. All electrical ports of the MID were wired with an Agilent 4395A Network/Spectrum/Impedance analyzer.

The manufactured impedance spectroscopy sensor was experimentally investigated with the mixture of DI water and ethanol. The possible sensor application as an in-line sensor for ethanol-containing fuel analysis should be initially characterized with liquids that demonstrate a predictable behavior at the defined composition. Due to the complexity of gasoline compositions, its analysis with different ethanol content can provide a non-linear sensor response, whose explanation is beyond the scope of the current contribution.

The impedance measurements were conducted utilizing an Agilent 4395A Network/Spectrum/Impedance analyzer in "impedance analyzer" mode together with the impedance measuring set. The experimental results were obtained in an analyzer multichannel mode where an impedance magnitude and a phase as well as real and imaginary parts of the impedance were measured in parallel.

In Figs. 8 and 9 the results of the complex impedance and phase measurements are shown for the frequency range of 10 kHz–10 MHz. All measurements were made at a constant temperature of 22 °C. After an initial test of the sensor behavior with air and deionized water to check the function of the system, pure ethanol was added to the DI water. In the concentration range of 2–10 % of ethanol mixed in deionized water, a sufficient sensitivity could be demonstrated with the presented microfluidic sensor system. The expected significant impedance and phase shift can be observed in the Cole–Cole and Bode plots, shown in Figs. 8 and 9.

5 Conclusions

The demonstrated technology process for the manufacturing of free-standing SU-8 structures provides the possibility of fabricating SU-8-based completely free-standing microfluidic sensors. As it was shown, the standard metal lift-off process opens the possibility of depositing sensor electrodes on the SU-8 surface on the wafer level. The technology of "buried" metal electrodes in SU-8 provides reliable structures with considerably low-cost processing. A control of the SU-8 process parameter makes it possible to achieve flexible and mechanically stable structures that can be used for the attachment of flexible sensor chips on 3-D molded interconnection devices. A releasing step completed with the standard Omnicoat™ process established the opportunity for a considerably clean wet SU-8 releasing even with relatively large-scale structures. As it was demonstrated on Fig. 5, the whole 4 in. wafer was released without any structural damage. The microfluidic multichannel sensor was assembled on the 3D-MID platform, which integrated all electrical and fluidic interconnections. The measurement results of the fabricated sensor system demonstrated a high sensitivity to the ethanol concentration by analyzing a water–ethanol mixture.

Acknowledgements. The authors would like to thank the Federal Ministry of Education and Research (BMBF) for funding of the MicroSens project and all employees of the MEMS cleanroom team of the IMOS in Magdeburg for their help and valuable contribution.

References

Agirregabiria, M., Blanco, F. J., Berganzo, J., Arroyo, M. T., Fulaondo, A., Mayoraa, K., and Ruano-Lópeza, J. M.: Fabrication of SU-8 multilayer microstructures based on successive CMOS compatible adhesive bonding and releasing steps, Lab on a chip, 5, 545–552, doi:10.1039/b500519a, 2005.

Barsoukov, E. and Macdonald, J. R.: Impedance Spectroscopy: Theory, Experiment and Applications, John Wiley & Sons, Inc, Hoboken, NJ, USA, 2005.

Despont, M., Lorenz, H., Fahrni, N., Brugger, J., Renaud, P., and Vettiger, P.: High-aspect-ratio, ultrathick, negative-tone near-uv photoresist for MEMS applications, in: IEEE The Tenth Annual International Workshop on Micro Electro Mechanical Systems, An Investigation of Micro Structures, Sensors, Actuators, Machines and Robots, Nagoya, Japan, 26–30 January 1997, 518–522, 1997.

Ezkerra, A., Fernández, L. J., Mayora, K., and Ruano-López, J. M.: Fabrication of SU-8 free-standing structures embedded in microchannels for microfluidic control, J. Micromech. Microeng., 17, 2264–2271, doi:10.1088/0960-1317/17/11/013, 2007.

Gawad, S., Schild, L., and Renaud, P. H.: Micromachined impedance spectroscopy flow cytometer for cell analysis and particle sizing, Lab on a chip, 1, 76–82, doi:10.1039/b103933b, 2001.

Gómez, R., Bashir, R., Sarikaya, A., Ladisch, M. R., Sturgis, J., Robinson, J. P., Geng, T., Bhunia, A. K., Apple, H. L., and Wereley, S.: Microfluidic Biochip for Impedance Spectroscopy of Biological Species, Biomed. Microdev., 3, 201–209, doi:10.1023/A:1011403112850, 2001.

Jo, B.-H., van Lerberghe, L. M., Motsegood, K. M., and Beebe, D. J.: Three-dimensional micro-channel fabrication in polydimethylsiloxane (PDMS) elastomer, J. Microelectromech. Syst., 9, 76–81, doi:10.1109/84.825780, 2000.

Leneke, T., Hirsch, S., and Schmidt, B.: A multilayer process for the connection of fine-pitch-devices on molded interconnect devices (MIDs), Circuit World, 35, 23–29, doi:10.1108/03056120910953286, 2009.

Lin, C.-H., Lee, G.-B., Chang, B.-W., and Chang, G.-L.: A new fabrication process for ultra-thick microfluidic microstructures utilizing SU-8 photoresist, J. Micromech. Microeng., 12, 590–597, doi:10.1088/0960-1317/12/5/312, 2002.

Liu, C.: Recent Developments in Polymer MEMS, Adv. Mater., 19, 3783–3790, doi:10.1002/adma.200701709, 2007.

Lorenz, H., Despont, M., Fahrni, N., LaBianca, N., Renaud, P., and Vettiger, P.: SU-8: A low-cost negative resist for MEMS, J. Micromech. Microeng., 7, 121–124, doi:10.1088/0960-1317/7/3/010, 1997.

Moser, D. and Krause, J.: 3-D-MID – Multifunctional Packages for Sensors in Automotive Applications, in: Advanced Microsystems for Automotive Applications 2006, edited by: Valldorf, J. and Gessner, W., VDI-Buch, Springer-Verlag, Berlin/Heidelberg, 369–375, 2006.

Oseev, A., Zubtsov, M., and Lucklum, R.: Gasoline properties determination with phononic crystal cavity sensor, Sensors and Actuators B: Chemical, 189, 208–212, doi:10.1016/j.snb.2013.03.072, 2013.

Pesantez, D., Amponsah, E., and Gadre, A.: Wet release of multipolymeric structures with a nanoscale release layer, Sensors and Actuators B: Chemical, 132, 426–430, doi:10.1016/j.snb.2007.10.060, 2008.

Schmidt, M.-P., Leneke, T., Hirsch, S., and Schmidt, B.: A novel injection molded fluidic interposer for microfluidic applications, 4th Electronic System-Integration Technology Conference (ESTC), Amsterdam, Netherlands, 17–20 September 2012, 1–4, doi:10.1109/ESTC.2012.6542059, 2012.

Schmidt, M.-P., Oseev, A., Engel, C., Brose, A., Aman, A., and Hirsch, S.: A Novel Design and Fabrication of Multichannel Microfluidic Impedance Spectroscopy Sensor for Intensive Electromagnetic Environment Application, Procedia Engineering, 87, 88–91, doi:10.1016/j.proeng.2014.11.272, 2014a.

Schmidt, M.-P., Oseev, A., Engel, C., Brose, A., and Hirsch, S.: Impedance spectroscopy microfluidic multichannel sensor platform for liquid analysis, 18th International Conference on Miniaturized Systems for Chemistry and Life Sciences, San Antonio, Texas, USA, 26–30 October 2014, 2137–2139, doi:10.13140/2.1.4254.8004, 2014b.

Vilares, R., Hunter, C., Ugarte, I., Aranburu, I., Berganzo, J., Elizalde, J., and Fernandez, L. J.: Fabrication and testing of a SU-8 thermal flow sensor, Sensors and Actuators B: Chemical, 147, 411–417, doi:10.1016/j.snb.2010.03.054, 2010.

Wang, P., Tanaka, K., Sugiyama, S., Dai, X., and Zhao, X.: Wet releasing and stripping SU-8 structures with a nanoscale sacrificial layer, Microelectronic Engineering, 86, 2232–2235, doi:10.1016/j.mee.2009.03.079, 2009.

Zhang, J., Tan, K. L., Hong, G. D., Yang, L. J., and Gong, H. Q.: Polymerization optimization of SU-8 photoresist and its applications in microfluidic systems and MEMS, J. Micromech. Microeng., 11, 20–26, doi:10.1088/0960-1317/11/1/304, 2001.

22

Shutter-less calibration of uncooled infrared cameras

A. Tempelhahn, H. Budzier, V. Krause, and G. Gerlach

Technische Universität Dresden, Electrical and Computer Engineering Department, Solid-State Electronics
Laboratory, Dresden, Germany

Correspondence to: A. Tempelhahn (alexander.tempelhahn@tu-dresden.de)

Abstract. Infrared (IR) cameras based on microbolometer focal plane arrays (FPAs) are the most widely used cameras in thermography. New fields of applications like handheld devices and small distributed sensors benefit from the latest sensor improvements in terms of cost and size reduction. In order to compensate for disturbing influences derived from changing ambient conditions, radiometric cameras use an optical shutter for online re-calibration purposes, partially also together with sensor temperature stabilization. For these new applications, IR cameras should consist only of infrared optics, a sensor array, and digital signal processing (DSP). For acceptable measurement uncertainty values without using an optical shutter (shutter-less), the disturbing influences of changing thermal conditions have to be treated based on temperature measurements of the camera interior. We propose a compensation approach based on calibration measurements under controlled ambient conditions. All correction parameters are determined during the calibration process. Without sensor temperature stabilization (TEC-less), the pixel responsivity is also affected by the camera temperature changes and has to be considered separately. This paper presents the details of the compensation procedure and discusses relevant aspects to gain low temperature measurement uncertainty. The residual measurement uncertainty values are compared to the shutter-based compensation approach.

1 Introduction

The evolution of infrared (IR) thermography including the discovery of new fields of applications is based on the development of small, power-efficient and, most of all, low-cost IR sensors. Microbolometer focal plane arrays (FPAs) are thermal sensors and are also called uncooled sensors because they work at room temperature without any cooling. This cost-reducing benefit compared to photon sensors is the reason why microbolometers are widely used in process control, fire prevention, fire protection and surveillance as well as for research and development tasks.

Each pixel of a microbolometer sensor array acts as a single sensor element. To ensure the required thermal isolation, such a microbolometer pixel consists of a semiconductor bridge and is coated with the bolometer material (Fig. 1). The cavity beneath the micro-bridge works as a $\lambda/4$-resonator/absorber. The entire sensor has to be under vacuum atmosphere for thermal isolation reasons. The temperature of the bolometer bridge changes due to the absorbed incident radiation. This temperature change affects the electrical resistance of the bolometer material due to its temperature coefficient of resistance (TCR). Common resistor materials are amorphous silicon (a-Si) and vanadium oxide (VOx) with a thin film TCR of about -0.03 and $-0.027\,\mathrm{K}$, respectively (Budzier and Gerlach, 2011). The change in electrical resistance is determined using a readout integrated circuit (ROIC) within the silicon substrate generating a radiation-proportional voltage signal for each pixel. The raw infrared data set comprises the signal voltages of the entire sensor array and contains temperature information from the observed scene but also from the camera interior due to the huge field of view (FOV) of the pixels.

The state-of-the-art semiconductor fabrication technology enables microbolometer sensors with $17\,\mu\mathrm{m}$ pixel pitch and a wide variety of spatial resolutions from about 80×80 to 1024×768 pixel (Tissot et al., 2013). This leads to sensor sizes from a few centimeters down to only a few millimeters depending on the pixel resolution of the sensor. Sensor manufacturers recently made a lot of efforts to improve

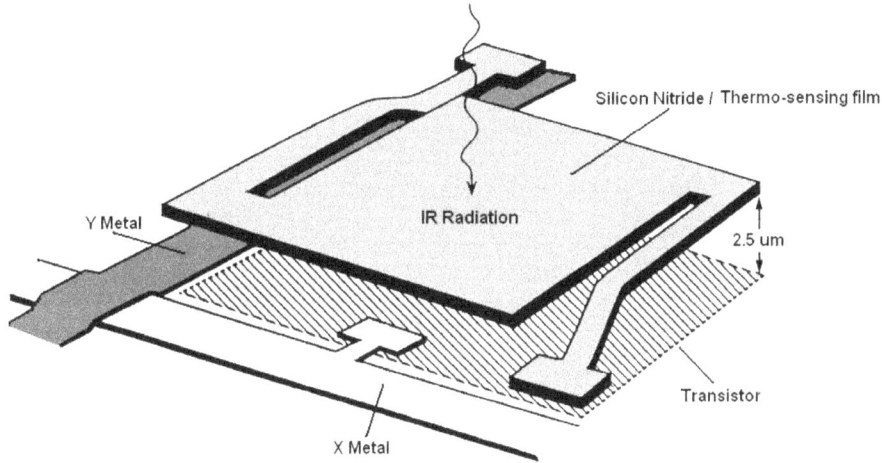

Figure 1. Bridge structure of a single microbolometer pixel (Moreno et al., 2012).

the vacuum packaging technology from common chip level packages (CLPs) via wafer level packaging (WLP) to pixel level packaging (PLP). This integration of vacuum packaging technology into the semiconductor processing will dramatically reduce the sensor cost and, therefore, will make microbolometer FPAs more attractive for lowest-cost applications like handheld devices and small distributed IR sensors (Hoelter et al., 2015; Takasawa, 2015).

IR cameras are used for radiometric measurement of temperature distributions. Here, two concerns mainly have to be taken into account: the sensor non-uniformity due to fabrication variations and the influences derived from changing ambient conditions, especially the ambient temperature. In the past such cameras used an integrated thermoelectric cooler (TEC) for keeping the temperature of the FPA constant and preventing the sensor parameters' offset voltage and responsivity to change according to the ambient temperature. But the high power consumption of this stabilization is a big drawback and is the decisive reason for using TEC-less microbolometer FPAs, e.g., for mobile IR devices or distributed sensor networks.

The correction approaches for infrared imagers and radiometric cameras differ in the required calibration effort (Tempelhahn et al., 2015). Radiometrically calibrated infrared cameras typically use optical shutters for runtime recalibration purposes in order to regularly correct thermal drift influences on the measurement. The calibration procedure for the shutter-based compensation approach is presented in detail in Budzier and Gerlach (2015). But the shutter is often the size-limiting component of an infrared camera because it has to cover the entire aperture. Another drawback of the shutter-based compensation approach is the interruption of the measurement during recalibration. Therefore, shutter-less infrared cameras are advantageous, especially for the new fields of application mentioned above. This paper presents a novel shutter-less compensation approach based on a cali-

bration procedure for determining correction parameters that allow one to compensate for the disturbing influences of a changing ambient temperature. Relevant aspects to gain low temperature measurement uncertainty are discussed and compared to the shutter-based compensation approach.

2 Radiation model

In the following, the setup of an IR camera as shown in Fig. 2 is considered. The microbolometer FPA consists of pixels located in line i and column j. The exchanged (incident minus emitted) radiant flux Φ_{ij} of each pixel (ij) is converted into the raw signal voltage $V_{\mathrm{raw},ij}$ corresponding to the linear relation between radiation and signal voltage,

$$V_{\mathrm{raw},ij} = R_{V,ij}\Phi_{ij} + V_{0,ij}, \qquad (1)$$

with the two sensor parameters responsivity $R_{V,ij}$ and offset voltage $V_{0,ij}$. Both of them vary over the sensor array due to process variations during the sensor fabrication. Furthermore, both also depend on the sensor temperature ϑ_{fpa}.

Previous investigations have shown that the pixel's field of view covers nearly the entire half space. Therefore, the corresponding pixel's projected solid angle ω_{pix} amounts almost to π (Tempelhahn et al., 2013). Infrared cameras usually comprise infrared optics with an f-number about unity. For that reason, the projected solid angle ω_{obj} related to the object irradiance E_{obj} is about $1/5$ of ω_{pix} (Budzier, 2014). The remaining projected solid angle ω_{cam} equals $4/5$ of ω_{pix} and is covered by the camera interior. Hence, each pixel also detects the irradiance E_{cam} emitted by the camera housing and depending on the camera temperature ϑ_{cam}. Furthermore, each pixel emits the radiant exitance M_{pix} into the environment due to the sensor temperature ϑ_{fpa}. The exchanged radiant flux Φ_{ij} from Eq. (1) can be written as a product of the pixel area A_{pix} and a linear combination of these three radiant den-

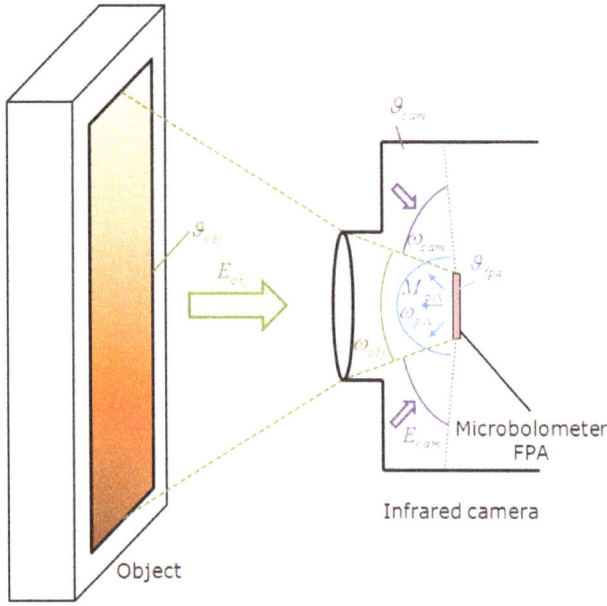

Figure 2. Radiation components in an infrared camera. E_{obj} radiation derived from the object depending on its temperature ϑ_{obj} covering the projected solid angle ω_{obj}, E_{cam} radiation from the camera interior related to the camera temperature ϑ_{cam} and the solid angle ω_{cam}, as well as the radiant exitance M_{pix} of the sensor depending on the sensor temperature ϑ_{fpa} emitted into the projected solid angle ω_{pix}.

sities with their corresponding projected solid angles:

$$\Phi_{ij} = A_{pix} \left[\omega_{obj,ij} E_{obj} \left(\vartheta_{obj} \right) + \omega_{cam,ij} \quad (2) \right.$$
$$\left. E_{cam} \left(\vartheta_{cam} \right) - \pi M_{pix} \left(\vartheta_{fpa} \right) \right].$$

Figure 2 illustrates this radiation composition and points out the influences of the three temperatures of the object, the camera and the sensor on the exchanged pixel radiant flux Φ_{ij}. The projected solid angles ω_{obj} and ω_{cam} of each pixel depend on the position within the FPA and are rotation-symmetrically distributed relating to the optical axis of the IR optics.

By replacing Φ_{ij} from Eq. (1) with Eq. (2), it is possible to separate E_{obj} containing the essential object temperature information ϑ_{obj} from all other disturbing influences:

$$V_{raw,ij} \left(\vartheta_{obj}, \vartheta_{cam}, \vartheta_{fpa} \right) = gain_{ij} \left(\vartheta_{fpa} \right) E_{obj,ij} \left(\vartheta_{obj} \right) \quad (3)$$
$$+ offset_{ij} \left(\vartheta_{cam}, \vartheta_{fpa} \right).$$

The slope $gain_{ij}$ of the pixel-specific linear equation comprises the pixel area A_{pix}, the object-projected angle $\omega_{obj,ij}$ and the sensor's temperature-dependent responsivity $R_{V,ij}$:

$$gain_{ij} \left(\vartheta_{fpa} \right) = A_{pix} \omega_{obj,ij} R_{V,ij} \left(\vartheta_{fpa} \right). \quad (4)$$

The intercept $offset_{ij}$ combines different disturbing parts: the offset voltage $V_{0,ij}$, the signal parts $V_{cam,ij}$ and V_{pix} de-

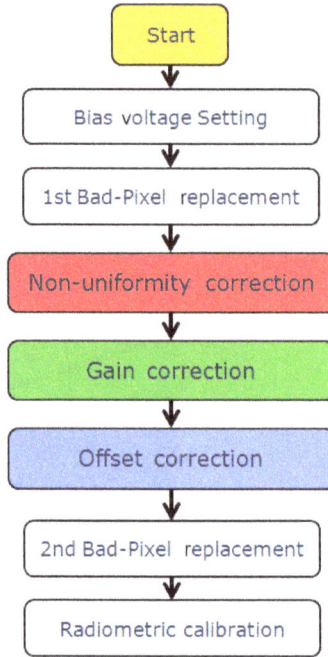

Figure 3. Correction procedure for shutter- and TEC-less IR cameras.

rived from camera radiation, and the pixel radiant exitance.

$$offset_{ij} \left(\vartheta_{fpa}, \vartheta_{cam} \right) \quad (5)$$
$$= V_{0,ij} \left(\vartheta_{fpa} \right) + V_{cam,ij} \left(\vartheta_{fpa}, \vartheta_{cam} \right) + V_{pix} \left(\vartheta_{fpa} \right)$$

3 Correction approach

The suggested approach adapts the shutter-based correction procedure presented in Budzier and Gerlach (2015) and considers the different disturbing influences separately. Figure 3 illustrates the flowchart of the calibration procedure to determine the necessary correction coefficients.

First, the bias voltages of the microbolometer have to be adjusted according to the required object and ambient temperature ranges. The first bad pixel replacement procedure determines pixels that show no response or have an abnormal behavior.

The main calibration steps are non-uniformity correction (NUC) as well as gain and offset correction. As mentioned before, each microbolometer FPA shows a pixel non-uniformity due to the semiconductor fabrication process. This can be corrected by applying the two-point non-uniformity correction (NUC) based on a slope coefficient a_{ij} and an intercept coefficient b_{ij}:

$$V_{nuc,ij} = a_{ij} V_{raw,ij} + b_{ij}. \quad (6)$$

The NUC is bound to a certain ambient temperature $\vartheta_{amb,ref}$. Changes in the ambient temperature ϑ_{amb} are transferred inside the camera housing due to heat conduction and convection, and change the camera's and sensor's temperature with specific delay times and time constants.

The sensor temperature dependency of the pixel responsivity $R_{V,ij}$ is determined using two switchable blackbodies at different constant temperatures. This ensures a constant incident radiant flux difference $\Delta\Phi$. Changes in $R_{V,ij}$ are measured using the signal voltage difference ΔV in relation to the sensor temperature ϑ_{fpa} via the relation

$$R_V = \frac{\Delta V}{\Delta\Phi}. \tag{7}$$

The regression function $g_{V,ij}$ is based on several supporting points at different ambient temperatures and, hence, different sensor temperatures. Using this pixel gain correction function, the NUC-corrected signal voltage is weighted:

$$V_{\mathrm{gain},ij} = \frac{V_{\mathrm{nuc},ij}}{g_{V,ij}\left(\vartheta_{\mathrm{fpa}}\right)}. \tag{8}$$

The offset correction considers the temperature influences in Eq. (5). The offset voltage $V_{0,ij}$ and the signal voltage part V_{pix} are compensated for in relation to ϑ_{fpa}. The signal voltage part $V_{\mathrm{cam},ij}$ corresponding to the disturbing camera radiant flux is estimated based on additional temperature information ϑ_{cam} from inside the camera using a polynomial regression function of the second order:

$$V_{\mathrm{cam},ij}\left(\vartheta_{\mathrm{cam}}\right) = c_{0,ij} + c_{1,ij}\vartheta_{\mathrm{cam}} + c_{2,ij}\vartheta_{\mathrm{cam}}^2. \tag{9}$$

The offset correction function $o_{V,ij}$ is pixel-specific and can be based on more than one camera temperature value, since the camera temperature measurement depends on the position of the thermometer:

$$V_{\mathrm{offset},ij} = V_{\mathrm{gain},ij} - o_{V,ij}\left(\vartheta_{\mathrm{fpa}}, \vartheta_{\mathrm{cam}}\right). \tag{10}$$

After this ambient temperature compensation, the signal voltage $V_{\mathrm{off},ij}$ should depend only on the object temperature ϑ_{obj}. Pixels that cannot be corrected appropriately are rejected due to the second bad pixel replacement procedure.

The conversion of the signal voltage level into real temperature values is the last step. This transfer function is based on either a second-order polynomial or a Planck-like regression function that has to be determined using a sufficient number of measurements of known temperatures as supporting data points (Budzier and Gerlach, 2015).

The proposed compensation approach can be easily adapted for infrared cameras using microbolometer FPAs with TEC. If the sensor temperature ϑ_{fpa} is stabilized, then the sensor parameters $R_{V,ij}$ and $V_{0,ij}$ stay constant and the intercept offset$_{ij}$ depends only on the camera temperature ϑ_{cam}. For that reason, the gain correction step can be skipped and the offset correction step can be simplified.

Table 1. Properties of the used infrared camera.

Manufacturer	ULIS, France
Sensor type	UL03162-028
TEC	w/o
Shutter	w/o
NETD	< 100 mK (F/1, 300 K, 50 Hz)
Resolution	384 × 288
Pixel pitch	25 μm
Uniformity (deviation)	< 1.5 %
Power consumption	< 100 mW
f-number	1.0
Focal length	18 mm

4 IR camera setup

The mentioned calibration procedure will be demonstrated for an IR camera comprising a microbolometer sensor array without temperature stabilization (Table 1).

The sensor temperature ϑ_{fpa} is provided by the detector itself and it is assumed that changes are uniformly distributed over the entire sensor array. The sensor specification describes the sensor temperature dependency of the pixel responsivity with a polynomial of the second order,

$$R_{V,ij}\left(\vartheta_{\mathrm{fpa}}\right) = r_{0,ij} + r_{1,ij}\vartheta_{\mathrm{fpa}} + r_{2,ij}\vartheta_{\mathrm{fpa}}^2, \tag{11}$$

and the sensor temperature dependency of the pixel offset voltage with a polynomial of the third order,

$$V_{0,ij}\left(\vartheta_{\mathrm{fpa}}\right) = v_{0,ij} + v_{1,ij}\vartheta_{\mathrm{fpa}} + v_{2,ij}\vartheta_{\mathrm{fpa}}^2 + v_{3,ij}\vartheta_{\mathrm{fpa}}^3. \tag{12}$$

The camera temperature is not homogeneously distributed inside the camera. Electrical losses derived from the digital signal processor (DSP) and the sensor array cause a temperature gradient between positions close to the sensor and in the inner side of the camera housing. For that reason, three temperature probes (LM61, Texas Instruments, USA) are placed inside the camera housing in different positions (Fig. 4). Each of the temperature probes TP$_1$–TP$_3$ shows different temporal step responses to changes in the ambient temperature ϑ_{amb}. The response time of the temperature probes, which indicates how fast temperature changes occur, varies from 39 s for TP$_1$ and 66 s for TP$_2$ to 105 s for TP$_3$. The same relations occur in terms of time constants that give information about the temperature settling time until the steady state is reached. The raw signal voltage $V_{\mathrm{raw},ij}$ responds to an ambient temperature change, even before TP$_1$ reacts to those temperature changes, and reaches the steady state together with TP$_3$. This demonstrates that each temperature probe provides different temporal information about occurring temperature changes, and should be used for compensation. The offset compensation based on different temperature probe values should benefit from having more information about temperature probes due to the possibility of capturing the cameras' temperature

Table 2. Properties of the used blackbodies.

Property	Water bath blackbody	RCN 300	4 element Peltier blackbody
Manufacturer	Self-built	HGH Infrared Systems, France	Self-built
Working principle	Water bath	Blackbody	Blackbody
Radiator size	350×350 mm	300×300 mm	$4 \times 35 \times 35$ mm
Material	Copper water tank	Aluminium block	Copper plate
Radiator size	Volume 20 L	Thickness 50 mm	Thickness 3 mm
Temperature range	0–80 °C	25–300 °C	-10–100 °C

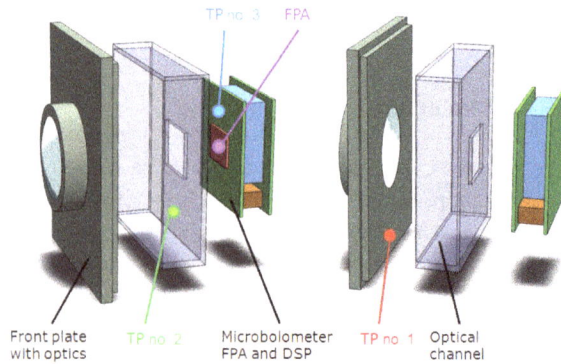

Figure 4. Positions of the temperature sensor sites inside the camera housing and the detector. TP_1 (red) is placed on the front plate carrying the optics. TP_2 (green) is located on the front side of the optical channel. TP_3 (blue) is placed close to the detector and the signal processing unit on the back side of the optical channel. The fourth site is located inside the sensor array itself (violet).

changes completely with respect to the direction of its propagation.

5 Calibration

In the following, the experimental results of the proposed calibration method using the IR camera from Table 1 will be presented. First, the calibration stand will be explained.

5.1 Calibration stand

The calibration stand was designed to meet the requirements of the proposed calibration procedure. The changing ambient temperature was simulated by using a heating chamber with the infrared camera positioned inside. A lateral opening allows the camera to be face panel blackbodies or testing scenes (Fig. 5).

For the NUC and the ambient temperature compensation steps, two panel blackbodies were used (Table 2): (1) RCN 300 from HGH Infrared Systems, France, and (2) a self-built water bath blackbody. A self-built four-element blackbody comprising four independently working small Peltier element radiators was used for radiometric calibration and testing purposes.

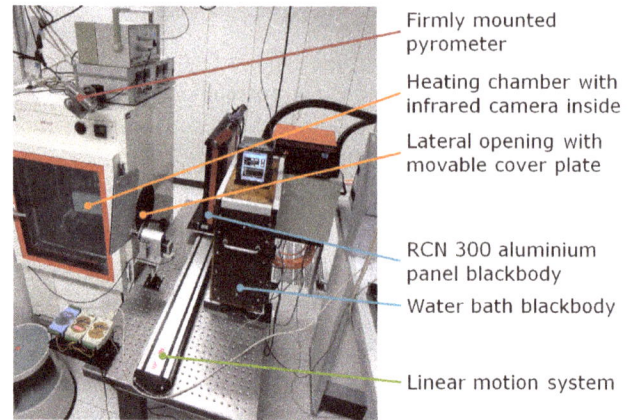

Figure 5. Photograph of the calibration setup.

A mechanically moved plate is controlled to open and close the lateral opening of the heating chamber in order to guarantee a homogeneous chamber temperature distribution and to reduce the convection between the heating chamber and the outside. Several thermo-elements were placed inside and outside the heating chamber to observe the temperature changes. The blackbody temperatures were captured using pyrometers.

5.2 Non-uniformity correction (NUC)

The NUC is based on raw infrared images of at least two different blackbody temperatures that should be spread over the targeted measurement range. The surface temperature should be uniformly distributed over the entire camera field of view. The pixel-dependent slope and intercept values of the pixel response curves from Eq. (3) are equalized, yielding the deviation ΔV_{ij}:

$$\Delta V_{ij} = V_{\mathrm{raw},ij} - V_{\mathrm{nuc}} = a_{ij} V_{\mathrm{raw},ij} + b_{ij}. \tag{13}$$

The resulting linear system of equations is solved in order to determine the pixel coefficients a_{ij} and b_{ij} (Fig. 6). After the NUC, each pixel shows the same behavior when looking at an object with the same surface temperature since the pixel responses follow the same so-called standard curve. However, this works only for certain ambient temperature conditions when the NUC images were taken. The accuracy of the

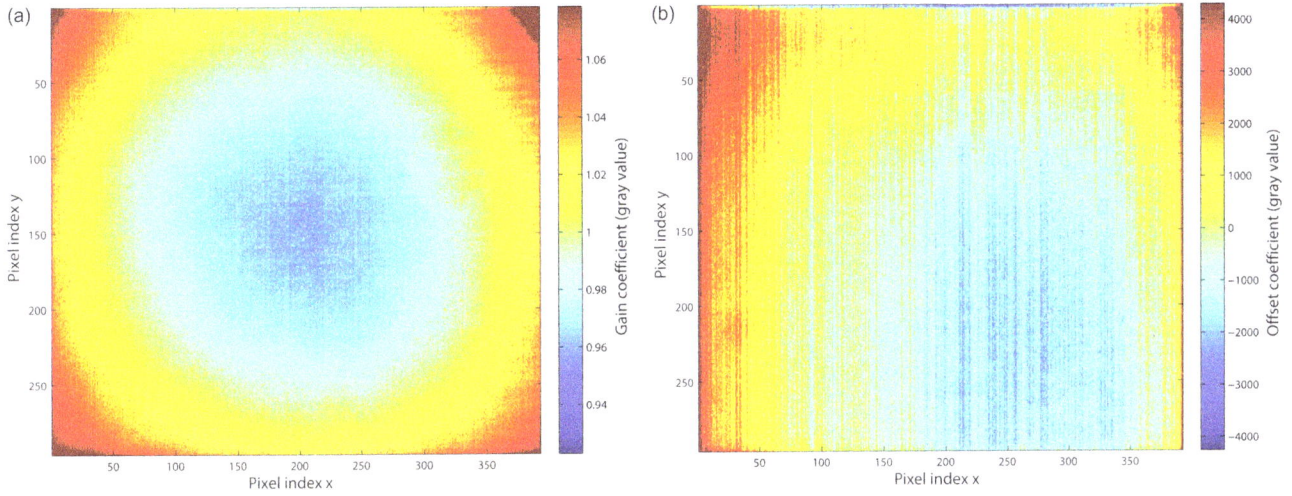

Figure 6. NUC coefficient matrices a_{ij} (**a**) and b_{ij} (**b**). The central symmetric shape of a_{ij} is due to the relation to the pixel-dependent projected solid angle $\omega_{\mathrm{fov},ij}$.

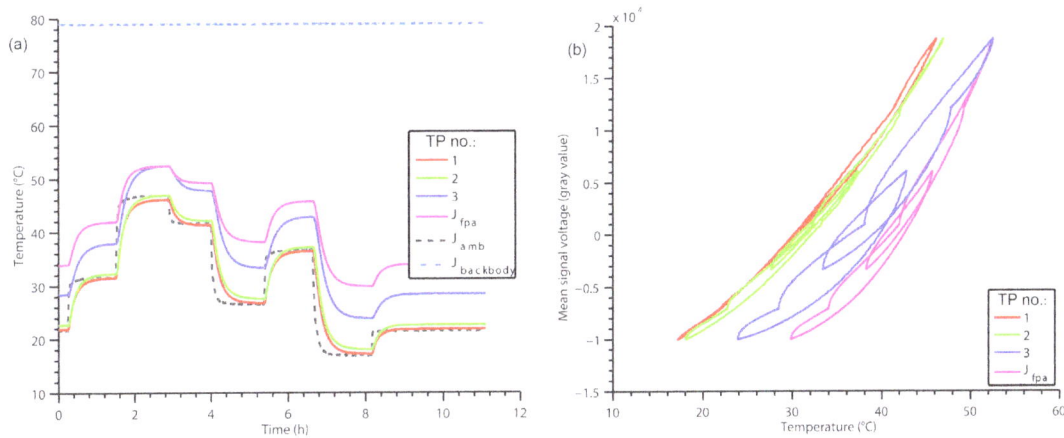

Figure 7. Temperature courses during the offset calibration regime: (**a**) temperature sensor signals; (**b**) mean signal voltage versus the four temperature inputs. The shape of the steady-state curves corresponds to a polynomial of the second or third order. The deviations from these steady-state functions are due to the transient behavior during the temperature change.

NUC can be improved by using more than two raw infrared images and applying the least-squares method to determine the NUC coefficients.

5.3 Gain correction

Here the relation between responsivity and sensor temperature pertaining to Eq. (11) is determined for each pixel and will be taken as the weighting function $g_{V,ij}$ in Eq. (8). According to Eq. (7), different infrared images of two blackbodies with constant temperatures of 20 and 80 °C were used. Since the responsivity depends only on the absolute sensor temperature ϑ_{fpa} and corresponds to a second-order polynomial, it is sufficient to use at least three pairs of infrared images taken at different ambient temperatures. Both blackbodies used (Table 2) cover the entire camera field of view.

5.4 Offset correction

The offset correction is based on a multivariate linear regression model comprising

- polynomials of the second order for each temperature probe inside the camera (Eq. 9), and

- one polynomial of the third order for the relation between offset voltage and sensor temperature (Eq. 12).

Since all temperature inputs are correlated with each other due to heat propagation and equalization processes, additional information on the temperature distribution is needed. It can be extracted from the cross-correlation of two or more temperature inputs.

The offset correction function $o_{V,ij}$ in Eq. (10) comprises for instance 17 coefficients if all temperature inputs and all

Table 3. Residual uncertainty after offset correction using different numbers and kinds of temperature inputs.

Temperature inputs	Residual temporal standard of deviation σ_t of the mean corrected pixel signal voltages (gray value)	Residual mean spatial standard deviation σ_S of the corrected pixel voltages (gray value)	Number of coefficients
ϑ_{fpa}	541.9	27.9	4
$\vartheta_{\text{TP}_1} - \text{TP}_3$	46.3	10.5	7
$\vartheta_{\text{fpa}}\vartheta_{\text{TP}_1} - \text{TP}_3$	39.8	9.8	10
$\vartheta_{\text{fpa}}\vartheta_{\text{TP}_m} \cdot \vartheta_{\text{TP}_n}$	37.8	9.3	14
$\vartheta_{\text{fpa}}, \vartheta_{\text{TP}_1} - \text{TP}_3$ $\vartheta_{\text{fpa}} \cdot \vartheta_{\text{TP}_1} - \text{TP}_3$ $\vartheta_{\text{TP}_m} \cdot \vartheta_{\text{TP}_n}$	36.1	9.2	17

possible corresponding cross-correlations are considered:

$$o_{V,ij} = \qquad (14)$$

$$c + \sum_{k=1}^{3} d_k \vartheta_{\text{fpa}}^{k} + \sum_{l=1}^{3} \sum_{m=1}^{2} e_{l,m} \vartheta_{\text{TP}_1}^{m} + \sum_{n=1}^{3} f_n \vartheta_{\text{fpa}} \cdot \vartheta_{\text{TP}_n}$$

$$+ g_1 \vartheta_{\text{TP}_1} \cdot \vartheta_{\text{TP}_2} + g_2 \vartheta_{\text{TP}_1} \cdot \vartheta_{\text{TP}_3} + g_3 \vartheta_{\text{TP}_2} \cdot \vartheta_{\text{TP}_3}$$

$$+ g_4 \vartheta_{\text{TP}_1} \cdot \vartheta_{\text{TP}_2} \cdot \vartheta_{\text{TP}_3}.$$

As mentioned earlier, all temperature inputs show different temporal behavior, which is used to separate them. An ambient temperature change regime with temperature steps from ±5 to ±20 K is applied to the camera inside the heating chamber, while the camera looks at a uniform blackbody with constant surface temperature. The camera temperature, the sensor temperature as well as other control temperatures were captured periodically. Figure 7 depicts the applied temperature change over time and illustrates the relation of the mean signal voltage in relation to the different temperatures. These relations contain all information about the offset changes due to the ambient temperature change. The coefficients of Eq. (14) were determined by applying the least-square fit to a polynomial (Bevington and Keith Robinson, 1992) in the offset calibration data.

Depending on the number and the kind of the used correction inputs, different correction uncertainty values in terms of absolute and spatial deviation can be achieved. Table 3 shows the comparison of different regression models based on the deviation of the mean absolute deviation and the mean spatial standard deviation of the pixel signal voltages over the sensor array. It should be noted that the more correction inputs are used, the lower measurement uncertainty values can be achieved. Figure 8 depicts the mean signal voltage and the spatial standard deviation versus time of the calibration using the offset correction function given by Eq. (14).

5.5 Radiometric calibration

The relation of the signal voltage $V_{\text{offset},ij}$ after ambient temperature compensation and the observed object temperature ϑ_{obj} can be estimated using a second-order polynomial.

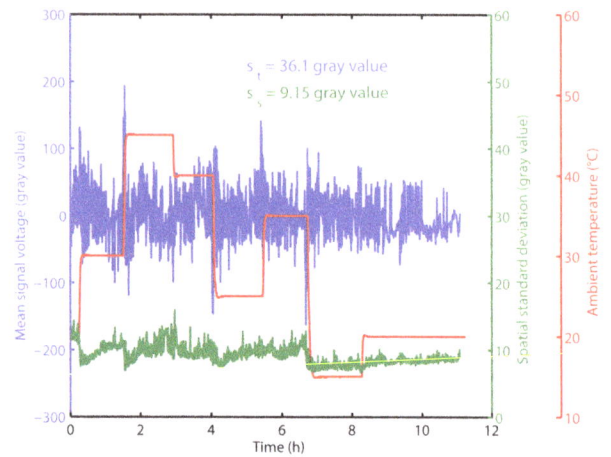

Figure 8. Mean signal voltage and spatial standard deviation of the entire sensor array after offset correction of the calibration data based on 17 correction coefficients. The mean value shows a constant level except for minor peaks when the ambient temperature changes. The spatial level changes slightly depending on the ambient temperature.

The uncertainty due to the regression can be reduced using a Planck-like approximation function based on four coefficients rbf and o,

$$V_{\text{offset},ij} = \frac{r}{e^{\left(\frac{b}{\vartheta_{\text{obj},ij} + 273.15}\right)} - f} + o, \qquad (15)$$

especially for interpolation outside the supporting points (Budzier and Gerlach, 2015). The inverted function defines the calculation rule to convert signal voltages into temperature values:

$$\vartheta_{\text{obj},ij} = \frac{b}{\ln\left(\frac{r}{V_{\text{offset},ij} - o}\right) + f} + 273.15. \qquad (16)$$

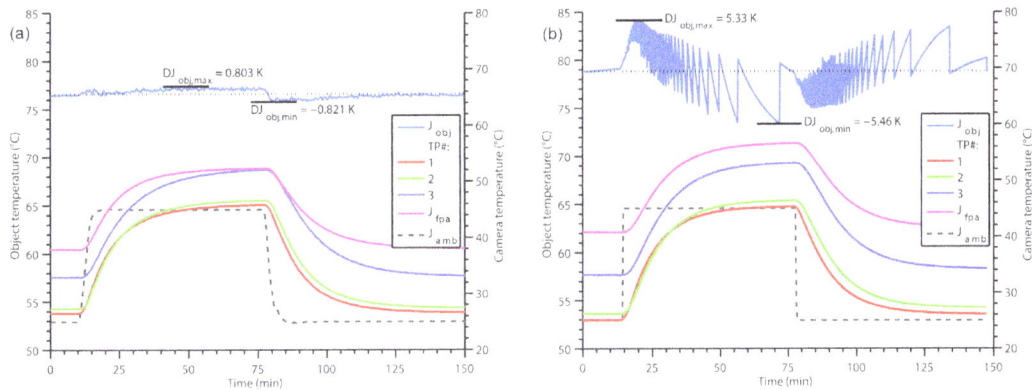

Figure 9. Comparison of the correction of (**a**) the shutter-less and (**b**) shutter-based IR cameras for ambient temperature jumps from 25 to 45 °C and back. The non-continuous shape of the shutter-based correction is due to the shutter procedure and the thermal drift occurring afterwards. The residual measurement uncertainty of the shutter-less approach is significantly lower due to the continuous online correction.

6 Comparison of shutter-based and shutter-less correction

The presented correction method for shutter-less IR cameras was compared to the common shutter-based approach using the calibrated IR camera from above and a second equivalent camera with a shutter. The ambient temperature was changed from 25 to 45 °C and back. The cameras were facing the same test scene one after another using the four-element blackbody. Figure 9 compares the results of the two measurements of one blackbody with a temperature of 76.6 °C. It should be noted that the residual deviation of the absolute temperatures of the shutter-less camera is significantly lower than that of the camera with a shutter. Furthermore, due to the fact that there is no interruption of the measurement, the temporal variation is also lower compared to the thermal drift after the shutter procedure.

7 Conclusions

The presented calibration and correction approach for shutter-less and TEC-less microbolometer-based infrared cameras shows very promising results, especially compared with the common shutter-based correction. The main drawbacks are the long calibration time and the huge number of pixel-dependent correction coefficients. This correction is only suitable for measurements with ambient temperature changes that affect the infrared camera almost uniformly. In the next study, the long-term stability of the calibration parameters will be studied.

Acknowledgements. This work was financially supported by the Deutsche Forschungsgemeinschaft under contract GE 779/26-1.

References

Bevington, P. R. and Keith Robinson, D.: Data Reduction and Error Analysis for the Physical Sciences, 2nd Edn., McGraw-Hill, New York, 1992.

Budzier, H.: Radiometrische Kalibrierung ungekühlter Infrarot-Kameras, Habilitationsschrift, Technische Universität, Fakultät Elektrotechnik und Informationstechnik, Dresdner Beiträge zur Sensorik, TUDpress, Dresden, Band 51, 2014 (in German).

Budzier, H. and Gerlach, G.: Thermal Infrared Sensors, Theory, optimisation and practice, Wiley, Hoboken, NJ., USA, Chichester, UK, 2011.

Budzier, H. and Gerlach, G.: Calibration of uncooled thermal infrared cameras, J. Sensors and Sensor Systems, 4, 187–197, 2015.

Hoelter, T., Kathman, A., Richards, A., and Walters, M.: Key Technology Trends and Emerging Applications for Compact Thermal Imagers, AMA Conferences 2015, Nürnberg, Germany, 19–21 May 2015, 938–943, doi:10.5162/irs2015/4.1, 2015.

Moreno, M., Torres, A., Kosarev, A., and Ambrosio, R.: Un-Cooled Microbolometers with Amorphous Germanium-Silicon (a-Ge$_x$Si$_y$:H) Thermo-Sensing Films, in: Bolometers, edited by: Unil Perera, A. G., INTECH Open Access Publisher, Chapter 2, doi:10.5772/32222, 2012.

Takasawa, S.: Uncooled LWIR imaging: applications and market analysis, Proc. SPIE, 9481, doi:10.1117/12.2183027, 2015.

Tempelhahn, A., Budzier, H., Krause, V., and Gerlach, G.: Modeling signal-determining radiation components of microbolometer-based infrared measurement systems, AMA Conferences 2013, Nürnberg, Germany, 14–16 May 2013, 100–104, doi:10.5162/irs2013/iP3, 2013.

Tempelhahn, A., Budzier, H., Krause, V., and Gerlach, G.: Development of a shutterless calibration process for microbolometer-based infrared measurement systems, The e-journal of Nondestructive Testing, 20, ISSN 1435-4934, 2015.

Tissot, J. L., Robert, P., Durand, A., Durand, S., Durand, E., and Crastes, A.: Status of uncooled infrared detector technology at ULIS, France, Defence Sci. J., 63, 545–549, 2013.

23

Joining technologies for a temperature-stable integration of a LTCC-based pressure sensor

J. Schilm[1]**, A. Goldberg**[1]**, U. Partsch**[1]**, W. Dürfeld**[2]**, D. Arndt**[2]**, A. Pönicke**[3]**, and A. Michaelis**[1]

[1]Fraunhofer IKTS, Fraunhofer Institute for Ceramic Technologies and Systems, Winterbergstraße 28, 01277 Dresden, Germany
[2]ALL IMPEX GmbH, Bergener Ring 43, 01458 Ottendorf-Okrilla, Germany
[3]Modine Europe GmbH, Arthur-B.-Modine-Straße 1, 70794 Filderstadt, Germany

Correspondence to: J. Schilm (jochen.schilm@ikts.fraunhofer.de)

Abstract. Besides the well-known application as circuit boards and housings, multilayer low-temperature co-fired ceramics (LTCC) offer a flexible and temperature-stable platform for the development of complex sensor elements. Commercial LTCC qualities are usually available with a matching set of metallization pastes which allow the integration of various electrical functions. However, for the integration of ceramic sensor elements based on LTCC into standardized steel housings it is necessary to compensate the mismatching thermal expansion behaviour. Therefore balancing elements made of Kovar® (Fe–29 wt% Ni–17 wt% Co) and alumina ceramic (Al_2O_3) can be used. These components have to be joined hermetically to each other and to the LTCC sensors. In this study, brazing experiments were performed for combinations of Kovar–Al_2O_3 and Kovar–LTCC with Ag–Cu–Ti- and Ag–Cu–In–Ti-based commercial braze filler metals, Cusil-ABA® and Incusil®-ABA, respectively. For both active braze filler metals, optimized processing parameters were investigated to realize hermetic Kovar–Al_2O_3 and Kovar–LTCC joints.

1 Introduction

Low temperature co-fired ceramics (LTCC) is a well-known technology for highly integrated, reliable, and high-temperature-stable microelectronic packages in mobile communication or for automotive, space, or medical applications (Peterson et al., 2008). Due to its linear stress/strain behaviour and its ability for integration of three-dimensional shapes like diaphragms, channels, and cavities, according to various authors, LTCC complies with all requirements for the integration of mechanical structures, e.g. for pressure sensors (Zarnik et al., 2010; Partsch et al., 2012; Fournier et al., 2010). A new piezo-resistive pressure sensor concept was developed by Partsch et al. (2007). Figure 1 shows a overview of a completely assembled LTCC-based pressure sensor with housing, pressure port and wiring. This new design principle allows the sensor cell to be fully mechanically decoupled and stress-free inside the sensor frame. Only thin LTCC cantilevers containing microchannels are used for the pressure connection of the sensor cell. Thick film resistors, screen-printed on the surface of the sensor cell, are connected to a Wheatstone bridge and measure the strain caused by the deflection of the LTCC diaphragm due to pressure differences in- and outside of the cell. Especially this sensor setup has been used for this study on a temperature-stable integration attempt. The LTCC technology easily enables variable sensor geometries, e.g. different diaphragm thicknesses for different pressure ranges by using different tape thicknesses and thus all types of pressure sensors (relative, absolute, differential) can be constructed. According to Fournier et al. (2010) in comparison with classic steel- or ceramic-based pressure sensors, such LTCC-based pressure sensors have many advantages like a very linear pressure-dependent signal behaviour together with an excellent high-temperature and long-term stability. Furthermore, LTCC is a high-volume technology which helps to produce sensor elements in a cost-effective manner as all components of the sensor system (sensor frame, electronic components) can be integrated in one LTCC-based multilayer substrate. However, for testing,

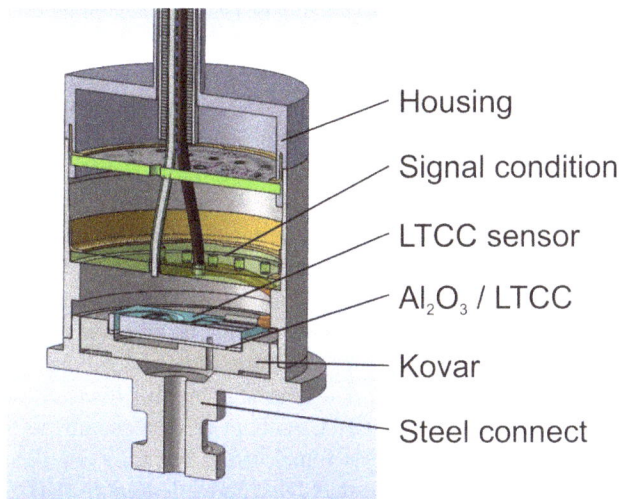

Figure 1. LTCC pressure transmitter and corresponding drawing of inner setup.

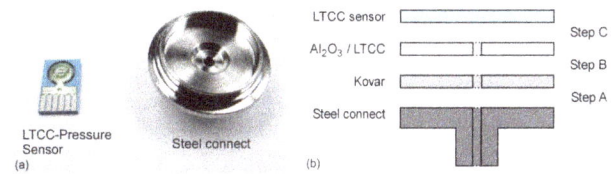

Figure 2. Single LTCC pressure and steel connect (**a**) and scheme for the stepwise integration of a LTCC-based sensor element into the steel connect (**b**).

calibration, and subsequent application, the LTCC sensor elements have to be connected gas-tight to the measuring equipment. In most cases standardized steel connectors are used to ensure a gas-tight connection of the LTCC sensors to the system. But, the integration of LTCC-based sensor elements for high-temperature applications requires suitable interconnect technologies. For example, if the LTCC sensors and the steel connectors are glued directly together, the maximum operation temperature of the sensor is limited to the glass transition temperature of the epoxy resin, which is in most cases below 190 °C. Also soldering offers no reliable option as the joint strength decreases rapidly at higher temperatures due to interdiffusion processes, which in turn results in a loss of gas-tightness. For an increased thermal stability other joining technologies like glass sealing or brazing can be used. However, in this case the mismatching coefficients of thermal expansion of LTCC and steel will limit the lifetime of the integrated sensor elements as thermal cycles will initiate cracks along the sealing or inside the sensor. To overcome these problems, a stepwise integration of the LTCC-based sensor elements into steel connectors was developed. This approach offers the opportunity to outbalance the different thermal expansion coefficients of LTCC and steel, and to increase the operation range of the sensor assembly to high temperatures up to 300 °C. The integration concept is schematically shown in Fig. 2. The integration approach of LTCC-based sensors can be divided into three steps. In step A a balancing element made of a nickel–cobalt ferrous alloy, i.e. Kovar® (Fe–29 wt% Ni–17 wt% Co) with a coefficient of thermal expansion that closely matches that of the ceramic materials at low temperatures, is brazed to the steel housing. Such a steel connection produced from cost-efficient and construction steel types are necessary for having a standard interconnection interface which can be processed easily by

electron beam welding to other devices or which may be a more complex device itself. The following step B is required in order to create a joinable surface for step C and will be described later in detail. In step B the bonding of a ceramic layer made from alumina (Al_2O_3) or LTCC offers the possibility to integrate the LTCC-based sensor elements in step C by sealing with a glass-based solder or other techniques. A convenient way for the implementation of step B is the so-called active metal brazing process.

1.1 Joining of ceramics to metals by brazing (step B)

Fernie et al. (2009) describes several direct bonding techniques to join ceramics and metals together for hermetic joints. Besides techniques without any liquid phase like diffusion bonding or friction welding, direct bonding methods utilizing a liquid phases based on adhesives, braze filler metals, or glass solders can be used. For the desired application, where in joining step B (Fig. 2) a planar joint between Kovar and Al_2O_3 or LTCC has to be realized which is suitable for operation temperatures up to 300 °C, brazing is the appropriate method. According to Nascimento et al. (2003) a common way to braze ceramics and metals is the metallization of the ceramic prior to the brazing process since then the metallized ceramics can be brazed to metals without any active braze filler metal. However, the metallization processes imply several individual process steps, which makes them complicated and expensive. In contrast to this quite old technology Walker and Hodges (2008) describe active metal brazing as a technique which allows brazing ceramics directly to metals or themselves without any additional metallization steps. Active brazing alloys are based on filler metals like Ag, Ag–Cu, or Au and contain low fractions of so-called active species (i.e. Ti, Zr, Hf). These elements enhance wetting of the ceramic surface during brazing in an oxygen-free environment using protective atmospheres or vacuum ($p < 10^{-4}$ mbar).

In the literature active metal brazing of Kovar to Al_2O_3 is much more extensively investigated than the brazing of Kovar to LTCC. A good overview of the problem can be found in Walker and Hodges (2013). When Al_2O_3 ceramics are brazed to themselves with Ag–Cu–Ti active braze filler metals, the joint microstructure shows a bi-layered reaction zone at the interface between the Al_2O_3 and the brazing alloy. This reaction layer consists of a Ti-rich oxide layer with a thickness below $1\,\mu m$ completely covering the alumina interface and a second, a few-microns-thick mixed oxide layer containing Ti, Cu, and Al (Stephens et al., 2003; Lin et al., 2014). Al_2O_3–Al_2O_3 joints prepared in this manner are hermetic and reach joint strengths of $>95\,\%$ of the strength of the base material. If now one of the Al_2O_3 pieces is substituted by Kovar, the results change. Hahn et al. (1998) demonstrated in their work that Kovar–Al_2O_3 joints achieved no high strength values and reached only $40\,\%$ of the initial Al_2O_3–Al_2O_3 joint strength. Additionally, Kovar–Al_2O_3 joints made by active metal brazing showed poor hermeticity. Stephens et al. (2000) revealed by microstructural analysis that in non-hermetic joints no continuous Ti_xO_y reaction layers were formed at the interface between Al_2O_3 and the brazing alloy. According to Vianco et al. (2003a) this phenomenon of titanium scavenging can be explained by the dissolution of nickel and iron from Kovar in the molten braze and the strong affinity of nickel to titanium. Arróyave and Eagar (2003) describe that during the brazing process nickel and titanium react to form intermetallic compounds while the activity of titanium in the melt is decreased and the formation of the necessary reaction zone at the Al_2O_3 interface is suppressed. To improve the joint strength of Kovar–Al_2O_3 and to prevent the formation of intermetallic compounds, different barrier layer concepts were developed. Mo, Ni, and Mo–Ni coatings on Kovar were tested by Hahn et al. (1998). They report an increased bending strength of the joints by more than $80\,\%$ of the uncoated base material. Vianco et al. (2003b) investigated the influence of Mo thickness and found that brazing of Kovar–Al_2O_3 with a $500\,\mu m$ thick Mo barrier layer yielded the best hermeticity performance and strength. Magnetron sputtering of titanium layers on Al_2O_3 was introduced by Zhu et al. (2014) as an alternative method to improve the joint strength and gas-tightness of Kovar–Al_2O_3 joints. The mechanical metallization of alumina surfaces was applied by Nascimento et al. (2007) to achieve a proper pre-metallization of the ceramic component. Besides introducing barrier layers, Wielage et al. (2012) showed that an improvement of the joints is possible if induction brazing with much shorter brazing times compared to conventional furnace brazing is applied. Microstructural analysis showed a remarkable reduction of intermetallic compounds and a decrease of the reaction layer thickness between Kovar and the brazing alloy. For the desired application and workflow in step B (Fig. 2) it seems to be interesting to use LTCC as a joining partner for Kovar instead of Al_2O_3. The main reason for the use of LTCC is given by minimized thermomechanical stresses in the case of using a LTCC balancing element because it has the same coefficient of thermal expansion (CTE) value as the sensor element itself. Another reason comes from the idea that it could be possible to braze the LTCC sensor directly on the Kovar element if the required brazing temperature is low enough to avoid changes in the microstructure of the complex LTCC sensor. However, in comparison to alu-

mina, less information on brazing of LTCC is available. One approach to braze LTCC is to use a metallized LTCC in combination with a non-active braze filler metal. For this purpose Keusseyan and Dilday (1993) investigated the brazeability of Cu-, Ag-, and Au-based thick film metallization layers and concluded that for LTCC–metal joints the brazing temperature should be limited to $500\,°C$ in order to minimize the mechanical stresses caused by the mismatching thermal expansion coefficients. Another approach, followed by Walker et al. (2006), was the investigation of PVD thin-film coatings like Ti–Au, Ti–Pt, and others. While brazing with a Ag–Cu–In braze filler metal, the Ti–Pt thin films yielded the best hermeticity performance and highest strength. However, on active metal brazing of LTCC without further metallization layers only one study was found in the literature. Furthermore Walker et al. (2006) stated that hermetic joints of LTCC and Kovar were only be possible if the LTCC was ground and re-fired prior to the brazing process. As shown, brazing of LTCC only is less investigated, and no systematic results are given in the literature which let one judge about the suitability of the active metal brazing technique for LTCC. In the case of Al_2O_3, besides Ag–Cu–Ti active braze filler metals, no brazing alloys with lower brazing temperatures were tested. Thus in the present study the commercially available active braze filler metal Incusil®-ABA is investigated to provide brazing parameters as a means to obtain hermetically brazed Kovar–Al_2O_3 and Kovar–LTCC joints. For comparison with the literature, joining of Al_2O_3 and LTCC to Kovar with Cusil-ABA® was investigated as well.

2 Methods and materials

2.1 Ceramic materials

The LTCC sensors are based on DuPont's GreenTape 951 system, as this material system is a fair compromise in comparison to other ceramic co-firing materials regarding Young's modulus and fracture strength. While the exact composition of this LTCC quality is not published by DuPont, within this work it is important to know that the main crystalline phase consists of Al_2O_3 grains which are bonded by a PbO-based glass frit. For joining experiments LTCC samples were made by laminating three tape layers followed by a firing process with a peak temperature of $850\,°C$ similar to the one described by Fournier et al. (2010). After firing the sintered LTCC had a thickness of $630\,\mu m$ and was cut into single samples (7×7 mm^2) with a dicing saw. The surface roughness of the as-fired LTCC was $Ra < 0.36\,\mu m$. The alumina ceramic was obtained in thick film standard quality with an Al_2O_3 content of $96\,\%$ (Rubalit® 708 S, CeramTec, Marktredwitz, Germany), a thickness of $250\,\mu m$, and a surface roughness of $Ra < 0.36\,\mu m$. Samples were prepared by laser scribing and subsequent breaking along the scribed lines. As metallic joining partners, balancing elements made of Kovar were prepared from a massive rod according to the

Table 1. Active metal brazing filler metals with their compositions and brazing temperatures.

Brazing filler metal	Ag (wt%)	Cu (wt%)	In (wt%)	Ti (wt%)	Brazing temperature
Incusil-25-ABA	43.6	29.1	24.3	3.0	650 °C
Incusil-ABA	59.0	27.25	12.5	1.25	755 °C
Cusil-ABA	63.0	35.25		1.75	810–850 °C

sensor setup indicated in Fig. 1. The bottom side of the element will be brazed to the steel interconnect and the cavity in the upper side a ceramic balancing substrate made of Al_2O_3 or LTCC will be brazed by an active metal brazing process. The Kovar components additionally contain a channel structure that allows access of the pressurized gas to the sensor membrane. The surfaces of all samples were degreased prior to assembly and brazing processes.

2.2 Brazing and glass sealing

In accordance with the previously described integration strategy, two brazing processes – one for joining the steel connect to Kovar and one for joining Kovar to Al_2O_3 – have been selected. A one-step brazing process appears to be more favourable but offers fewer opportunities for a process control regarding the hermeticity of the different interfaces. For this reason the brazing processes were separated. For the first brazing joint between steel and Kovar, a nickel-based brazing foil (MBF-20 from Metglas Inc., Conway, SC, USA) is applied which has a liquidus temperature well above the active filler metal brazes used for the second brazing process. For this purpose the MBF-20 brazing foil was cut by a laser process into shapes matching strictly to the joined side of the components. The foils were placed in between both components adjusted with an additional load on top of the arrangement and brazed in vacuum ($<1 \times 10^{-5}$ mbar) with the following brazing cycle: from room temperature at 5 K min^{-1} to 940 °C with a hold for 15 min in order to achieve a homogenous furnace temperature and than again with 5 K min^{-1} up to the brazing peak temperature of 1055 °C with an additional hold time of 15 min. Cooling down to room temperature was conducted at 3 K min^{-1}.

For joining of the ceramics (Al_2O_3 and LTCC) to Kovar, three types of braze filler metals provided by Wesgo Metals (Hayward, CA, USA) were used, which are listed in Table 1. These alloys were applied in the form of laser cut foils with a thickness of 50 µm. Brazing was carried out in a full-metal vacuum furnace with molybdenum heating elements at a pressure $<1 \times 10^{-5}$ mbar. The following process cycle was used for brazing: heating from room temperature to 550 °C (for Incusil-ABA) or 700 °C (Cusil-ABA) at a rate of 10 K min^{-1}, holding for 20 min to obtain a temperature homogenization inside the furnace, further heating up to the desired brazing temperature (Table 1), holding of brazing temperature for 10 min, and then cooling down to room

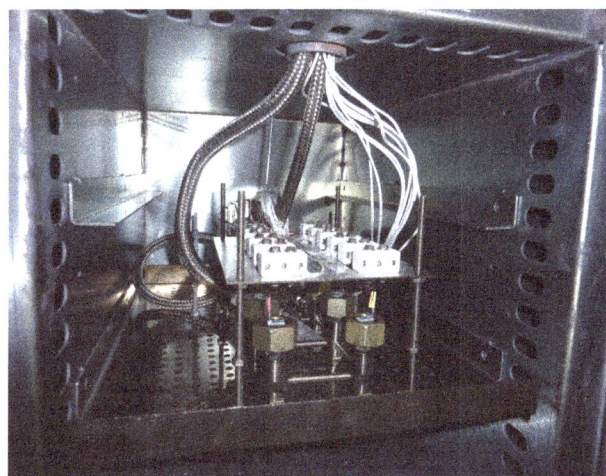

Figure 3. High temperature pressure measurement station for characterizing sensors up to 200 bar and 650 °C.

temperature at a rate of 5 K min^{-1} to 400 °C with subsequent furnace cooling. The brazing temperatures were varied between 810 and 850 °C for Cusil-ABA in order to investigate the influence of brazing temperature on gas-tightness and microstructure of the joined assemblies. For the final integration of the LTTC sensor into this prepared steel connector, a commercially available, lead-free sealing glass from ASAHI (4115DS-NY01) supplied as a ready-to-use paste with an appropriate firing profile having a peak temperature of 500 °C for use in muffle furnace was screen-printed on the back side of the LTCC sensor element. For the joining process the sensor was placed on the ceramic balancing element together with a mechanical load of 10 g.

2.3 Characterization

Upon brazing, the gas-tightness of each brazed assembly in each integration step was measured using a helium leak detector (Phoenix XL30, Oerlikon Leybold Vacuum, Cologne, Germany). A joint with a helium leak rate of 1×10^{-8} mbar s^{-1} or better was considered as hermetic. Microstructural analysis of the joints was performed on polished cross-sectioned samples using a scanning electron microscope (abbreviation: SEM; NVision 40, Carl Zeiss SMT, Oberkochen, Germany). The micrographs shown in this paper were recorded in the element specific back-scattered electron mode. Additionally, the scanning electron microscope

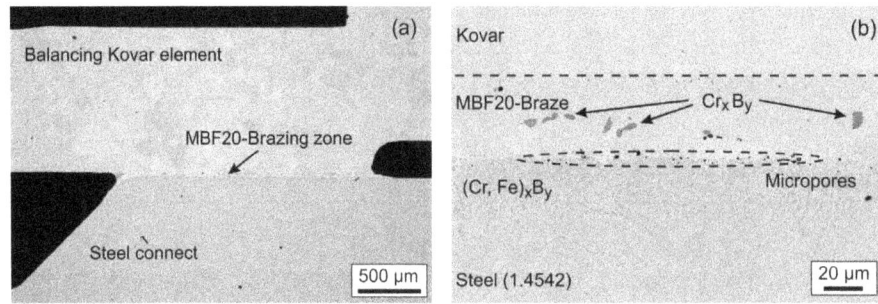

Figure 4. Scanning electron micrographs of steel (1.4542)–Kovar joints brazed with MBF-20 showing an overview in (**a**) and details of the microstructure in (**b**).

is equipped with an energy dispersive X-ray analysis system (abbreviation: EDX; Inca x-sight, Oxford Instruments, Abingdon, England), which allows for a quantitative detection of elements.

The characteristic of the pressure sensor is performed by a newly developed pressure measurement system. The sensors can be measured in a special chamber oven KU70/07-A of the company THERMCONCEPT using a high-temperature-capable pressure rail and a ceramic-insulated electrical wiring for a temperature range of 25–650 °C. At the same time, the sensors were applied by means of pressure controller PACE 5000 of the company GE Measurement & Control in the range of 0–200 bar with and the characteristic curve is measured with a computer-controlled system. Figure 3 shows the inner setup of this newly developed measurement device which allows the simultaneous characterization of 6 sensors at maximum in the range of 25–600 °C.

3   Results and discussion

As claimed in the Introduction and with respect to the chosen integration strategy, especially the joining process of metals to ceramics represent a challenge. Active metal brazing of Kovar and similar alloys to Al_2O_3 by active metal brazes is not generally a new topic, but in terms of reduced brazing processing temperatures the use of indium-containing active metal brazes appears attractive. In the case of the SiO_2- and PbO-containing LTCC material it can be supposed that the active component titanium in the active metal brazing alloy will undergo a redox reaction with these oxides. Possible reaction products could be titanium silicides of titanium–lead intermetallic compounds. Especially the formation of silicides with different components of high-temperature-stable brazing alloys is described by McDermid and Drew (1991) or Liu et al. (2009). Such intermetallic phases have a brittle character and can have disadvantageous effects on the adhesion of the braze at the ceramic surfaces. For these reasons special interest must be paid to the interfacial reactions between the different brazing alloys and joint materials in order to identify proper brazing alloys and brazing conditions.

3.1   Brazing of Kovar and steel

Brazing of the balancing Kovar element into the steel housing is the first step of the integration procedure. In accordance with brazing temperatures which are required for active filler braze between 750 and 850 °C, it is necessary to perform this brazing step at a higher temperature which lies well above the formerly mentioned one. One must take care on stability of the Kovar alloy and potential reactions with the filler braze. Based on these boundary conditions MBF-20, an amorphous nickel braze filler metal was chosen. The brazing process, was performed at temperatures between 1040 and 1060 °C. The micrograph in Fig. 4a shows an overview of the brazing zone indicating a good and pore-free adhesion of both components. A closer look at Fig. 4b reveals the formation of darker chromium borides which are brittle intermetallic phases in the brazing alloy. This indicates that chromium from the steel slightly dissolves into the molten brazing alloy MBF-20, which also contains small amounts of boride for reduced melting temperatures. Without going into much into detail we can say that it was possible to achieve hermetic dense joints with this materials and the SEM investigations gave no hints for significant interfacial reactions. With these results the CTE adjusted steel connector for the further integration of the ceramic components is available.

3.2   Brazing KOVAR to ceramic interlayer

3.2.1   Brazing with Incusil-ABA

Active metal brazing of LTCC and Al_2O_3 to Kovar with Incusil-ABA at 755 °C for 10 min yields in both cases to hermetically sealed assemblies. Surprisingly, we were able to realize hermetic joints of as-fired LTCC and Kovar, which is in sharp contrast to the results of Walker et al. (2006). SEM images of the microstructures of Kovar–Al_2O_3 and Kovar–LTCC joints brazed with Incusil-ABA are shown in Fig. 5a and b, respectively. These micrographs show the typical structure of the Ag–Cu–In eutectic with a Ag-rich phase (white regions, with dissolved In and Cu) and a Cu-rich phase (grey regions) together with enclosed intermetallic phases in the brazing alloy (dark grey regions) and reac-

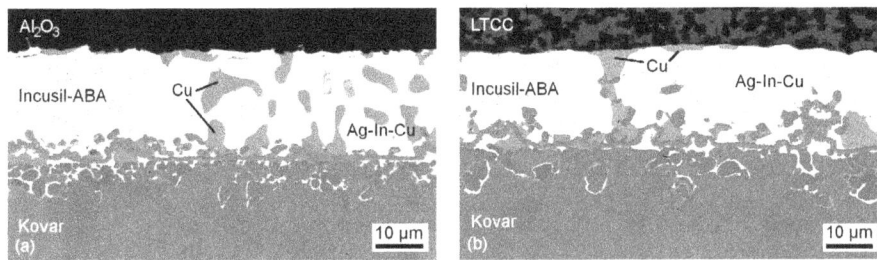

Figure 5. SEM images of Kovar–Al$_2$O$_3$ (**a**) and Kovar–LTCC (**b**) joints brazed with Incusil-ABA at 755 °C for 10 min.

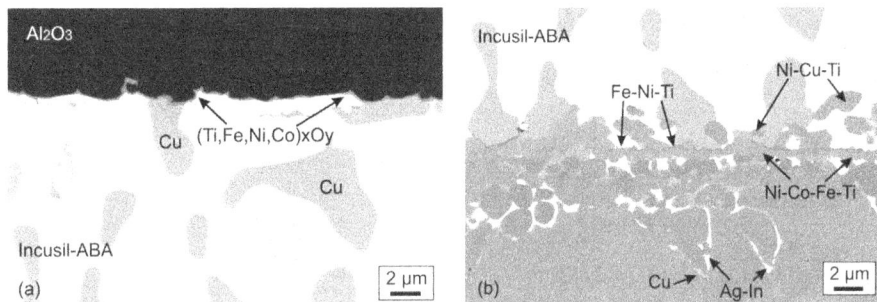

Figure 6. Enlarged SEM images of Fig. 5a showing the Al$_2$O$_3$/Incusil-ABA (**a**) and the Kovar/Incusil-12.5-ABA interfaces (**b**).

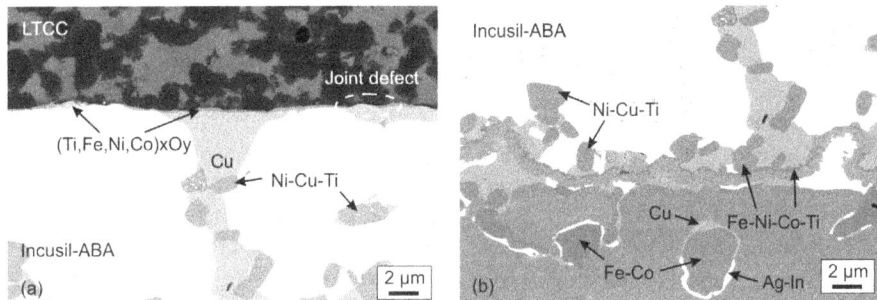

Figure 7. Enlarged SEM images of Fig. 5b showing the LTCC/Incusil-ABA (**a**) and the Kovar/Incusil-12.5-ABA interfaces (**b**).

tion layers on both interfaces. Figure 6a and b are the enlarged images from Fig. 5a displaying the reaction layers at the interface between Al$_2$O$_3$ and the brazing alloy, and between Kovar and the brazing alloy, respectively. At the interface between Al$_2$O$_3$ and the brazing alloy a very thin reaction layer with submicron thickness was formed. This reaction layer completely covers the alumina interface, yielding a mean helium lake rate of 6×10^{-10} mbar s^{-1}. The main constituents of the reaction layer are Ti and O, but also elements of the Kovar, i.e., Ni, Fe, and Co, are detected. This suggests that the constituents of the Fe–Ni–Co alloy show a strong affinity to Ti even at lower temperatures than in the publications of Stephens et al. (2000) and Vianco et al. (2003a). The strong reactivity of Fe, Ni, and Co with Ti shaped the interface between Kovar and the brazing alloy as several intermetallic compounds like (Fe,Ni,Co)$_x$Ti$_y$ with a high amount of Ni (abbreviation: Ni–Co–Fe–Ti) or Fe (abbreviation: Fe–Ni–Co–Ti) are observed. These intermetallic phases form a

small band which meanders parallel to the Kovar surface. Further away from the interface in the brazing alloy Ni-Cu-Ti compounds are visible. In addition, down to a depth of 25 μm, Ag, In, and Cu from the brazing alloy are found at the grain boundaries of the Kovar and along Fe–Co grains which are depleted of Ni. In comparison with Kovar–Al$_2$O$_3$ joints, the microstructure of Kovar–LTCC joints with Incusil-ABA looks similar. Figure 7a and b are the enlarged images from Fig. 5b showing the interfaces between LTCC and the brazing alloy, and between Kovar and the brazing alloy, respectively. Again, through diffusion of Fe, Ni, and Co and their reaction with the active element Ti, intermetallic compounds were formed in the brazing alloy and along the interface to Kovar. However, in the micrographs two differences in comparison with Kovar–Al$_2$O$_3$ joints are found. Firstly, the very thin (Ti,Fe,Ni,Co)$_x$O$_y$ reaction layer at the interface LTCC/brazing alloy contains traces of Si and Pb, the main constituents of the glass phase of the LTCC. Secondly,

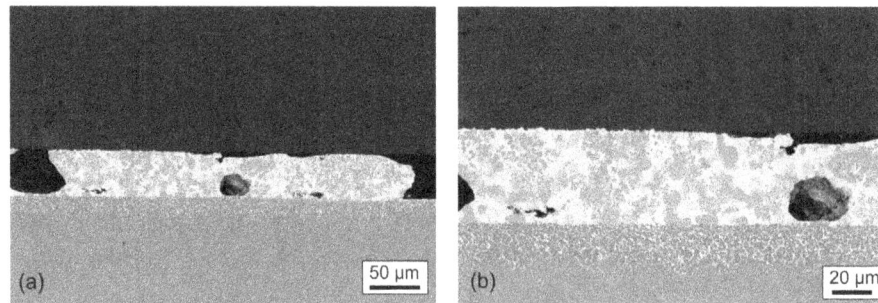

Figure 8. SEM images of Kovar–Al$_2$O$_3$ joints brazed with Incusil-25-ABA at 650 °C for 10 min.

Table 2. Hermeticity after brazing with Cusil-ABA as a function of brazing parameters.

Joint setup	Peak process temperature	Hermetic joints (fraction)
Al$_2$O$_3$–Kovar	810 °C	2/6
Al$_2$O$_3$–Kovar	830 °C	12/12
Al$_2$O$_3$–Kovar	850 °C	6/6
LTCC–Kovar	810 °C	2/6
LTCC–Kovar	830 °C	6/6
LTCC–Kovar	850 °C	Not tested

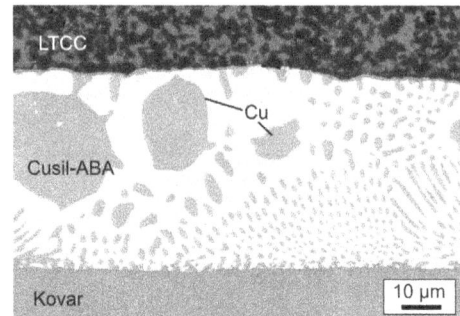

Figure 9. SEM image of a Kovar–LTCC joint brazed with Cusil-ABA at 810 °C for 10 min.

the reaction layer is non-continuous with some pores where the brazing alloy was not able to wet the LTCC completely. However, all brazed assemblies were hermetic with an average helium leak rate of 4×10^{-9} mbar s^{-1}.

Further experiments were conducted by using the Incusil-25-ABA brazing paste. An apparent advantage of the brazing alloy is the lower processing temperature between 640 and 680 °C. Within the scope of this study it was not possible to achieve hermetically brazed joints between Al$_2$O$_3$–Kovar and LTCC–Kovar while using the Incusil-25-ABA brazing alloy. The wetting of the braze on the Kovar surface was excellent, which in turn led to spreading of the melt out of the brazing gap all over the Kovar surface. As a consequence the brazing joints contained numerous and quite large pores, and a porous microstructure was formed as seen in Fig. 8a. Also the reason for the excellent wetting can be taken form the SEM images in Fig. 8. A strong interaction between the active metal braze and the Kovar alloy leads to the destruction of the microstructure beneath the Kovar surface and is quite more pronounced than is the case for the Incusil-ABA braze. This strong reactivity enables the wetting of the molten brazing alloy on the Kovar. Also here the dissolution of the Kovar into the brazing melt results in the formation of nickel–titanium-based phases in the brazing alloy, which can be recognized as the darker disperse phase in the brazing zone. In accordance with the enhanced dissolution of the Kovar, this phase formation seems to be more pronounced. However Incusil-25-ABA contains more titanium than Incusil-ABA,

which may also be a reason for the stronger phase formation. This reaction captures at least a fraction of the active-phase titanium from the brazing alloy which is necessary to enable a wetting process on ceramic surfaces. Thus due to insufficient brazing results no further experiments were performed with this brazing alloy containing a high percentage of indium. The presented results showed that brazing of Kovar with indium-containing active braze filler metals leads to considerable destruction of the Kovar microstructure. An optimization of the brazing cycle could help to minimize this behaviour. However this was beyond the scope of the present study and will be addressed in the future. As explained in the next section, the indium-free Cusil-ABA braze filler metal leaves the microstructure of the Kovar nearly intact.

3.2.2 Brazing with Cusil-ABA

Active metal brazing of Al$_2$O$_3$ and LTCC to Kovar with Cusil-ABA was performed at three different brazing temperatures for a minimum of six samples for each brazing condition. Table 2 summarizes the obtained hermeticity data. While brazing of Al$_2$O$_3$ yielded to hermetically sealed assemblies in most cases, brazing of LTCC gave rather different results. After brazing of LTCC–Kovar joints at a temperature of 810 °C, only a few of the assemblies were hermetic (Fig. 9). Microstructural investigation showed that the interface between LTCC and the brazing alloy is weakly bonded

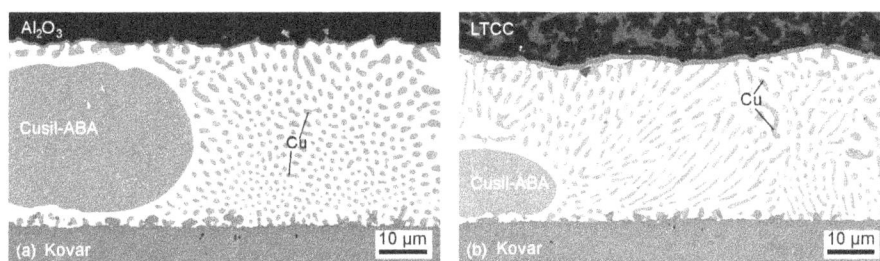

Figure 10. SEM images of Kovar–Al$_2$O$_3$ (**a**) and Kovar–LTCC (**b**) joints brazed with Cusil-ABA at 830 °C for 10 min.

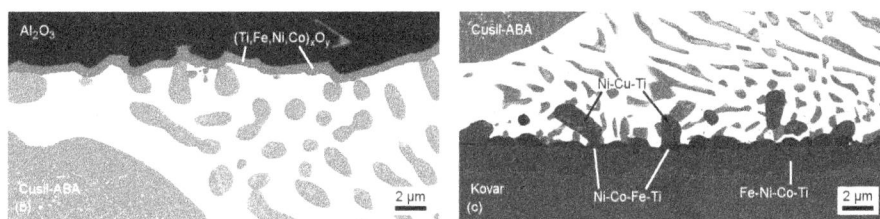

Figure 11. Enlarged SEM images of Fig. 10a showing the Al$_2$O$_3$–Cusil-ABA (**a**) and the Kovar–Cusil-ABA interfaces (**b**).

because it seems that only a non-continuous and thin reaction layer was formed. Thus at this brazing temperature no reliable joining was possible. When the brazing temperature was increased to 830 °C, all brazed LTCC–Kovar joints showed gas-tightness due to the formation of a continuous reaction layer at the interface between LTCC and Kovar, which is shown later in detail. Due to the fact that this brazing temperature is close to the sintering temperature of the LTCC tape, it was initially assumed that brazing is not possible because of softening of the residual glassy phase in the LTCC. This was not confirmed, and the results showed strong and hermetic bonding. However, in contrast to Al$_2$O$_3$, brazing of LTCC at 850 °C was not tried as the LTCC is sintered at this temperature and the stability of the ceramic material is limited. In fact a repeated heating of the DuPont 951 tape up to the processing temperature is possible without any degradation of the microstructure. Scanning electron micrographs of the microstructures of Kovar–Al$_2$O$_3$ and Kovar–LTCC joints brazed with Cusil-ABA at 830 °C are shown in Fig. 10a and b, respectively. In these micrographs the brazing alloy displays the typically structure of the Ag–Cu eutectic with a Ag-rich phase (white region) and a Cu-rich phase (grey regions). Furthermore, the formation of reaction layers on both interfaces is visible.

These reaction layers are shown in more detail in Fig. 11a and b. The active element titanium formed a continuous reaction layer with a thickness of 0.7–1 μm bordering the interface between Al$_2$O$_3$ and the brazing alloy. The reaction layer consists of titanium and oxygen with minor amounts of Ni, Fe, and Co. The elemental composition is the same as observed for Kovar–Al$_2$O$_3$ joints brazed with Incusil-ABA. At the interface between Kovar and the brazing alloy an up to 3 μm thick reaction layer with multiple phases was formed.

The main phase comprises a Fe-rich intermetallic compound (Fe–Ni–Co–Ti) that covers the interface of the Kovar completely. Adjacent to the Fe–Ni–Co–Ti phase a second Ni-rich phase (Ni–Co–Fe–Ti) was found. A third intermetallic phase composed of Ni, Cu, and Ti is observed in the brazing alloy. The microstructural analysis of Kovar–Al$_2$O$_3$ joints brazed with Cusil-ABA and Incusil-ABA showed that during the active metal brazing processing similar phases were formed in the brazing seam. However, in the case of Cusil-ABA the intermetallic compounds are located near the Kovar surface (Fig. 11b), whereas in the case of Incusil-ABA a lacework phase was formed (Fig. 6b). Additionally, while brazing with Cusil-ABA no penetration of the Kovar along the grain boundaries by the brazing alloy was observed (Fig. 10a and b). This leads to the conclusion that the reactivity of Fe, Ni, and Co with the active element titanium in the braze filler metals is enhanced because of the presence of indium or of the lower melting temperature of the Ag–Cu–In eutectic. The observation that the Incusil-ABA braze filler metal with the higher indium content leads to a stronger destruction of the Kovar microstructure along the grain boundaries lets one assume that especially the grain boundary phases of the Kovar consist of a alloy composition which forms low melting compositions with indium. A look at the binary phase diagrams iron–indium, cobalt–indium, and nickel–indium reveals that nickel and cobalt can form low melting phases under brazing conditions (Okamoto, 1997, 2003). In the case of iron this behaviour is shifted to higher temperatures and should not be pronounced below 800 °C (Okamoto, 1990). On the other side according to Berry (1987) the grain boundary phases of Kovar-based alloys tend to form oxide-rich phases, which makes them susceptible to corrosion processes. So we should

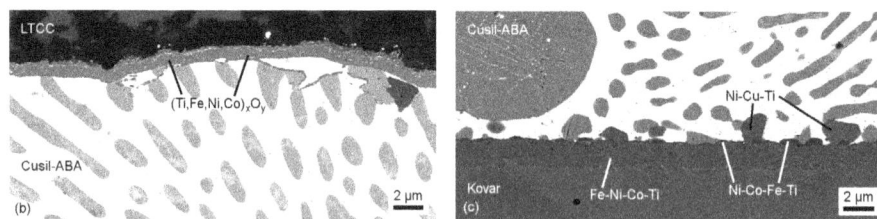

Figure 12. Enlarged SEM images of Fig. 10b showing the LTCC–Cusil-ABA (**a**) and the Kovar–Cusil-ABA interfaces (**b**).

Table 3. Compositions of interfacial reaction layers Cusil-ABA–LTCC at 830 °C for 10 min and Incusil-ABA–LTCC after brazing at 755 °C for 10 min.

Element Ma. %	Cusil-ABA–LTCC	Incusil-ABA–LTCC
O	10.2	16.5
Al	0.6	1.6
Si	3.8	4.5
Ti	44.8	36.4
Fe	6.7	10.1
Co	–	3.9
Ni	8.2	10.1
Cu	25.8	13.0
In	–	–
Ag	–	–

Table 4. Compositions of interfacial reaction layers Cusil-ABA–LTCC and Cusil-ABA–Al_2O_3 after brazing at 850 °C for 10 min.

Element Ma. %	Cusil-ABA–LTCC	Cusil-ABA–Al_2O_3
O	14.3	12.5
Al	1.2	4.14
Si	4.7	
Ti	32.9	36.6
Fe	11.6	12.6
Co	5.6	4.7
Ni	24.0	25.0
Cu	4.5	4.3
Ag	1.0	1.3

note that a more detailed investigation of the grain boundary phase is necessary in order to clarify this behaviour.

The microstructure of Kovar–LTCC joints brazed with Cusil-ABA is similar to the microstructure of Kovar–Al_2O_3 joints. Figure 12a and b are the enlarged images from Fig. 10b displaying the reaction layers at the interface between LTCC and the brazing alloy, and between Kovar and the brazing alloy, respectively. At the interface between LTCC and the brazing alloy a nearly 1 μm thick reaction layer was formed (Fig. 12a). Besides the main constituents of titanium and oxygen, the EDX analysis revealed the presence of Fe, Ni, and Co from Kovar and of minor traces of silicon and lead from the glass phase of the LTCC. It is noteworthy that the compositions of the interfacial reaction layers bordering the LTCC interface are quite similar to the one found after brazing of LTCC and Kovar with Incusil-ABA as seen by EDX data in Table 3, which compares the compositions of the interfacial layers brazed with Cusil-ABA at 830 °C and with Incusil-ABA at 755 °C. A large difference is only recognized for the copper content. However we should not forget the small thickness of the reaction layer in the case of Incusil-ABA which adds an uncertainty to the spectral data. So additionally in Table 4 similar results are presented for compositions of two interfacial layers resulting from brazing both LTCC and Al_2O_3 with Cusil-ABA at 850 °C for 10 min. However, the higher brazing temperature for Cusil-ABA yielded a much thicker reaction layer than for samples

brazed with Incusil-ABA. The comparison of the interface between Kovar and the brazing alloy after brazing to Al_2O_3 (Fig. 11b) and LTCC (Fig. 12b) showed no difference in microstructural appearance like thickness, phases formed, or elemental composition. Based on these results it was decided to use the Cusil-ABA braze filler metal with a brazing temperature of 830 °C for the construction of the complete sensor as shown in the next section.

3.2.3   Joining of sensor and electrical connection

In accordance with the integration procedure the last step involves the soldering of the sensor element by a glass paste which was screen-printed on the back side on the sensor and fired at maximum temperature of 550 °C. Joining and sealing processes for packaging of ceramic-based sensor elements are established for quite a long time, and so numerous qualified glass solders are available for this task. Figure 13a illustrates the final assembling steps of the sensor element into the steel connector with both brazed balancing elements made of Kovar and alumina. In Fig. 13b a SEM image shows the joining zones Kovar–Al_2O_3 and Al_2O_3–LTCC sensor of a completely assembled sensor. It can be seen that both zones contain only few pores and are well attached to each other. After the soldering step of the sensor a final measurement of the helium leakage rate was performed. Completely assembled sensors were mounted and characterized in the aforementioned pressure rail at temperatures up to 300 °C. As an example, Fig. 14 displays a set of characteristic curves of

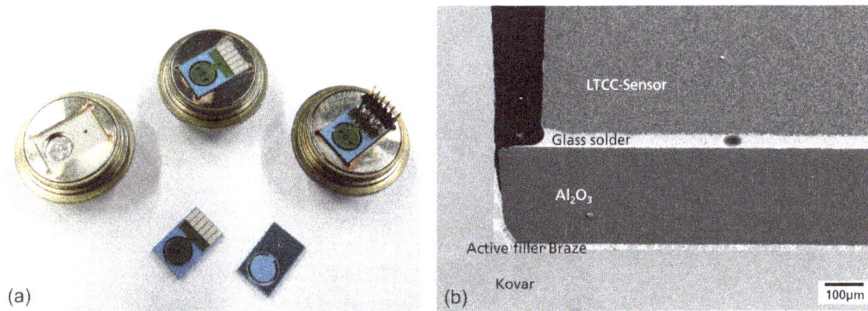

Figure 13. Joint components showing the stepwise integration procedure (**a**) and a SEM image of a cross section of a packaged LTCC pressure sensor (**b**).

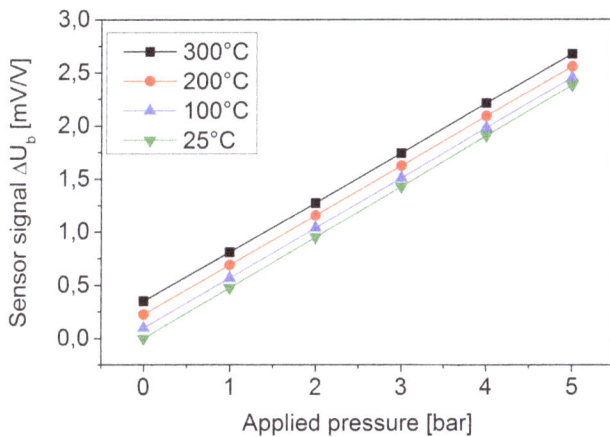

Figure 14. Temperature-dependent pressure-signal characteristics of a completely packaged LTCC sensor.

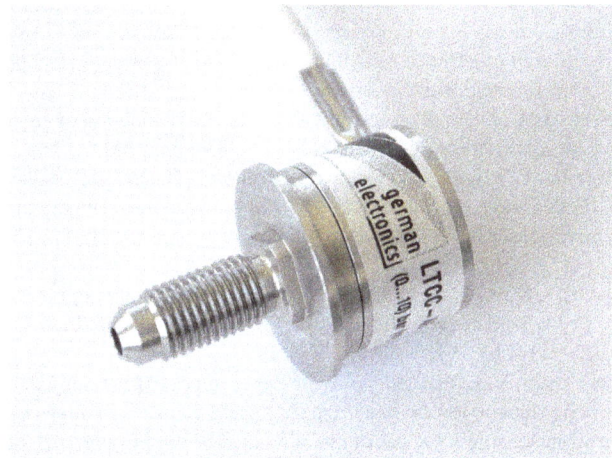

Figure 15. Completely assembled sensor with steel connect screw and welded steel housing with wiring.

a completely assembled sensor based on an applied pressure (bar) and the corresponding sensor signal (ΔmV / V) between 25 and 300 °C. The sensor signal shows a good linearity in the investigated pressure and temperature range. The sensitivity remains unaffected, and the particular curves are only shifted by a small offset, which can be compensated by an accompanied temperature measurement.

4 Conclusions

The present work focused on the joining process of Kovar with alumina and LTCC as part of on approach to integrate LTCC-based sensors into steel connects. While using commercially available active braze filler metals (Cusil-ABA, Incusil-ABA, Incusil-25-ABA) under certain conditions, both ceramic types were hermetically sealed to Kovar. Hermetic joining of Al_2O_3 to Kovar was possible with Incusil-ABA and with Cusil-ABA for all investigated temperatures. Additionally, brazing of LTCC to Kovar was possible and shown for the first time. At 755 °C with Incusil-ABA hermetic LTCC–Kovar joints were realized. The higher indium content of Incusil-25-ABA would enable lower brazing temperatures, but the strong interaction with the Kovar metal and the porous brazing seams result in unreliable joints. With Cusil-ABA, joints were hermetically sealed at brazing temperatures > 810 °C. In all cases microstructural analysis revealed the development of intermetallic compounds that might be brittle, but their influence on the joint strength is unclear and will be investigated in the future. The combination of this metal-to-ceramic brazing step with additional joining processes allows the hermetic integration of a ceramic LTCC pressure sensor into steel housing with an adapted standardized thread (Fig. 15).

Acknowledgements. The authors thank Felix Köhler, Birgit Manhica, Maria Striegler, and Sabine Fischer for sample preparation, helium leak rate measurement, and scanning electron microscopy.

References

Arróyave, R. and Eagar, T. W.: Metal substrate effects on the thermochemistry of active brazing interfaces, Acta Mater., 51, 4871–4880, 2003.

Berry, K. A.: Corrosion Resistance of military microelectronics package at the lead-glass interface, in: Proceedings of the ASM's 3rd Conference on Electronic Packaging: Materials and Corrosion in Microelectronics, Minneapolis, MN, Materials Park, OH, ASM International, 28–30 April 1987, 55–61, 1987.

Fernie, J. A., Drew, R. A. L., and Knowles, K. M.: Joining of engineering ceramics, Int. Mater. Rev., 54, 283–331, 2009.

Fournier, Y., Maeder, T., Boutinard-Rouelle, G., Barras, A., Craquelin, N., and Ryse, P.: Integrated LTCC pressure flow temperature multisensor for compressed air diagnostics, Sensors, 10, 11156–11173, 2010.

Hahn, S., Kim, M., and Kang, S.: A study of the reliability of brazed Al_2O_3 joint systems, IEEE Trans. Comp. Pack. Manuf. Tech. C., 21, 211–216, 1998.

Keusseyan, R. L. and Dilday, J. L.: Development of brazing interconnection to low thermal expansion glass-ceramics for high performance multichip packaging, Proc. of the 43rd Conference on Electronic Components and Technology, Orlando, FL, USA, 1–4 June 1993, 896–903, 1993.

Lin, K.-L., Singh, M., and Asthana, R.: Interfacial characterization of alumina-to-alumina joints fabricated using Silver-Copper-titanium interlayers, Mater. Char., 90, 40–51, 2014.

Liu, Y., Huang, Z. R., and Liu, X. J.: Joining of sintered silicon carbide using ternary Ag-Cu-Ti active brazing alloy, Cer. Int., 35, 3479–3484, 2009.

McDermid, J. R. and Drew, R. A. L.: Thermodynamic brazing alloy design for joining silicon carbide, J. Am. Cer. Soc., 74, 1855–1860, 1991.

Nascimento, R. M., Martinelli, A. E., and Buschinelli, A. J. A.: Review Article: Recent advances in metal-ceramic brazing, Cerâmica, 49, 178–198, 2003.

Nascimento, R. M., Martinelli, A. E., Buschinelli, A. J. A., and Sigismund, E.: Interface microstructure of alumina mechanically metallized with Ti brazed to Fe-Ni-Co using different fillers, Mater. Sci. Eng. A, 466, 195–200, 2007.

Okamoto, H.: Co-In (Cobalt-Indium), J. Phase Equilib., 18, p. 315, 1997.

Okamoto, H.: Fe-In (Iron-Indium), Binary Alloy Phase Diagrams, 2nd Edn., edited by: Massalski, T. B., 2, 1712–1714, 1990.

Okamoto, H.: In-Ni (Indium-Nickel), J. Phase Equilib., 24, p. 379, 2003.

Partsch, U., Gebhardt, S., Arndt, D., Georgi, H., Neubert, H., Fleischer, D., and Gruchow, M.: LTCC based sensors for mechanical quantities, Proceedings of the 16th European Microelectronics and Packaging Conference & Exhibition, Oulu, Finland, 17–20 June 2007, 381–388, 2007.

Partsch, U., Lenz, C., Ziesche, S., Lohrberg, C., Neubert, H., and Maeder, T.: LTCC-based sensors for mechanical quantities, Informacije MIDEM, 42, 260–271, 2012.

Peterson, K. A., Knudson, R. T., Garcia, E. J., Patel, K. D., Okandan, M., Ho, C. K., James, C. D., Rohde, S. B., Rohrer, B. R., Smith, F., Zawicki, L. R., and Wroblewski, B. D.: LTCC in microelectronics, microsystems, and sensors, Proceedings of the 15th International Conference on Mixed Design of Integrated Circuits and Systems, Poznan, Poland, 19–21 June 2008, 23–37, 2008.

Stephens, J. J., Vianco, P. T., Hlava, P. F., and Walker, C. A.: Microstructure and performance of Kovar/alumina joints made with Silver-Copper base active metal braze alloys, in: Advanced brazing and soldering technologies, edited by: Vianco, P. T. and Singh, M., ASM International (Materials Park), 240–251, 2000.

Stephens, J. J., Hosking, T. J., Headly, P. F., Hlava, P. F., and Yost, F. G.: Reaction layers and mechanisms for a Ti-activated braze on sapphire, Metall. Mater. Trans. A, 34, 2963–2972, 2003.

Vianco, P. T., Stephens, J. J., Hlava, P. F., and Walker C. A.: Titanium scavenging in Ag-Cu-Ti active braze joints, Weld. J., 82, 268S–277S, 2003a.

Vianco, P. T., Stephens, J. J., Hlava, P. F., and Walker C. A.: A barrier layer approach to limit Ti scavenging in FeNiCo/Ag-Cu-Ti/Al_2O_3 active braze joints, Weld. J., 82, 252s–262s, 2003b.

Walker, C. A. and Hodges, V. C.: Comparing metal-ceramic brazing methods, Weld. J., 87, 43–50, 2008.

Walker, C. A. and Hodges, V. C.: Metal-nonmetal brazing for electrical, packaging and structural applications, in: Advances in Brazing: Science, Technology and Applications, edited by: Sekulić, D. P., Woodhead Publishing Oxford, 498–524, 2013.

Walker, C. A., Uribe, F., Monroe, S. L., Stephens, J. J., Goeke, R. S., and Hodges, V. C.: High-temperature joining of low-temperature co-fired ceramics, Proceedings of the 3th International Brazing and Soldering Conference, San Antonio, TX, USA, 24–26 April 2006, 54–59, 2006.

Wielage, B., Hoyer, I., and Hausner, S.: Induction brazing of alumina and zirconia with various metals, Proc. 5th International Brazing and Soldering Conference, Las Vegas, NV, USA, 22–25 April 2012, 101–108, 2012.

Zarnik, M. S., Mozek, M., Macek, S., and Belavic, D.: An LTCC-based capacitive pressure sensor with a digital output, Informacije MIDEM, 40, 74–81, 2010.

Zhu, W., Chen, J., Jiang, C., Hao, C., and Zhang, J.: Effects of Ti thickness on microstructure and mechanical properties of alumina-Kovar joints brazed with Ag-Pd/Ti filler, Ceram. Int., 40, 5699–5705, 2014.

24

Paradigm change in hydrogel sensor manufacturing: from recipe-driven to specification-driven process optimization

M. Windisch[1], K.-J. Eichhorn[2], J. Lienig[1], G. Gerlach[3], and L. Schulze[1]

[1]Institute of Electromechanical and Electronic Design, Dresden University of Technology,
01062 Dresden, Germany
[2]Leibniz Institute of Polymer Research Dresden, Hohe Str. 6, 01069 Dresden, Germany
[3]Solid-State Electronics Laboratory, Dresden University of Technology, 01062 Dresden, Germany

Correspondence to: M. Windisch (markus.windisch@tu-dresden.de)

Abstract. The volume production of industrial hydrogel sensors lacks a quality-assuring manufacturing technique for thin polymer films with reproducible properties. Overcoming this problem requires a paradigm change from the current recipe-driven manufacturing process to a specification-driven one. This requires techniques to measure quality-determining hydrogel film properties as well as tools and methods for the control and optimization of the manufacturing process. In this paper we present an approach that comprehensively addresses these issues. The influence of process parameters on the hydrogel film properties and the resulting sensor characteristics have been assessed by means of batch manufacturing tests and the application of several measurement techniques. Based on these investigations, we present novel methods and a tool for the optimization of the cross-linking process step, with the latter being crucial for the sensor sensitivity. Our approach is applicable to various sensor designs with different hydrogels. It has been successfully tested with a sensor solution for surface technology based on PVA/PAA hydrogel as sensing layer and a piezoelectric thickness shear resonator as transducer. Finally, unresolved issues regarding the measurement of hydrogel film parameters are outlined for future research.

1 Introduction

Stimuli-sensitive hydrogels are swellable polymer networks. They respond to changes in the pH value and the concentration of certain kinds of ions or organics, respectively, with well-defined, reversible shifts of their swelling degree. This chemo-mechanical transducer effect together with the wide diversity of available polymers qualifies them as versatile sensing layers for a multiplicity of measurement tasks (Tokarev and Minko, 2009). However, despite more than 1000 scientific publications in the last decade and numerous promising designs (Bashir et al., 2002; Richter et al., 2004; Trinh et al., 2006), hydrogel sensors have still not reached the commercial mass product level. One reason for that is the lack of a well-studied, controllable and scalable manufacturing process for thin hydrogel layers in the lower micrometre

or sub-micrometre range. Such thin films are necessary in order to meet the requirement of industrial sensing applications for response times of less than a minute.

This lack of knowledge can be vividly illustrated by comparing the design and manufacturing of hydrogel sensors with that of traditional mechanical sensors. The outcome of the design process of mechanical systems is usually a set of physically defined and measureable parameters with certain tolerances for each part. Typical examples are length specifications, standardized fits or values for the surface roughness. Furthermore, the design and the manufacturing technique need to be matched. Injection moulding, for example, imposes certain restrictions on the design of the part geometry. Conversely, design specifications, like very narrow tolerance ranges, may require certain manufacturing techniques (e.g. honing). This specification-driven procedure ensures the interchangeability of parts or manufacturers as well as the functionality of the product. Function-based declarations ("Part A needs to fit in part B with a friction force of 10 N.") and recipes ("Polish with 400 grit sandpaper for 10 min") are usually avoided.

The current state of the design and manufacturing of thin-film hydrogel sensors contradicts these principles. Specifications are often function-based, such as "The hydrogel should have a sufficient swelling degree." This is due to a lack of proven measurement techniques for the quality-determining physical parameters and, as a result of this, a knowledge gap regarding the optimal values and acceptable tolerance ranges of these parameters. The manufacturing is mostly recipe-driven ("Spin-coating at 3000 rpm for 3 min."), because the precise quantitative interrelations between the controllable process parameters and the resulting values of the quality-determining film parameters are in most cases unknown.

It was our objective to address these shortcomings by performing the first process-oriented study of the manufacturing of thin hydrogel layers using the example of a sensor solution for monitoring the concentration of industrial cleaners (main component: sodium pyrophosphate) in surface technology. The outcome, presented in this paper, contributes to the following research questions, affecting a wide range of hydrogel sensors for industrial, biomedical and other applications.

– What is a suitable versatile manufacturing process for thin hydrogel films?

– Which measurement techniques are feasible for the monitoring of quality-determining film parameters?

– Which process parameters are crucial for the quality of the manufactured sensors and which methods exist for their sensor-specific optimization?

2   Initial hydrogel sensor solution

This paper is structured in three main sections, according to these issues.

2.1   Sensor principle and motivation

The sensor, previously presented in Windisch and Junghans (2013), consists of a piezoelectric thickness shear resonator (TSR, AT-cut quartz crystal) as a highly sensitive transducer with a PVA/PAA (polyvinyl alcohol/polyacrylic acid) hydrogel coating as a sensing layer. These two coupled elements form an electro-mechanical vibration system. It shifts its frequency-dependent impedance if the hydrogel changes its swelling degree together with its mass and complex shear modulus. The sensor is excited with an AC voltage in a frequency range of ± 20 kHz around the resonance frequency (about 10 MHz) and the corresponding current is measured. The subsequent pre-processing calculates

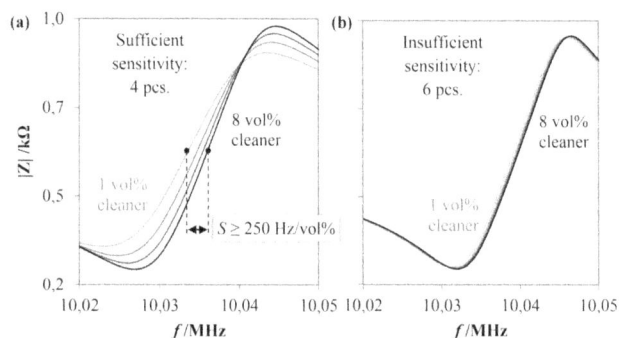

Figure 1. Variance of sensor sensitivity in the sample batch of 10 PVA/PAA-coated TSRs: **(a)** four sensors have a sufficient sensitivity, and **(b)** six sensors have an insufficient sensitivity.

the electrical impedance and converts it to the frequency domain. Figure 1 shows examples of the resulting impedance spectra, from whose changes the measured concentration is finally computed through further data processing steps (see Windisch and Junghans, 2014, for details).

The impedance spectrum of a TSR with a thin viscoelastic hydrogel film depends on four parameters (Bruenig, 2011):

– thickness d,

– density ρ,

– storage modulus G', and

– loss modulus G''.

The density of a swollen hydrogel varies at the utmost between 1 g cm^{-3} (water) and 1.3 g cm^{-3} (dry polymer) and can therefore be regarded as nearly constant for a given swelling degree. The thickness d and the complex shear modus $G^* = G' + i\,G''$ strongly depend on the manufacturing process. They must be kept within narrow tolerance ranges in order to assure an optimal and reproducible sensor function.

The quality criterion for the sensor is its sensitivity, which is defined as the concentration-dependent frequency shift of the inflection point of the impedance spectra $S = \partial f_{ip}/\partial c$ (Fig. 1a). Other deviations, such as offsets of the frequency or the impedance, can be corrected with appropriate measurement-processing algorithms and calibration procedures (Windisch and Junghans, 2014). However, an insufficient sensitivity derogates the signal-to-noise ratio and, consequently, directly limits the achievable measurement accuracy.

A sample batch with a volume of ten pieces was produced and characterized in order to investigate the manufacturability of the previously developed sensor. Figure 1 shows examples of the different sensor characteristics that occurred. The selected application requires a sensitivity of at least 250 Hz vol%$^{-1}$ in order to achieve the necessary measurement accuracy for the cleaner concentration. Using this criterion, the yield of the batch was 40 %, which corresponds to our experiences from previous manufacturing tests. This low yield more than doubles the effective sensor cost, which puts them out of economical range for many industrial applications, for example in surface technology.

Hence, it is necessary to study the interrelation between process parameters and the function determining hydrogel film properties in order to optimize the manufacturing technique and improve the yield. Furthermore, research and, subsequently, volume production require suitable measurement techniques for these parameters.

2.2 Manufacturing process for thin hydrogel layers

Spin-coating was chosen as a process step for creating thin hydrogel layers on substrates. It produces homogenous films with a controllable thickness and can be performed on cost-saving standard equipment. In exchange for these advantages, it also imposes three restrictions for the overall process and the hydrogel synthesis.

– All hydrogel components must be soluble in the same solvent.

– The viscosity of the solution must be well defined and constant (for a given shear rate s).

– The cross-linking has to take place in dry polymer, since the solvent must vaporize during the spin-coating in order to obtain uniform and stable films.

One consequence of these restrictions is that chemical cross-linking in liquid or semi-liquid state is not compatible with spin-coating. Even long gelation times of up to 40 min lead to a constant drift of the solution viscosity, which prevents the batch coating of films with a constant and reproducible thickness.

The low molecular mobility in dry polymers severely limits the cross-linking rate and may even prevent a sufficient network formation. Some publications show that radiation-chemical cross-linking is in principle feasible for dry films (Buller et al., 2013; Hegewald et al., 2005). However, these methods impose additional technological restrictions or specific requirements, like the co-polymerization of special cross-linker molecules. This is contrary to the goal of a simple and inexpensive manufacturing technology.

Another approach is to apply a cross-linking temperature above the glass transition temperature T_G, where the molecular mobility is significantly increased. The thermal cross-linking of PVA and PAA starts at about 120 °C (Arndt et al., 1999). Both polymers are in a liquid-like rubbery state at this temperature and provide a sufficient chain mobility for the cross-linking reaction. PVA and PAA are thermally cross-linked for these reasons and the simplicity of this method.

Based on these preliminary considerations, the following manufacturing process was used for the subsequent studies.

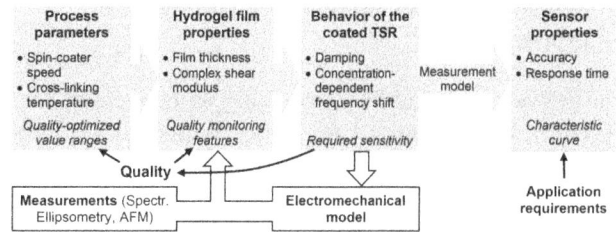

Figure 2. Methodology of the manufacturing process with a comprehensive optimization approach.

1. Deposition of mercaptoundecanoic acid from a 1 mM ethanol solution as an adhesion promoter between the gold electrode of the TSR and the hydrogel.

2. Polymer film formation through spin-coating (see Table A1 for speed profile) of an aqueous solution of PVA (PolySciences, MW $\approx$ 125 000 kg, 88 mol% hydrol., $T_G = 85$ °C) and PAA (PolySciences, MW $\approx$ 450 000 kg, $T_G = 106$ °C) in the ratio 8 : 1.

3. Thermal cross-linking of the two polymers through the formation of anhydrides and ester bonds using the hydroxyl groups and the carboxylic acid groups (Arndt et al., 1999).

The thickness d and homogeneity $s^2(d)$ of the hydrogel film are predefined through both the interrelation of the polymer solution concentration and the spin-coating speed. In the subsequent thermal cross-linking step, the temperature profile determines the complex shear modulus G^* and the range of its concentration-dependent changes. This allows the separate optimization for both parameters without disturbing interrelations.

Figure 2 shows the process-oriented approach for the application-specific sensor optimization. The goal is defined by the application, which requires certain sensor properties in order to fulfil the measurement task. The measurement model matches the characteristic curve to the requirements, provided that the coated TSR has a sufficient sensitivity.

Considering these interdependencies, the ultimate goal of the process optimization is to find the value ranges of the process parameters that ensure the desired film properties and, consequently, a sufficient sensitivity. In a first step, this requires a detailed understanding of both the interrelations between the hydrogel film properties and the resulting behaviour of the coated TSR. From this, target values for d, G' and G'' can be defined and used as quality monitoring features. The second step is then an experimental investigation of the interrelation between the process parameters and these monitoring features. This concept for the process optimization requires measurement techniques for the thickness and the mechanical properties of thin hydrogel films.

3 Measurement techniques for hydrogel film properties

The relevant hydrogel film properties can be measured using either direct or indirect procedures. We have investigated both approaches for parameters d and G'. The loss modulus G'' of films in the sub-micrometre range is difficult to measure. Only the tested indirect model-based measurements (see Sect. 3.2) provide values for G'' (or the viscosity η) of the hydrogel. The lack of a second, independent measurement technique for the verification of these values prevents a profound evaluation of their accuracy. Furthermore, the qualitative and quantitative interrelations between the process parameters and the loss modulus are still widely unknown. Therefore, the measurement of G'' is not covered in this paper and will be subject to future research.

3.1 Direct measurements

Direct measurements are the first option for investigating the properties of thin hydrogel films. Amongst various tested techniques for measuring the thickness of dry and swollen hydrogel films (see Table A2), spectroscopic ellipsometry and atomic force microscopy (AFM) provided the best results.

Ellipsometry as a measurement principle is based on an optical model, which computes the film thickness from the measured amplitude component Ψ and the phase difference Δ of light reflected by the sample. By reason of operating completely in the optical domain and providing the thickness as a well-defined output parameter, it is regarded as a direct method in the context of this paper, although ellipsometry is – in a narrower sense – an indirect measurement method.

A variable angle spectroscopic ellipsometer (M-2000 by J. A. Woollam Co.) was used for measuring the average optical thickness $\overline{d}$ and refractive index n of hydrogel layers on TSRs simultaneously. The latter parameter provides additional information on the optical properties, which is useful for evaluating the chemical homogeneity and, to some extent, the material density of the hydrogel (Bittrich et al., 2014; Ogieglo et al., 2015).

AFM was used for measuring Young's modulus E and the storage modulus G' (using the relation $E = 3G'$ for $\upsilon = 0.5$) of a hydrogel film (Domke and Radmacher, 1998; Markert et al., 2013). However, the effective glass transition temperature of polymers, and therefore G', strongly depends on the measurement frequency (Williams et al., 1955). Consequently, the measured effective values of G' differ, since the cantilever frequency of an AFM is usually in the range of 10^1–10^2 kHz, while the TSR works at 10 MHz. A conversion is currently not possible, because the necessary interrelations are still unknown for the specific PVA/PAA hydrogel. Therefore, the directly measured material properties can currently only be used for relative comparisons.

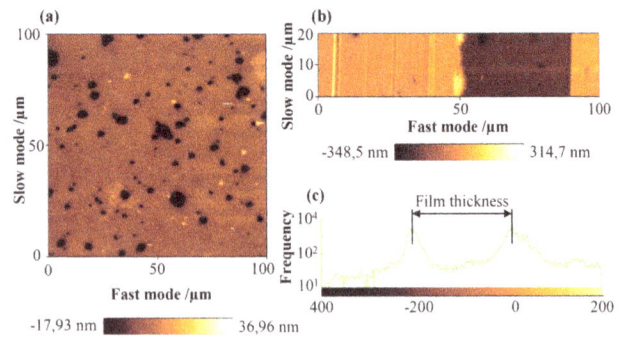

Figure 3. AFM image of (**a**) the hydrogel surface and (**b**) a scratch for the determination of the layer thickness with (**c**) the corresponding height histogram.

In addition to the elastic moduli, AFM can simultaneously measure the thickness of a hydrogel film. Such measurements require a sharp step between the substrate and the film surface, which is generated by scratching the hydrogel film. Figure 3b and c show an image of such a scratch and the corresponding height histogram, respectively. The two peaks in the histogram indicate the most often occurring height values. They represent the hydrogel surface and the TSR surface (scratch), respectively. Hence, the height difference between both is the thickness of the hydrogel layer. This direct measurement procedure allows the determination of the mechanical film thickness at a single position $d(x, y)$.

Gesang et al. (1995) have shown that for polymer films thicker than 10 nm, ellipsometrically measured thickness values d_{ellips} are significantly larger than the corresponding values d_{mech} obtained by AFM measurements. With respect to the mechanical (acoustic) transducer principle, d_{mech} seems more suitable as a quality-determining film parameter than d_{ellips}. However, the AFM measurements are destructive (scratch is necessary) and very limited in the lateral range. Therefore, they are not feasible for the quality monitoring of the manufacturing process. For this reason, only variable angle spectroscopic ellipsometry (VASE) was used for measuring the film thickness in the subsequent experimental section.

3.2 Indirect measurements

Two different approaches were tested for the indirect measurements. The first one is based on the relation between the swelling degree and the storage modulus, given by Eq. (1) (Arndt et al., 2009; Philippova and Khokhlov, 2012).

$$G' = k_{\text{B}} T \iota \left(\frac{V}{V_{\text{dry}}} \right)^{-\frac{1}{3}}, \tag{1}$$

where k_{B} denotes the Boltzmann constant.

Since the thin hydrogel film is confined in the x and y directions by the TSR, it can only swell in the z direction (thickness d). Therefore, G' depends only on the cross-linking density ι and the measured thickness d (the temperature T was kept constant at 22 °C):

$$G' = k_B T \iota \left(\frac{x_{dry} y_{dry} d}{x_{dry} y_{dry} d_{dry}} \right)^{-\frac{1}{3}} = k_B T \iota \left(\frac{d}{d_{dry}} \right)^{-\frac{1}{3}}. \qquad (2)$$

The values d_{ellips} and G'_{AFM}, measured in 1 vol% cleaner concentration, were used for eliminating the unknown value of ι from the equation. This yields the storage modulus at 8 vol%:

$$G'_{cal} = G'_{AFM, 1\_vol\%} \left(\frac{d_{ellips, 8\_vol\%}}{d_{ellips, 1\_vol\%}} \right)^{-\frac{1}{3}}. \qquad (3)$$

A second option for indirectly measuring the hydrogel film properties is the exploitation of the coated TSR itself as a quartz crystal microbalance (QCM). In this approach, an analytical electromechanical model of the quartz crystal and the attached visco-elastic layer is used for simulating its behaviour in the frequency domain. By matching the simulated spectra to the measured ones, the corresponding values for thickness and the mechanical properties of the hydrogel layer can be extracted from the model. The following explanations provide an overview of the application of this complex method. An in-depth description of the complete procedure is provided by Bruenig (2011).

The basis of the model-based approach is the 1-D solution of the wave differential equation for loaded TSRs (Weihnacht et al., 2007):

$$Z = \frac{i \zeta C_0 2 v_q / d}{1 - K^2 \frac{\tan(\zeta)/\zeta}{1 + \frac{iZ_l}{2Z_q} \tan(\zeta)/\left(1 - \frac{iZ_l}{2Z_q \tan(\zeta)}\right)}} + i \omega C_s. \qquad (4)$$

In this equation ζ denotes the normalized frequency $\zeta = \omega d_q / (2 v_q)$ with the angular frequency ω, the shear wave velocity v_q and the thickness d_q of the quartz crystal. C_p is the ideal electrical capacitance of the plate capacitor formed by the resonating area of the TSR electrodes and C_s is its stray capacitance. K and Z_q are the piezoelectric coupling factor and the acoustic impedance of the quartz crystal, respectively. All of the aforementioned parameters (except for ω) are transducer-specific constants for a given TSR. Z_l is the load impedance representing the varying mechanical properties of the hydrogel and the surrounding liquid. The latter were modelled as a viscoelastic layer (hydrogel) and a Newtonian half space (water). The Kelvin–Voigt model was used as a simplified representation of the mechanical hydrogel properties. This simplification is necessary, because more sophisticated modelling approaches for hydrogels, like e.g. Burger's model (Gerlach et al., 2005), contain too many free parameters for a unique solution of the equation system. The application of Mason's transmission line model (Bruenig, 2011; Mason, 1950) yields for the described load case

$$Z_l = \frac{i \tan\left(\omega d_{sim} \sqrt{\frac{\rho}{G^*}}\right) \sqrt{\rho G^*} + \sqrt{\rho_w \cdot i \omega \eta_w}}{1 + \frac{i \tan\left(\omega d_{sim} \sqrt{\frac{\rho}{G^*}}\right) \sqrt{\rho_w \cdot i \omega \eta_w}}{\sqrt{\rho G^*}}}, \qquad (5)$$

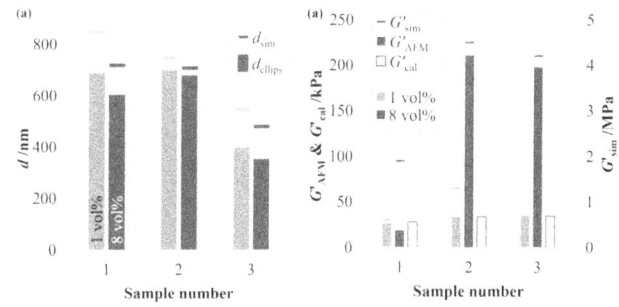

Figure 4. Comparison of directly (measured) and indirectly (simulated, calculated) determined film parameters in 1 and 8 vol% cleaner concentration: (**a**) thickness; (**b**) storage modulus.

with $G^* = G' + i \omega \eta$ for the hydrogel and the index $_w$ denoting the properties of the surrounding liquid half space.

The model-based indirect measurement approach was evaluated in comparison to the direct measurement of the thickness (using VASE) and the storage modulus (using AFM) of three PVA/PAA-coated TSRs in two different cleaner concentrations. Adapted proprietary software (provided by Bruenig, 2011) was used for the numerical simulation of the model and its matching to measured spectra.

3.3 Results

Figure 4 illustrates the main outcome of the experimental investigation of the different measurement techniques. The complete results are summarized in Table A3.

All ellipsometrically measured values (d_{ellips}) are systematically smaller than the indirectly measured ones (d_{sim}). A possible explanation of this effect could be the little pores, which had been found with an AFM scan of the surface of a dry hydrogel film (Fig. 3a). These pores reduce the ellipsometrically determined mean thickness, since they optically behave like the surrounding solution they are filled with. In the case of the acoustic measurement principle the liquid in the pores is moved together with the hydrogel film. They behave like part of the gel due to the very similar densities of hydrogel and liquid.

The systematic deviation between the absolute values d_{ellips} and d_{sim} is largely cancelled out for the relative values of the thickness change caused by shifts of the cleaner concentration (see row "Relative changes from 1 to 8 vol% cleaner" in Table A3). This means that both measurement approaches are potentially suitable for assessing the swellability, which is related to the cross-linking density of the hydrogel film.

The indirectly measured storage moduli are more than 1 order of magnitude higher than the ones measured with the AFM. This outcome is in qualitative agreement with the expected stiffening of the polymer network at high frequencies (Lucklum et al., 1997). The value of G'_{AFM} of the less swollen gel at 8 vol% cleaner concentration is peculiar for

Table 1. Ellipsometry measurements of PVA/PAA films with different coating parameters.

No.	No. of TSRs	c_p (wt.%)	ω (min^{-1})	η (Pa · s) for different shear rates s (s^{-1})				$\overline{n}$	$X(\overline{d})$ (nm)	$s^2(\overline{d})$ (nm)	$X[s^2(d)]$ (nm)
				1	10	100	1000				
1	4	2.5	3900	0.0823	0.0778	0.0620	0.0404	1.529	159.3	1.7	6.9
2	4	3.5	3750	0.2332	0.2028	0.1530	0.0949	1.548	323.4	2.6	5.8
3	5	4.5	3350	0.4920	0.4253	0.3008	0.1773	1.563	525.5	17.0	7.9
4	3	5.5	2500	1.3960	1.0842	0.7155	0.4119	1.565	1011.2	16.2	7.0

TSR$_1$, because it is smaller than the corresponding value of the highly swollen gel at 1 vol%. This is contrary to the results for TSR$_2$ and TSR$_3$ and to the softening with increasing swelling degrees described in the literature (Arndt et al., 1999; Philippova and Khokhlov, 2012). The comparably high mean variance (TSR$_1$: 16.09 kPa^2; TSR$_2$: 2.96 kPa^2; TSR$_3$: 3.11 kPa^2) of the 225 single measurement points indicates a possible measurement deviation for TSR$_1$. Therefore, the values G'_{AFM} and $\Delta G'_{AFM}$ for TSR$_1$ (values in brackets in Table A3) were excluded from the further evaluation of the results.

The remaining relative values $\Delta G'_{AFM}$ and $\Delta G'_{sim}$ do not allow profound conclusions, since they differ significantly for TSR$_2$, while they are in good agreement for TSR$_3$.

The interrelation between storage moduli and swellability is in qualitative agreement with the expected behaviour. This means that the gel with the largest change in the thickness (high swellability, low cross-linking degree) has the lowest storage modulus and vice versa. However, the quantitative agreement is poor. The calculation of storage moduli G'_{cal} using Eq. (3) yielded systematically much lower values than the measurements. This holds true for the application of Eq. (3) to the directly measured values as well as to the ones determined with the model-based approach (the results of the latter are not presented in Table A3). These results indicate that Eq. (1) is not valid for the high frequencies of the AFM and the QCM measurement.

In the aggregate, the results infer that both tested thickness measurement techniques are in principle applicable for the monitoring of quality-determining film properties. Future AFM-based thickness measurements could contribute to the understanding of the observed systematic deviation between the two techniques and help to develop rules for the mathematical correction of these deviations.

Further experiments and a broader database are necessary for the evaluation of the tested measurement techniques for the storage modulus. The approach based on the interrelation between the swelling degree and the storage modulus is – in the investigated form – not applicable for high measurement frequencies and is therefore not further pursued.

4 Investigation and optimization of the manufacturing process

4.1 Spin-coating of homogenous polymer films

Research objectives for the optimization of the spin-coating step are (i) the creation of uniform hydrogel films with a reproducible thickness within a batch and (ii) the development of a mathematical expression for the interrelations between the process parameters and the resulting thickness. To acquire the necessary experimental data, 16 TSRs were coated using different polymer concentrations and rotation speeds according to Table 1. Subsequently, the film thickness was measured with the VASE M-2000 ellipsometer in a dry state. In addition, the viscosity of the polymer solutions was measured at room temperature for different shear rates (Anton Paar Physica MCR 301; cone plate ⌀ 50 mm, angle 1°).

The following statistical parameters were calculated from the ellipsometry results.

- $\overline{d}$: average thickness within one film

- $X(\overline{d})$: average of $\overline{d}$ within the batch

- $s^2(\overline{d})$: variance of the average film thickness within the batch

- $s^2(d)$: thickness variance within one film

- $X[s^2(d)]$: average of these variances within the batch

The values of these parameters in Table 1 illustrate the good homogeneity and reproducibility of the films. In agreement with qualitative simulation results of the electromechanical model, it can be concluded that such small deviations of the thickness are not the reason for the observed variance of the sensor sensitivity.

The second objective of the experimental investigation of the spin-coating step was the development of a relationship for pre-calculating the film thickness from the process parameters. Based on the works of Meyerhofer (1978) and Spangler et al. (1990), Eq. (6) was chosen as a general relation:

$$d = C \frac{(\eta_0 / Pa \cdot s)^a}{(\omega / \text{min}^{-1})^y}, \tag{6}$$

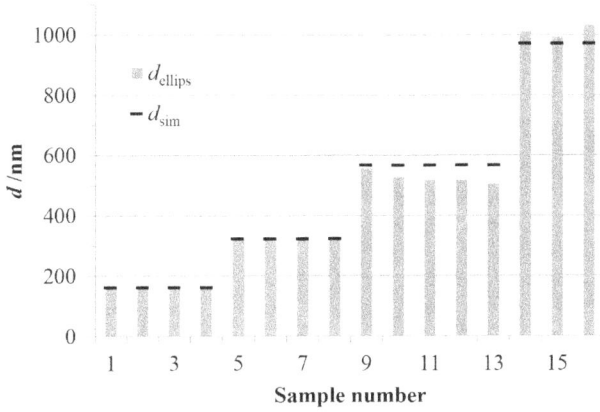

Figure 5. Measured and calculated film thickness.

where C is a polymer-specific parameter, which depends on various physical polymer properties like the molecular weight.

Deviating from the aforementioned publications, the polymer concentration c_p is used as a parameter instead of the initial solution viscosity η_0, based on the approximation given in Eq. (7) (Spangler et al., 1990).

$$\eta_0 = \eta_{solvent} + \eta_{solid}c^\gamma \approx \eta_{solid}c_p^\gamma \tag{7}$$

for $\eta_{solvent} \ll \eta_{solid}c_p^\gamma$.

The motivation for this approximation is the pronounced shear-thinning behaviour of the polymer solution (Table 1). Its viscosity decreases with increasing shear rates s. Since s is unknown and not constant in the polymer solution during the spin-coating, it is ambiguous at which shear rate the value η_0 is to be measured. Therefore, c_p is used as a more feasible parameter.

The product $C \cdot \eta_{solid}^a$ was summarized in a single polymer-specific constant k. Writing the resulting exponent of c_p as $x = a \cdot \gamma$ leads to the quantity Eq. (8), which was used as a mathematical model for the spin-coating.

$$d = k \frac{(c_p/\text{wt.\%})^x}{(\omega/\text{min}^{-1})^y} \tag{8}$$

Factor k must be determined for the PVA/PAA solution. Furthermore, different values had been reported for exponents x and y (Schubert and Dunkel, 2003; Jung et al., 2010). Consequently, all three parameters were determined from the experimental results by minimizing the least mean square error between the calculated and measured thicknesses. With the resulting values $k = 1600$ nm, $x = 1.99$ and $y = 0.5$, the average relative error amounted to 3.8 % (Fig. 5). Hence, Eq. (6) describes the interrelation between the process parameters and the hydrogel film properties for the spin-coating step sufficiently well. The excellent conformity with Meyerhofer's experimental results is remarkable ($x = 2$ and $y = 0.5$), although the shear-thinning phenomenon had not been taken into account for the underlying theoretical considerations.

4.2 Thermal cross-linking

A batch of 13 TSRs was spin-coated with equal parameters ($c_p = 2.5$ wt%; $\omega = 2500$ min^{-1}) and evenly distributed in a laboratory oven (Binder VD 23) for the thermal cross-linking at 160 °C for 20 min. The subsequently measured concentration curves showed large sensitivity differences of the sensors towards the selected proprietary industrial cleaner. These differences could be correlated with the position in the oven (Fig. 7). Considering the direct interrelation between Young's modulus (or shear modulus) and the cross-linking degree (Gerlach et al., 2009), this indicates a non-uniform cross-linking of the hydrogel layers in the oven.

The lack of an approved method for reliable measurements of the storage modulus (see Sect. 3) currently prevents quantitative studies of the interrelations between the cross-linking degree and the sensor sensitivity according to the methodology outlined in Fig. 2. However, the current state of research gives a reason for the hypothesis that the process parameters cross-linking temperature and time have an optimum with respect to the sensor sensitivity (for $d =$ const.). Arndt et al. (1999) found a strong non-linear interrelation between these parameters and the cross-linking degree of PVA/PAA. Higher temperatures or longer cross-linking times lead to more rigid (increased storage modulus, decreased loss modulus) and less swellable films. These mechanical properties affect the energy dissipation within the hydrogel film. The vibrational amplitude can be described as a function of the distance z from the TSR surface (Mecea, 1994; Landau and Lifschitz, 1989):

$$A(z) = A_0 e^{-2\omega\sqrt{\frac{\rho}{2G''}}\cdot z}, \tag{9}$$

where A_0 is the vibration amplitude at the TSR surface.

Equation (9) shows that the dissipation is lower in rigid films ($\downarrow G''$). This means in the inverse that the penetration depth of the acoustic wave in soft rubbery films is limited by the higher dissipation. Based on these results, the following extreme cases can be anticipated for hydrogel-coated TSRs.

Very weakly cross-linked films become necessarily quite thick in liquid environments due to their high swelling degree, while the penetration depth of the acoustic wave is small. If this depth is smaller than the film thickness, the TSR cannot sense thickness changes, since the wave does not reach the gel surface. This means that the influence of d_{sim} on Z_l (Eq. 5) vanishes and only the concentration-dependent change in G^* contributes to the sensor signal. The result is a reduced sensitivity. The other extreme case results in highly cross-linked films. Their very low swelling capability also limits the sensitivity of the sensor. Consequently, an optimum can be expected to exist somewhere between these two extremes. An empirical approach was pursued in order to

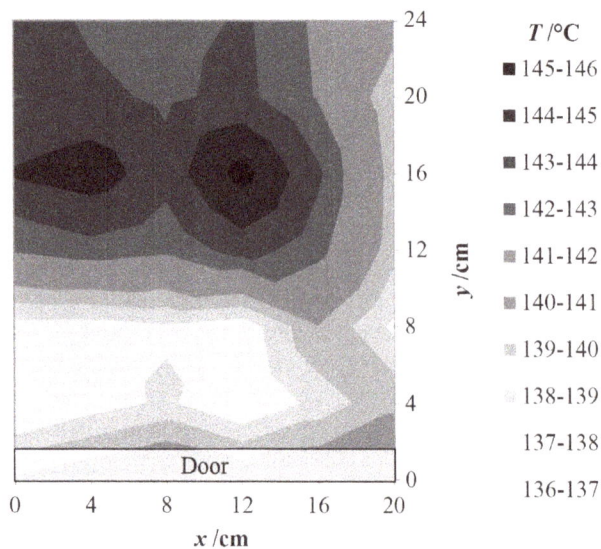

Figure 6. Temperature distribution in cross-linking oven after inserting the coated TSRs.

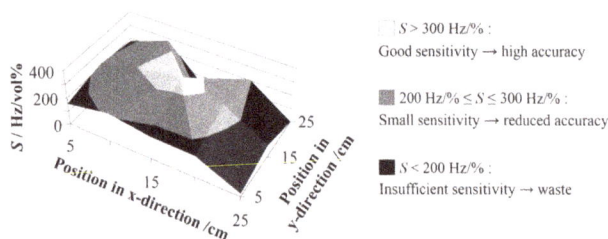

Figure 7. Effect of the inhomogeneous cross-linking temperature distribution in the oven on the resulting sensitivity and functionality of the manufactured sensor.

determine this optimum for the given sensor design without measurements of the mechanical film parameters.

The temperature field in the cross-linking oven was measured with a matrix of PT100 temperature sensors and an in-house developed signal processing device (max. measurement deviation in the range $20 < T < 180\,°C$: $\Delta T \leq 0.5\,K$). Figure 6 shows the inhomogeneous temperature distribution with local differences of up to $10\,K$ after inserting the coated TSRs into the preheated oven. The open door generates an air swirl, which cools the outer parts and causes a hot spot in the middle during the insertion time (about $30\,s$). After closing the door, it takes about 4 min to restore the original temperature (with a maximum deviation $1\,K$) in all areas of the oven. Afterwards, the temperature remains constant for the rest of the cross-linking step.

The similarities between the spatial distribution of temperature and sensitivity in Figs. 6 and 7 suggest the following explanation for the observed sensitivity deviations: the hot spot leads to a higher cross-linking degree of the hydrogel films in the middle of the oven. A consequence of that is

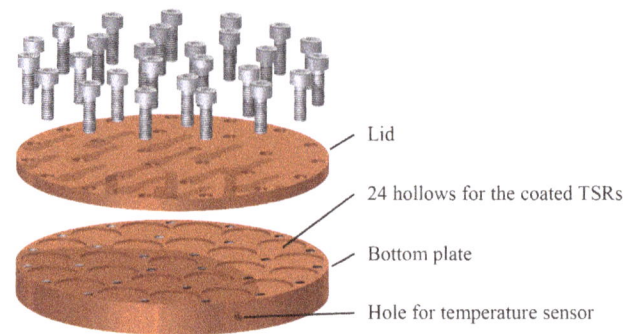

Figure 8. Cross-linking tool.

a higher Young's modulus of these hydrogels, which corresponds to a higher sensitivity, as the AFM measurements indicate. A possible reason why higher cross-linked – and thus less swellable – hydrogels exhibit a better sensitivity relates to the transducer principle: overly soft hydrogels do not sufficiently follow the shear oscillation of the TSR and therefore have less influence on the vibration behaviour of the system, even if they strongly swell.

4.2.1 Development and investigation of a batch manufacturing tool

A batch manufacturing tool for 24 TSRs made of massive copper was designed in order to unify and control the cross-linking temperature (Fig. 8). It reduces the spatial temperature differences to less than $0.3\,K$ and has an integrated sensor for real-time temperature monitoring. One application-driven motivation for developing this tool was to provide defined and controllable thermal conditions in – compared to sophisticated thermal curing devices of the semiconductor industry – inexpensive lab ovens.

The temporal temperature profile in the tool was monitored, together with the spatial temperature distribution in the oven, during the cross-linking process step. Figure 9 shows the distribution of the temperature sensors on the shelf panel in the oven. In order to provide identical conditions for the heating and the cooling process, the tool was placed on an identical shelf panel outside of the oven for cooling down.

The measured profiles (Fig. 10) allow several conclusions. First of all, the steady-state temperature, the intensity of the cooling effect due to the open door as well as the time for recovering the initial temperature after closing the door vary with the position in the oven. For example, sensor 7, which was placed right behind the door, is generally cooler than the other sensors because of the heat loss through the door. This position is most distant to the heat sources and has therefore the longest recovery time. The opposite holds true for the sensors 5 and 6, which are closest to the heat source (mantle heating) and do, therefore, hardly change their temperature.

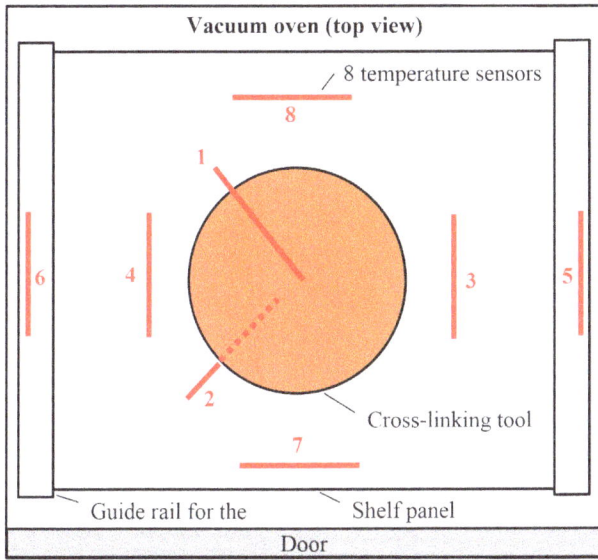

Figure 9. Sensor position in the cross-linking oven.

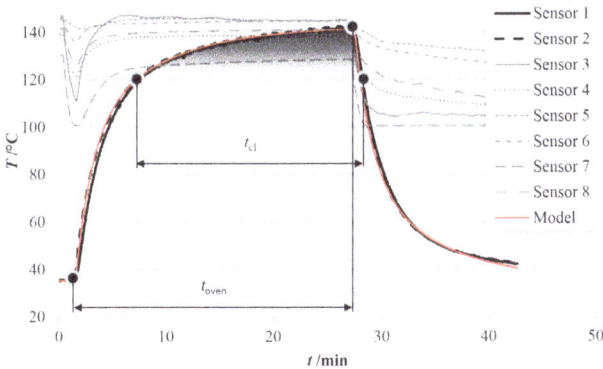

Figure 10. Temperature profile during the cross-linking process.

As a second outcome, Fig. 10 shows that the heat distribution within the cross-linking tool is sufficiently even. Sensor 1 was placed on the lid, while sensor 2 was placed in the hole in the bottom plate of the tool. The temperature of both sensors (bold curves) is almost equal.

The important manufacturing parameter is the cross-linking temperature profile $T_{cl}(t)$. Arndt et al. (1999) have shown that the cross-linking rate reaches a significant level for temperatures above approximately 120 °C. It increases strongly non-linearly with rising temperatures up to about 160 °C. This means that the cross-linking time t_{cl} is the duration at temperatures above 120 °C. The resulting cross-linking degree ι and, therefore, the swellability and storage modulus of the gel, depend on the actual temperature profile within t_{cl}. This area is grey-shaded in Fig. 10, where the colour gradient qualitatively indicates the non-linearly increasing cross-linking speed. The interrelation between the dwell time in the oven t_{oven} (as the actual controllable pro-

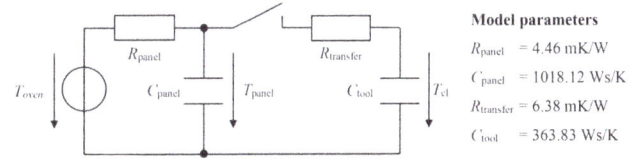

Figure 11. Equivalent thermal circuit.

cess parameter) and the manufacturing parameters $T_{cl}(t)$ and t_{cl} is determined by the thermal time response of the system. Hence, this system behaviour must be sufficiently well known in order to enable the forward-control of the cross-linking process. This means in practice that values of the process parameters t_{oven} and the temperature T_{oven}, at which the oven is pre-heated, can be pre-calculated for any given manufacturing parameters t_{cl} and $T_{cl}(t)$.

The thermal system behaviour of the cross-linking tool in the oven was modelled using the method of electro-thermal analogies and thermal networks. Figure 11 shows the simplified equivalent circuit of the set-up. C_{tool} is the heat capacitance of the copper tool. The tool is placed in the pre-heated oven on a thick (for better heat conductance) aluminium shelf panel with the heat capacitance C_{panel}. Between the two elements is (due to surface roughness) a small air gap of a few micrometres that causes the transfer resistance $R_{transfer}$. The heat conduction resistance of the shelf panel R_{panel} limits the heat flow from the mantle of the oven to the cross-linking tool. The switch indicates the coupling of the pre-heated shelf panel and the cross-linking tool upon its insertion in the oven.

The system behaviour of the equivalent circuit given in Fig. 11 is described by

$$T_{oven} = A \frac{d^2 T_{cl}(t)}{dt^2} + B \frac{dT_{cl}(t)}{dt} + T_{cl}(t) \qquad (10)$$

with $A = R_{panel} C_{panel} R_{transfer} C_{tool}$ and
$\quad B = R_{panel} C_{panel} + R_{panel} C_{cl} + R_{transfer} C_{tool}$.

The general solution of this differential equation is given by Eq. (11).

$$T_{cl}(t) = c_1 e^{\frac{t(-\sqrt{B^2-4A}-B)}{2A}} + c_2 e^{\frac{t(\sqrt{B^2-4A}-B)}{2A}} + T_{oven} \qquad (11)$$

The two constants c_1 and c_2 are calculated for the heating step using the boundary conditions $T_{cl}(0) = T_{cl,0}$ and $dT_{cl}(0)/dt = (T_{oven} - T_{cl,0})/(C_{tool} \cdot R_{transfer})$, with $T_{cl,0}$ being the initial temperature of the tool (e.g. room temperature). Denoting the two exponents in Eq. (11) as $t \cdot X$ and $t \cdot Y$, re-

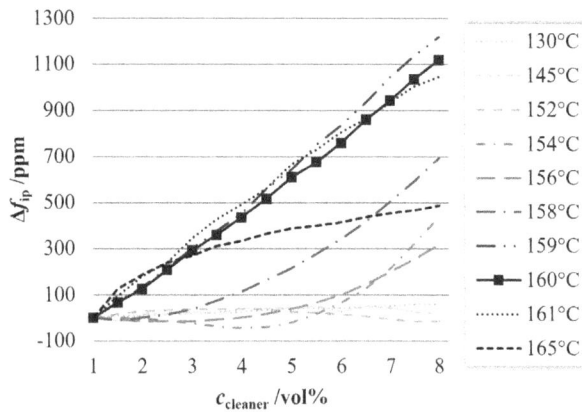

Figure 12. Normalized sensor sensitivity for different cross-linking temperatures at a given time $t_{\text{oven}} = 25\,\text{min}$ (see Table A4 for the corresponding t_{cl} values).

spectively, the analytical model is given by Eq. (12).

$$T_{\text{cl}}(t) = \left(T_{\text{oven}} - T_{\text{cl,0}}\right)\left[\frac{Y + 1/(C_{\text{tool}} \cdot R_{\text{transfer}})}{X - Y}e^{t \cdot X}\right.$$
$$\left. - \frac{X + 1/(C_{\text{tool}} \cdot R_{\text{transfer}})}{X - Y}e^{t \cdot Y}\right] \quad (12)$$

The constants for the cooling step (starting at t_1 with the temperature $T_{\text{cl},1} = T_{\text{cl}}(t_1)$) can be calculated the same way with the boundary conditions $T_{\text{cl}}(t_1) = T_{\text{cl},1}$ and $dT_{\text{cl}}(t_1)/dt = -(T_{\text{oven}} - T_{\text{cl,0}})/(C_{\text{tool}} \cdot R_{\text{transfer}})$. It is assumed that the cool shelf plate outside of the oven, on which the hot cross-linking tool is put, has room temperature ($T_{\text{cl,0}}$).

In order to apply Eq. (12) for controlling cross-linking processes, the values of the elements in the equivalent circuit need to be determined. The modelling of the cross-linking tool itself as concentrated element matches its real physical properties very closely. Its high heat conductivity eliminates any temperature differences within the copper body. Therefore, the value $C_{\text{tool}} = 363.83\,\text{Ws}\,\text{K}^{-1}$ was calculated analytically from its mass (945 g) and the specific thermal heat capacity of copper (385 Ws (kg · K)$^{-1}$).

The heat capacitance and conduction resistance of the panel could not be calculated analytically because of the complex geometry of a special clamping mechanism inside and under the panel. Furthermore, the insertion of the cold cross-linking tool causes spatial temperature differences within the shelf plate. Hence, C_{panel} and R_{panel} can be regarded as virtually concentrated elements, which do not completely represent the real physical properties, but are suitable for describing the average heat conduction behaviour of the shelf panel. The value of R_{transfer} cannot be pre-calculated analytically either, because it depends on the unknown microstructure of the contact surfaces. Hence, the values of C_{panel}, R_{panel} and R_{transfer} were determined by fitting (least squares method) the analytical model to the measured thermal time response of the cross-linking set-up (results denoted

in Fig. 11). With these values, the heating and cooling steps were simulated (red graph in Fig. 10). They match the measured temperature profile very closely. This proves that the simplified model is suitable for the pre-calculation of cross-linking temperature profiles.

The developed cross-linking tool was used for the investigation of the temperature influence on the sensor sensitivity. A batch of 20 TSRs was coated with a PVA/PAA layer ($c_{\text{p}} = 2.5\,\text{wt\%}$, spin-coater speed profile according to Table A1). The cross-linking was carried out with ten different temperatures T_{oven}. Two coated TSRs were placed in the cross-linking tool at each temperature. The process parameter t_{oven} was kept constant to 20 min. Equation (12) was used for calculating the resulting values for t_{cl} and $T_{\text{cl,max}}$ (Table A4). Subsequent to the manufacturing, the sensitivity of the sensors was investigated by measuring the change in their inflection point frequency (Fig. 1) for the application-relevant concentration range of 1–8 vol% industrial cleaner.

4.2.2 Results

Figure 12 shows the results of the investigation of the influence of the cross-linking temperature on the sensor sensitivity. The curves are the mean frequency change of the two samples for each temperature. The relative deviation between every two samples was within a maximum range of 12 % for all samples. This indicates that the cross-linking tool significantly improved the reproducibility. The best sensitivity (about 1.6 kHz vol%$^{-1}$) was achieved in the range of 159–161 °C. Temperatures above 160 °C lead to a decreased sensitivity. Therefore, the process parameters $T_{\text{oven}} = 160\,°$C and $t_{\text{oven}} = 20\,\text{min}$ were chosen as a robust optimum. The corresponding parameters t_{cl} and $T_{\text{cl,max}}$ are printed in bold type in Table A4.

Another five sensors, which were manufactured as one batch with the cross-linking tool and the optimized parameters, exhibited sensitivities in the range of 1.3–1.6 kHz vol%$^{-1}$. This represents a yield of 100 % for the selected industrial application.

5 Conclusions

The optimized manufacturing of hydrogel sensors requires specifications for measurable design parameters and knowledge about their interrelation to the controllable process parameters of the manufacturing technique. We have presented an acoustic sensor principle for industrial applications together with a basic, versatile coating technique for the manufacturing of thin hydrogel layers. The proposed process is easily scalable and very controllable because of the separation of the film formation and the cross-linking step. Due to these advantages, we suggest using it as a manufacturing standard and adapt the hydrogel synthesis, if necessary (and not vice versa). For example, Hirata et al. (2004) showed that a previously only chemically (in liquid state) cross-linkable hydrogel can be thermally cross-linked as well.

Several measurement techniques were tested for the quality-determining hydrogel film parameters storage modulus G' and thickness d. The thickness can in principle be measured with variable angle spectroscopic ellipsometry as well as with a model-based approach that uses the coated TSR itself for determining the film parameters. A systematic deviation between the thickness values obtained with these two techniques was observed. An in-depth study of this deviation with additional measurements using a third technique, such as AFM, is subject to future research.

The same holds true for the measurement of the storage modulus. Major deviations occurred between the values of G' measured with the model-based approach and the AFM, respectively. Further investigations are necessary in order to evaluate the obtained results. Furthermore, Eq. (3) seems to be not applicable for high frequencies. The swelling-dependent changes in the storage moduli calculated from the thickness changes are 2 orders of magnitude smaller than the ones obtained by AFM and the model-based approach.

The investigation of the manufacturing process showed that the film thickness is sufficiently reproducible. Simulations using Eqs. (4) and (5) indicate that the remaining deviations within a batch do not cause the initially observed strong sensitivity variance (Fig. 1). Furthermore, an existing analytical model for the interrelation between the process parameters spin-coater speed ω, polymer concentration c_p and resulting film thickness d was successfully adapted to the manufacturing process of thin hydrogel films.

The strong influence of the cross-linking time t_cl and the temperature profile $T_\mathrm{cl}(t)$ on the sensor sensitivity (Fig. 12) suggests that the temperature deviations within the oven are the main reason for the initial sensitivity variance within a batch of sensors. A cross-linking tool was developed, which works as a heat spreader and reduces the spatial temperature deviations. Furthermore, an analytical model for the thermal system response of the cross-linking set-up was established and successfully tested. It interrelates the controllable process parameters t_oven and T_oven to the manufacturing parameters t_cl and $T_\mathrm{cl}(t)$. The application of the tool and the model for the optimization of the thermal cross-linking process step yielded a quintupled sensitivity and a significantly reduced variance.

The objective of providing optimal values and acceptable tolerance ranges for d and G' could not be completely accomplished for the reason of lacking reliable measurement techniques. VASE, AFM and the model-based approach are promising candidates for this purpose. However, they could not be sufficiently validated within the scope of this paper to be regarded as proven techniques. Concepts for addressing this shortcoming in subsequent studies were pointed out. Finally, there is still a need for a second measuring technique to validate the values for G'' (respectively η), which will also be subject to future research.

Appendix A: Additional data of the experimental part

Table A1. Speed profile for spin-coating.

Time (s)	Speed (rpm)
0	0
5	250
35	250
60	2500 (or according to Table 1)
180	2500 (or according to Table 1)
195	500
205	500
210	0

Table A2. Tested measurement techniques and their applicability for thin hydrogel layers.

Measurand	Measurement technique	Instrument	Remarks on applicability
$\bar{d}$, n	Variable angle spectroscopic ellipsometry (VASE)	M-2000, J. A. Woollam Co.	Applicable
$d(xy)$	Interference measurement	ETA-CSS-BID, ETA-Optik GmbH	Not applicable: reflectivity of the hydrogel surface too low (compared to reflectivity of the underlying gold electrode)
$\bar{d}$	X-ray reflectometry (XRR)	In-house development of the Institute of Structural Physics (TU Dresden)	Not applicable: hydrogel layers are too thick; no measurements in swollen state possible
$d(xy)$	Confocal microscopy	NanoFOCUS μScan AF2000	Not applicable: overly high transmissivity of the hydrogel prevents the detection of the gel surface.
E $d(xy)$	Atomic force microscopy (AFM)	Nanowizard II, JPK Instruments	Applicable with limitations: G' is not measured directly, but calculated: $G' = \frac{E}{2(1+v)}$; measurement of $d(xy)$ requires scratching the gel layer

Table A3. Parameter determination of the model-based indirect measurement approach.

	Parameter	TSR_1	TSR_2	TSR_3	Determination of parameter value
TSR	K v_q (m s^{-1}) Z_q (10^6 kg (m^2 s)$^{-1}$)		0.0893 3321.2 8.795		Material constants from the literature (Bruenig, 2011)
	d_q (μm) C_p (pF)	165.31 6.6	165.17 6.52	165.51 6.22	Extracted from model by matching the measured and simulated spectrum of the unloaded TSRs (prior to hydrogel coating)
	C_s (pF)	7.35	6.21	7.97	Extracted from model (spectrum of loaded TSR outside of the resonance range)
Liquid half space	$\rho_{w,1vol\%}$ (g cm^{-3}) $\rho_{w,8vol\%}$ (g cm^{-3})		1.015 1.065		Measured with hydrometer (uncertainty: ±0.005 g cm^{-3})
	$\eta_{w,1vol\%}$ (mPa s^{-1}) $\eta_{w,8vol\%}$ (mPa s^{-1})		1.8 1.9		Extracted from model (uncoated TSR with liquid half space as load)
Dry film	d_{ellips} (nm) n ($\lambda = 632.8$ nm)	208.3 1.514	231.4 1.515	137.3 1.489	Measured with VASE
Hydrogel film in aqueous solution with a cleaner concentration of 1 vol%	ρ (g cm^{-3})	1.11	1.12	1.12	Calculated from thickness change
	d_{ellips} (nm) n ($\lambda = 632.8$ nm)	690.4 1.391	701.9 1.398	401.7 1.389	Measured in situ with spectroscopic ellipsometry at an angle of incident of 70°
	d_{sim} (nm)	850	750	550	Extracted from model (PVA/PAA-coated TSR with liquid half space)
	G'_{AFM} (kPa)	27	33.7	33.3	Measured in situ with AFM (average of altogether 225 measurement points in three areas of $100\,\mu$m $\times\,100\,\mu$m each)
	G'_{sim} (kPa) η_{sim} (mPa s^{-1})	600 59	1300 57	700 37	Extracted from model (PVA/PAA-coated TSR with liquid half space)
Hydrogel film in aqueous solution with a cleaner concentration of 8 vol%	ρ (g cm^{-3}) d_{ellips} (nm) n ($\lambda = 632.8$ nm) d_{sim} (nm) G'_{AFM} (kPa) G'_{sim} (kPa)	1.12 606.5 1.4 720 (19) 1900	1.12 681.1 1.408 710 211 4500	1.14 355.8 1.401 480 198 4200	Same as parameter determination for 1 vol%
	G'_{cal} (kPa)	28.2	34	34.7	Calculated using Eq. (3)
	η_{sim} (mPa s^{-1})	64	65	40	Same as parameter determination for 1 vol%
Relative changes from 1 to 8 vol% cleaner	Δd_{ellips} (%) Δd_{sim} (%) $\Delta G'_{AFM}$ (%) $\Delta G'_{sim}$ (%)	12.2 15.5 (-29.6) 216.7	3 5 526.1 246.2	11.4 12.7 495.6 500	Calculated from absolute values measured at 1 and 8 vol% cleaner concentration
Sensor sensitivity	S (Hz vol%$^{-1}$)	245	268	136	Average concentration-dependent shift of the inflection point frequency between 1 and 8 vol% cleaner concentration (calculated from spectra measured with Agilent R3765CG VNA)

Table A4. Pre-calculated cross-linking parameters (for $T_{cl,0} = 22\,°C$ and $t_{oven} = 25\,\min$). The bold font indicates the optimal parameter set.

T_{oven} (°C)	t_{cl} (min)	$T_{cl,max}$ (°C)
130	14.20	128.67
145	19.63	143.48
152	20.85	150.40
154	21.11	152.37
156	21.35	154.35
158	21.61	156.32
159	21.70	157.31
160	**21.81**	**158.30**
161	21.93	159.28
165	21.97	163.24

Acknowledgements. This research was funded by a PhD scholarship of the German National Academic Foundation. Previous financial support of R&D project Prozessmesstechnik zur Badueberwachung in der Oberflaechentechnik mittels Hydrogelsensoren by the EU (ERDF, European Regional Development Funds) and the Free State of Saxony is gratefully acknowledged. The authors also thank the former project partner SITA Messtechnik GmbH for the supplied material and S. Abril-Guevara, A. Gehre, L. Guenther, J. Nowak and Y. Wang for supporting the experimental section with measurements and test rigs. Furthermore, the authors are indebted to Jens Friedrichs (Max Bergmann Center) for the AFM measurements and R. Schulze (Leibniz Institute of Polymer Research Dresden) for the VASE measurements.

References

Arndt, K.-F., Richter, A., Ludwig, S., Zimmermann, J., Kressler, J., Kuckling, D., and Adler, H. J.: Poly(vinyl alcohol)/poly(acrylic acid) hydrogels: FT-IR spectroscopic characterization of crosslinking reaction and work at transition point, Acta Polym., 50, 383–390, 1999.

Arndt, K.-F., Krahl, F., Richter, S., and Steiner, G.: Swelling-Related Processes in Hydrogels, in: Hydrogel Sensors and Actuators, edited by: Gerlach, G. and Arndt, K.-F., Vol. 6, Springer Science & Business Media, Heidelberg, Germany, 69–136, 2009.

Bashir, R., Hilt, J. Z., Elibol, O., Gupta, A., and Peppas, N. A.: Micromechanical cantilever as an ultrasensitive pH microsensor, Appl. Phys. Lett., 81, 3091–3093, 2002.

Bittrich, E., Uhlmann, P., Eichhorn, K.-J., Hinrichs, K., Aulich, D., and Furchner, A.: Polymer Brushes, Hydrogels, Polyelectrolyte multilayers: Stimuli responsivity and control of protein adsorption, in: Ellipsometry of Organic Surfaces and Thin Films, edited by: Hinrichs, K. and Eichhorn, K.-J., Springer-Verlag, Berlin-Heidelberg, 79–105, 2014.

Bruenig, R.: Modellierung von akustischen Dickenscherschwingern im Frequenzbereich, Dr. Hut, Munich, Germany, 2011.

Buller, J., Laschewsky, A., and Wischerhoff, E.: Photoreactive oligoethylene glycol polymers–versatile compounds for surface modification by thin hydrogel films, Soft Matter, 9, 929–937, 2013.

Domke, J. and Radmacher, M.: Measuring the elastic properties of thin polymer films with the atomic force microscope, Langmuir, 14, 3320–3325, 1998.

Gerlach, G., Guenther, M., Sorber, J., Suchaneck, G., Arndt, K.-F., and Richter, A.: Chemical and pH sensors based on the swelling behavior of hydrogels, Sensor Actuat. B-Chem., 111, 555–561, 2005.

Gesang, T., Fanter, D., Hoeper, R., Possart, W., and Hennemann, O. D.: Comparative film thickness determination by atomic force microscopy and ellipsometry for ultrathin polymer films, Surf. Interface Anal., 797–808, 1995.

Hegewald, J., Schmidt, T., Gohs, U., Günther, M., Reichelt, R., Stiller, B., and Arndt, K.-F.: Electron beam irradiation of poly(vinyl methyl ether) films: 1. Synthesis and film topography, Langmuir, 21, 6073–6080, 2005.

Hirata, I., Okazaki, M., and Iwata, H.: Simple method for preparation of ultra-thin poly(n-isopropylacrylamide) hydrogel layers and characterization of their thermo-responsive properties, Polymer, 45, 5569–5578, doi:10.1016/j.polymer.2004.06.015, 2004.

Jung, J. Y., Kang, Y. T., and Koo, J.: Development of a new simulation model of spin coating process and its application to optimize the 450 mm wafer coating process, Int. J. Heat Mass Tran., 53, 1712–1717, 2010.

Landau, L. and Lifschitz, E.: Mécanique des fluides, Éditions Mir, Paris, vol. 6., 752 pp., 1989.

Lucklum, R., Behling, C., Cernosek, R. W., and Martin, S. J.: Determination of complex shear modulus with thickness shear mode resonators, J. Appl. Phys., 30, 346–356, 1997.

Markert, C. D., Guo, X., Skardal, A., Wang, Z., Bharadwaj, S., Zhang, Y., Bonin, K., and Guthold, M.: Characterizing the micro-scale elastic modulus of hydrogels for use in regenerative medicine, J. Mech. Behav. Biomed., 27, 115–127, 2013.

Mason, W. P.: Piezoelectric crystals and their application to ultrasonics, D. van Nostrand, van Nostrand, USA, 1950.

Mecea, V. M.: Loaded vibrating quartz sensors, Sensor Actuat. A-Phys., 40, 1–27, 1994.

Meyerhofer, D.: Characteristics of resist films produced by spinning, J. Appl. Phys., 49, 3993–3997, doi:10.1063/1.325357, 1978.

Ogieglo, W., Wormeester, H., Eichhorn, K.-J., Wessling, M., and Benes, N. E.: In situ ellipsometry studies of thin swollen polymer films, a review, Prog. Polym. Sci., 42, 42–48, 2015.

Philippova, O. E. and Khokhlov, A. R.: Polymer Gels, in: Polymer Science: A Comprehensive Reference, edited by: Matyjaszewski K. and Moeller, M., Elsevier, Amsterdam, 339–366, 2012.

Richter, A., Bund, A., Keller, M., and Arndt, K.-F.: Characterization of a microgravimetric sensor based on pH sensitive hydrogels, Sensor Actuat. B-Chem., 99, 579–585, 2004.

Schubert, D. W. and Dunkel, T.: Spin coating from a molecular point of view: its concentration regimes, influence of molar mass and distribution, Mater. Res. Innov., 7, 314–321, 2003.

Spangler, L. L., Torkelson, J. M., and Royal, J. S.: Influence of Solvent and Molecular Weight in Thickness and Surface Topography of Spin-Coated Polymer Films, Polym. Eng. Sci., 30, 644–653, 1990.

Tokarev, I. and Minko, S.: Stimuli-responsive hydrogel thin films, Soft Matter, 5, 511–524, doi:10.1039/B813827C, 2009.

Trinh, Q. T., Gerlach, G., Sorber, J., and Arndt, K.-F.: Hydrogel-based piezoresistive pH sensors: design, simulation and output characteristics, Sensor Actuat. B-Chem., 117, 17–26, 2006.

Weihnacht, M., Bruenig, R., and Schmidt, H.: More accurate simulation of quartz crystal microbalance (QCM) response to viscoelastic loading, IEEE Ultrasonics Symposium, New York, NY, 28–31 October 2007, 377–380, 2007.

Williams, M. L., Landel, R. F., and Ferry, J. D.: The temperature dependence of relaxation mechanisms in amorphous polymers and other glass forming liquids, J. Am. Chem. Soc., 77, 3701–3707, 1955.

Windisch, M. and Junghans, T.: Hydrogel Sensors for Process Monitoring, Adv. Sci. Tech., 77, 71–76, doi:10.4028/www.scientific.net/AST.77.71, 2013.

Windisch, M. and Junghans, T.: Innovative Hydrogel Sensor Solution for Process Monitoring, Sci-eConf 2014 conference proceedings, 2014.

25

Investigation of low-temperature cofired ceramics packages for high-temperature SAW sensors

Jochen Bardong[1], Alfred Binder[1], Sasa Toskov[2], Goran Miskovic[2], and Goran Radosavljevic[2]

[1]Carinthian Tech Research, Europastraße 4/1, Villach, Austria
[2]Institute of Sensor and Actuator Systems, Gusshausstrasse 27–29, Vienna, Austria

Correspondence to: Jochen Bardong (jochen.bardong@ctr.at)

Abstract. Surface acoustic wave (SAW) temperature sensor devices have been developed for operating temperatures up to and above 1000 °C. A challenging task to make these devices available on the market is to develop an appropriate housing concept. A concept based on low-temperature cofired ceramics (LTCC) has been investigated and tested under elevated temperatures up to 600 °C. The devices showed promising results up to 450 °C. Thorough analysis of the possible failure mechanisms was done to increase the maximum temperature above this limit in further production cycles.

1 Introduction

As surface acoustic wave (SAW) devices are widely spread as frequency filters and resonators for frequencies above 100 MHz, sensing applications are distributed in niche fields where silicon-based sensors cannot be applied. For wireless temperature measurements, SAW sensors have been shown to be capable to withstand temperatures up to and above 1000 °C, but with bare crystals without housing. For easy handling and to protect the crystal, housing is needed. As packaging materials, two main families can be identified: metal housings with glass or glass-ceramic wire feedthrough for signal routing and ceramic housings with filled vias and contact pads. In this paper, the latter types will be examined in a first approach to verify if the whole assembly concept is feasible. Low-temperature cofired ceramics (LTCC) allows for integration of embedded antenna structures on the outside of the housing, so this material has additional advantages over metal housings, which always need to be contacted to an antenna structure for radio frequency (RF) interrogation and tend to become leaky or break at the feedthrough. As test device, a reflective SAW delay line on Langasite (LGS) with aluminum–platinum (Al–Pt) based metallization is chosen due to its robustness (Bardong, 2013).

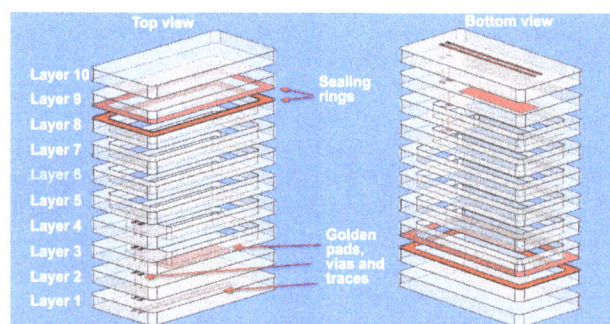

Figure 1. 3-D model of LTCC housing: top and bottom view, split up in the described layers.

2 Materials and methods

LTCC technology offers versatile structuring of complex channels and cavities within a single ceramic layer due to its possibility for selection of various tapes with different thicknesses and physical characteristics. This technology is suitable for applications in harsh and corrosive environments due to excellent hermeticity as well as thermal and chemical resistance of ceramic material.

The designed LTCC housing is made out of several tape layers. Some carry vias for signal routing, some are gold

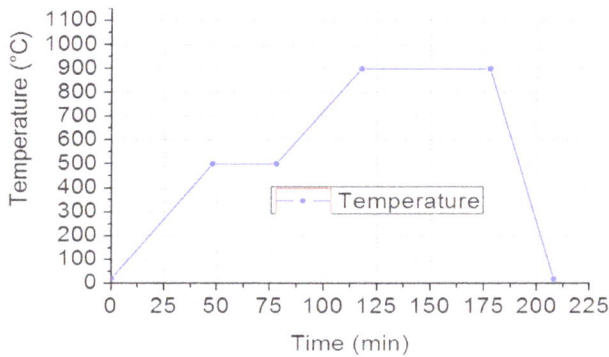

Figure 2. Sintering phase 1: Two-step sintering profile of LTCC CeramTec GC tape.

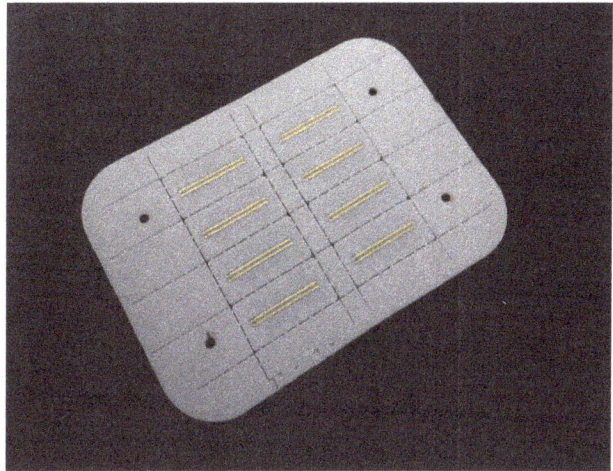

Figure 3. Fabricated housings after two-step sintering profile of LTCC CeramTec GC tape: top view.

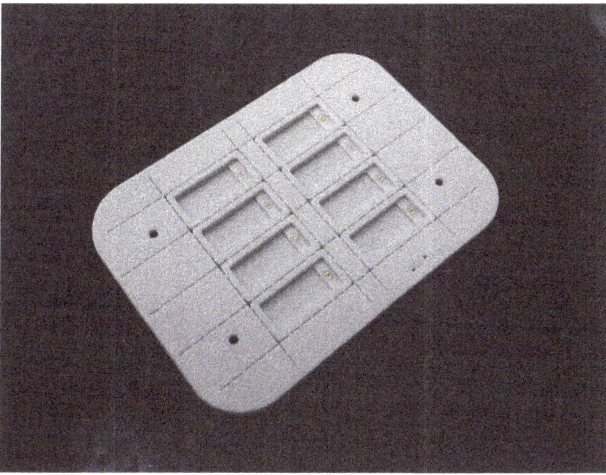

Figure 4. Fabricated housings after two-step sintering profile of LTCC CeramTec GC tape: bottom view.

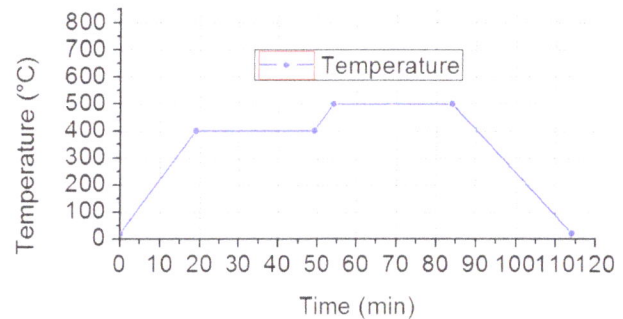

Figure 5. Sintering phase 2: Two-step ESL 4031-B glass sealing paste burn off and sintering.

plated via screen printing (Fig. 1). The top layer has a glass sealing ring on it to apply the lid.

The total outer dimensions of the housing are $7 \times 13.8 \times 2.4\,\text{mm}^3$ for width, length, and height, respectively. The structuring is done in the green state of the material by an Nd:YAG laser.

For this housing, the standardized process to develop 3-D structures with cavities has to be modified to achieve defect-free devices; as the cavities are quite large, making a defect-free lamination during the isostatic pressing process is a challenging task.

The LTCC housing is fabricated using commercially available CeramTec GC LTCC tape. This tape shows pronounced reliability and stability in high-temperature environments (Toskov et al., 2013). Commercially available and compatible ESL 8881 gold conductive paste has been selected to screen print the conductive pads and traces and ESL 8835 VF gold conductive paste for the via filling. The drying process of the gold conductive pastes was carried out at 125 °C for 15 min. Due to the large cavity in layers 3, 4, and 5–8, the

lamination process has been executed in four separate steps. In the first three steps, layers 1–2, 3–4, and 5–8 were laminated to ensure lamination integrity of each of these parts, respectively. In each step, single ceramic layers were centred and stacked together in a steel compression mold and thermally compressed in a uniaxial press for 3.5 min, at 75 °C, and a pressure of 64 bar. The final step was dedicated to the lamination of all the previously laminated subgroups.

The sintering process was carried out in three separate phases devoted to: (1) LTCC sintering in a Linn box furnace (Fig. 2), (2) burn off and sintering of screen-printed glass sealing rings, and (3) finally sealing of the entire LTCC housing after chip mounting and wire bonding.

Fabricated structures after the first sintering phase are presented in Figs. 3 and 4.

Right after the first sintering phase, the glass seals were screen printed at the top of the eighth layer stack and the bottom side of the lid (layer 9). Figure 5 shows the corresponding heat treatment profile to burn out organic compounds of the paste. In Fig. 6, the housing parts after these steps are presented.

Figure 6. Fabricated LTCC housing after ESL 4031-B glass sealing paste burn off and sintering.

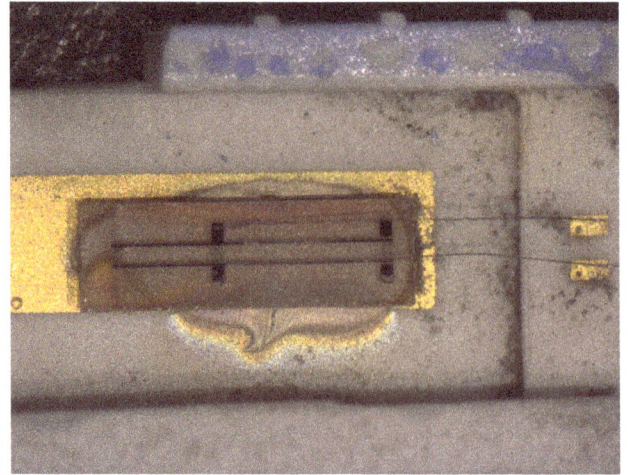

Figure 8. Reopened LTCC housing after heat treatment. Langasite crystal, contacting wire bonds, and nanosilver paste can be identified. Residual grinding dust from the opening process is visible between crystal and bonding pads of the LTCC.

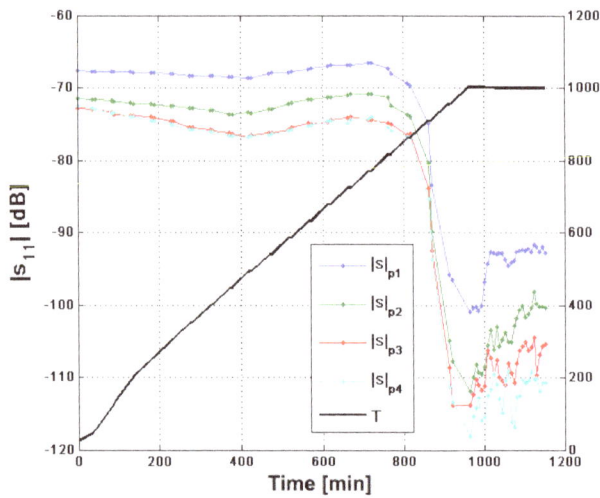

Figure 7. Bare device impulse amplitudes during annealing up to 1000 °C in Argon. The black line indicates the linear temperature rise and corresponds with the abscissa at the right. The device readings are stable up to 850 °C.

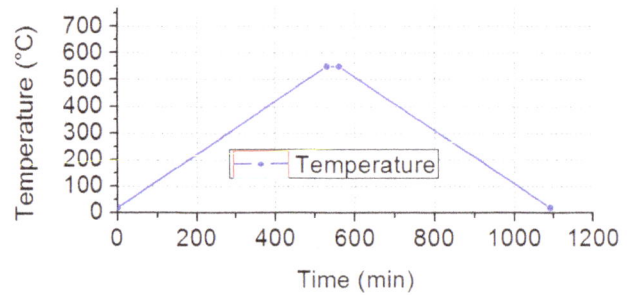

Figure 9. Temperature profile for the ESL 4031-B sealing paste with Langasite SAW sensor inside the housing.

To establish electrical contacts, the dies were bonded with Pt wires to the feed-through gold pads. In Fig. 8, a reopened setup is shown after heat treatment.

Finally, the lid was closed using a slow heating and cooling rate to avoid thermally induced stress (Fig. 9).

The closed devices were contacted and fixed in place with several platinum wires to a device carrier (Fig. 10), which was then mounted into a sample holder described in Bardong et al. (2009) and processed under the temperature setting described in the following part of this text. The atmosphere was set to 1 hPa of nitrogen in a tube furnace.

3   Measurements and signal processing

For heat treatment, the furnace was programmed to hold temperature levels from 250 to 600 °C in steps of 50 K. To get a first impression about the long-term stability of the housed device, each temperature level was kept for 20 h with an RF interrogation cycle every 15 min.

For the test, Langasite reflective delay lines with a Pt–Al stack of 105 nm thickness were chosen to be mounted into the housing (da Cunha et al., 2007; Bardong et al., 2010). The dies alone showed signals up to 900 °C (Fig. 7) (Bardong, 2013).

Therefore, the dies were placed on a dispensed amount of nanosilver paste and sintered onto the gold area inside the housing. Sintering was performed at 280 °C for 40 min on a hotplate.

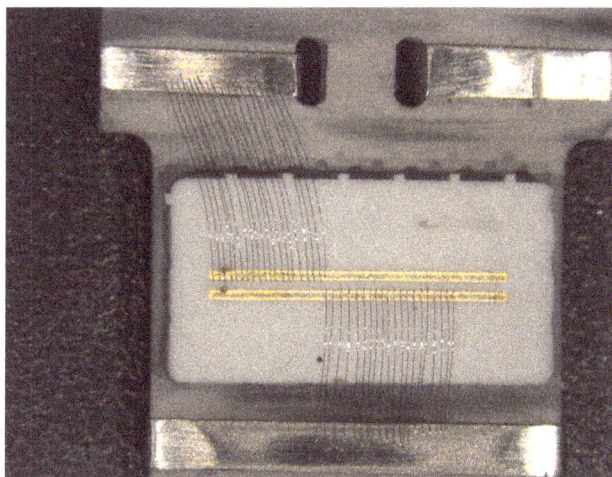

Figure 10. Closed LTCC device, mounted on a carrier. Twenty bonding wires of 25 µm in diameter hold the device in position and feed the signal in and out of the housing.

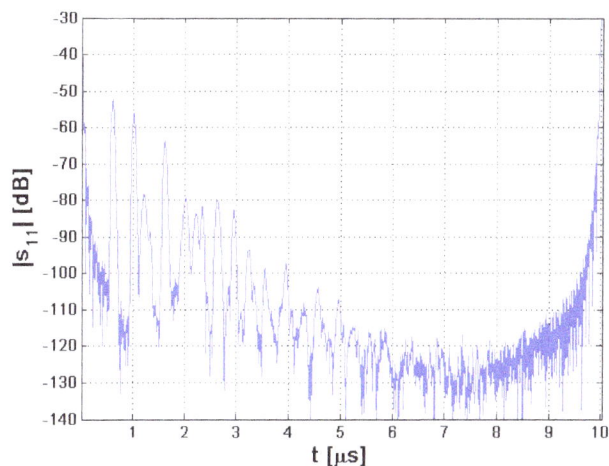

Figure 12. Impulse response of the measured signal from Fig. 11. Aside from the three impulses corresponding to the device design, there is a multitude of weaker peaks indicating echoes of the surface acoustic wave running back and forth on the crystal, leading to these artifacts.

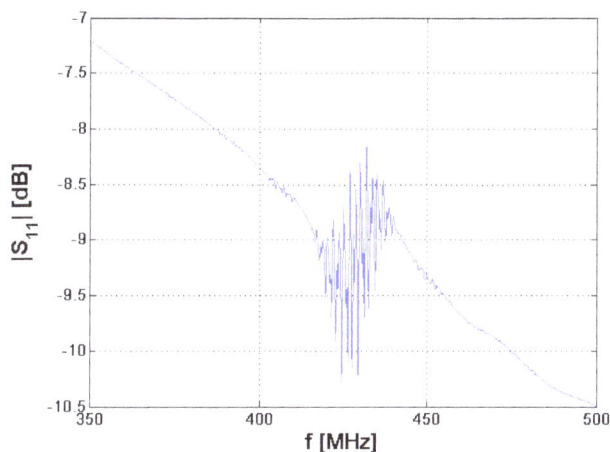

Figure 11. Device response to an interrogating sweep between 350 and 450 MHz. Scales are in decibels to enhance signal details. The blurred part at around 425 MHz is the device response – a superposition of three impulses with slightly different centre frequencies resulting in this interference pattern. Electrical crosstalk is distributed all over the signal resulting in this warped line.

RF interrogation was performed via reflective measurements of RF scattering parameters (S_{11}) from 370 to 470 MHz in 100 kHz steps using a vector network analyser (VNA). By doing this, the device is interrogated with a sine-shaped signal of a dedicated frequency and its response is compared to the reference. The differences between received interrogation signal and reference are calculated and stored. The next frequency (step) is tested likewise until all frequency steps have been interrogated. The achieved data matrix consists of frequency data points and corresponding, complex-valued device response data points, so-called scattering parameters, whose short form "S-parameters" is usu-

ally used. Depending on the number of applied connectors of the VNA, called "ports", the S-parameters are indexed (Hiebel, 2006; Zinke and Brunswig, 1999); in this case, port 1 is used as a sending and receiving port, and the short form of the only resulting S-parameter is written as "S_{11}". These data are always recorded in the frequency domain. An exemplary plot is shown in Fig. 11.

From this signal, only qualitative assumptions can be made considering the functionality of the device. The main response energy is located at 425 MHz, as can be seen in Fig. 11. The blurred interference pattern is the result of the superposition of three impulses with different time delays and slightly different centre frequencies. This signal is processed as follows: a window is applied over its whole length to force the starting and ending points to the same value, preferably zero. This is needed to apply the inverse fast Fourier transform (IFFT) algorithm which can only be applied to periodic signals. The windowed signal is interpreted as one full period of the signal to be transformed, which allows the operation (Butz, 2003). The signal, in time domain, is shown in Fig. 12.

Figure 13 shows an enhanced area of Fig. 12 in which the main impulses of the device design are located.

Each one of these impulses is identified and analysed considering delay time, amplitude, and phase value at the maximum. This is possible as the measurement is complex valued and therefore includes amplitude and phase information of the signal. The impulses are multiplied each with a separate window. Thus, the signal becomes reduced to one single peak with surrounding zeroes which is transformed back to the frequency domain. This has to be done for each of the impulses separately. In Fig. 14, the result is shown after processing the data or one device. The algorithm is named time gating as a

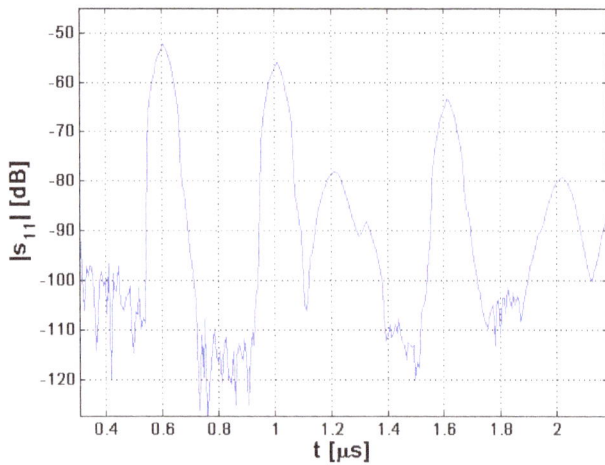

Figure 13. Detail view of Fig. 12. The main impulses show their dedicated maximum at around 0.6, 1, and 1.6 µs, respectively, and have the strongest amplitude in that area.

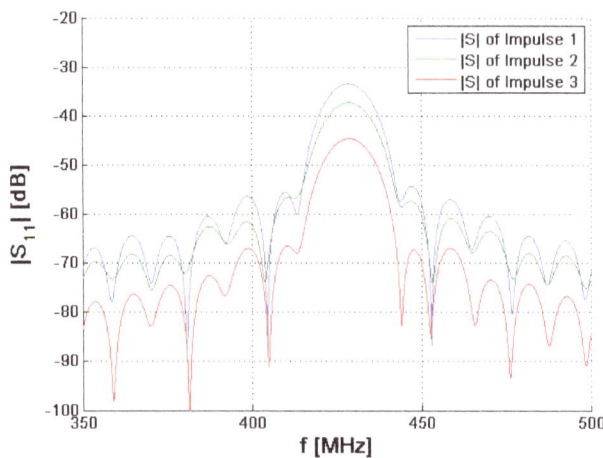

Figure 14. Transfer function of each impulse after time gating each single impulse and applying FFT.

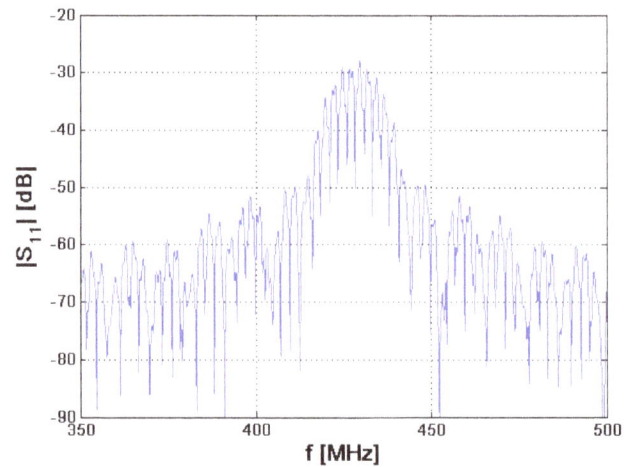

Figure 15. Interference pattern after superpositioning the TFs of all three main impulses. The difference to Fig. 11 is removed electrical crosstalk in this figure.

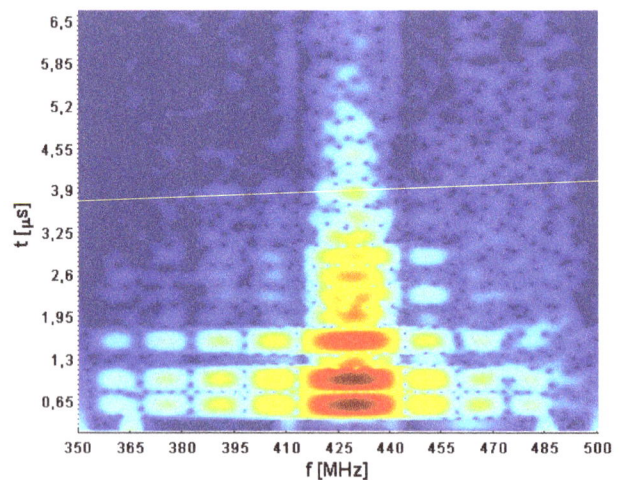

Figure 16. Spectrogram of the measured signal from Fig. 11.

window (gate) is applied in time domain. Again, the maximum is analysed regarding centre frequency, amplitude, and phase position.

Here, the transfer functions (TF) of the device for each single impulse are shown. The slight differences between these TFs result in an interference pattern, if superpositioned, as can be seen in Figs. 11 and 15.

All signal representations discussed previously are one-dimensional projections to the time or frequency axis, respectively. If the signal energy distribution is of interest – and sometimes, it is impossible to make out the signal from artifacts – a spectrogram can be used to analyse this energy distribution in the time–frequency plane (Boashash, 2015). Figure 16 shows the energy distribution of the tested device in this plane. The signal energy is concentrated in the main peaks (see Fig. 12) at around 430 MHz each. Sidelobes can

be determined for each of the main peaks as separated circular areas in parallel to the frequency axis. Echoes from the chip edges and inter-IDT reflections can be seen at 430 MHz in between peaks 2 and 3 and after peak 3. As these artifacts show to have the same frequency as the SAW main signals, they are most likely SAW-based artifacts. Otherwise, the IDT would act like a frequency filter and suppress them.

After analysis of a measurement file, one single measurement point is done. Each of the dots in e.g. Fig. 17 is linked to one measurement. Thus, the whole measurement series is then condensed to line graphs showing the behaviour of an aspect of the device under the test circumstances, be it the centre frequency, the delay, or the corresponding amplitude and phase values.

It is expected that the analysed values change between temperature plateaus and stabilize during the time of 20 h in

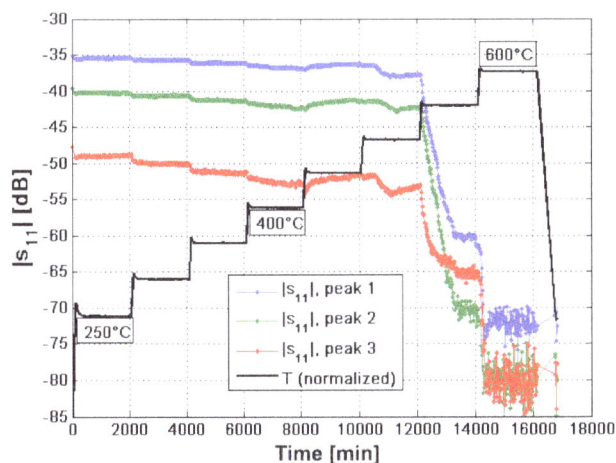

Figure 17. Amplitude values of the three impulses shown in Fig. 11 during the heat treatment. At 550 °C, amplitudes drop considerably and do not recover, indicating device failure.

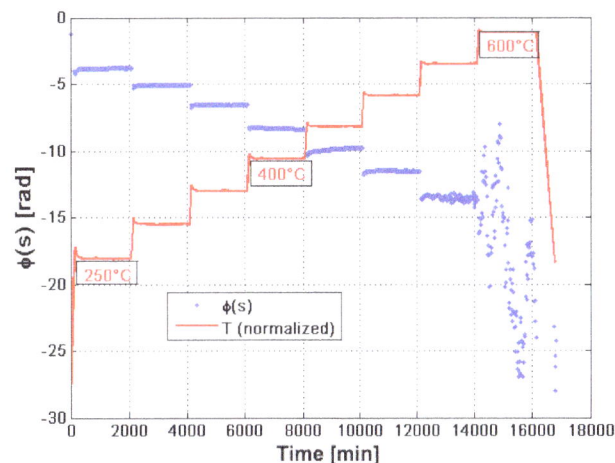

Figure 18. Phase response (dots) of a device's first impulse (see Fig. 12 or 13 for a complete impulse response) during the temperature treatment (line).

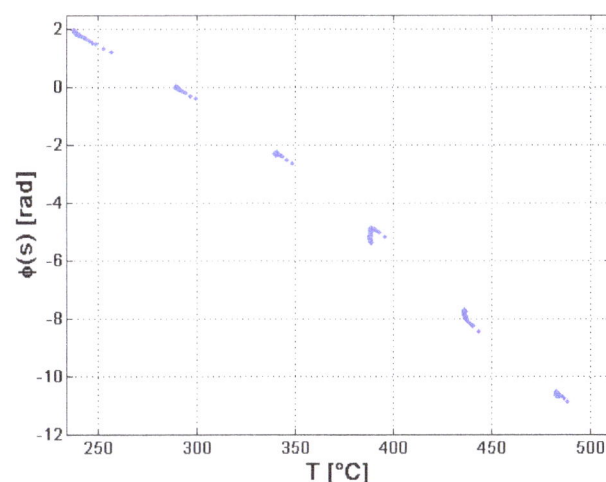

Figure 19. Phase response plotted over temperature.

which one such plateau is kept. Amplitude values might vary slightly by several decibels due to diffusion-driven resistivity changes of the platinum–aluminum thin film from which the electrodes are fabricated (Bardong, 2013; Bardong et al., 2010; Wall et al., 2015). Frequency, delay, and their corresponding phase values should stay constant on each individual temperature level and change due to the temperature coefficient of delay when a new temperature level has been set. As the phase value can be interpreted as the derivative of the observed main value (time or frequency), it is more sensitive to small changes and will therefore be preferably investigated (Smith, 2013).

Figure 17 shows the amplitude values of three impulses after evaluation of a complete measurement campaign. The normalized temperature levels are shown with the black line. The first level on the left indicates 250 °C, rising in 50 K steps to 600 °C on the right. Each vertical point triplet corresponds to a single measurement discussed earlier. The amplitude values do not change much due to temperature changes, but the temperature levels can be recognized as small step-like signal drops. At the second to last temperature level, the device shows amplitude degradation from −65 to −70 dB for all three impulses, which indicates a considerable degradation of the transducer structure leading to device failure by almost 21 dB (factor 128).

Whereas the amplitude holds the information of what happens to the metal layer during temperature treatment (Fig. 17), the phase values show shifts corresponding to temperature changes and indicate, with growing noise, device failure. Figure 18 shows the phase values of the first peak (blue line in Fig. 17) over time with a labelled, normalized temperature plot for comparison. The phase shifts can clearly be distinguished from one temperature level to the next. As the strong amplitude degradation occurs, the noise

of the phase readings rises slightly, but still allows it to match the temperature level of 550 °C with a device response. At 600 °C, device failure is indicated by the wide-spread phase values.

To test if the crystal itself behaves as estimated, the phase values of the first impulse maximum are plotted over the applied temperature. As expected, a slight parabolic behaviour of this crystal cut can be observed in this temperature regime (Hornsteiner, 1999; Naumenko and Solie, 2001; Malocha et al., 2002).

4 Results

During annealing, the devices showed good signals up to 500 °C (Figs. 17 and 18). Above this temperature, degradation occurred. This was not expected as non-housed de-

Figure 20. Detail of Fig. 8, centre structure, 200 times enhanced. Several dots on the finger structures indicate metal droplets leading to electrical shortcuts on the structure.

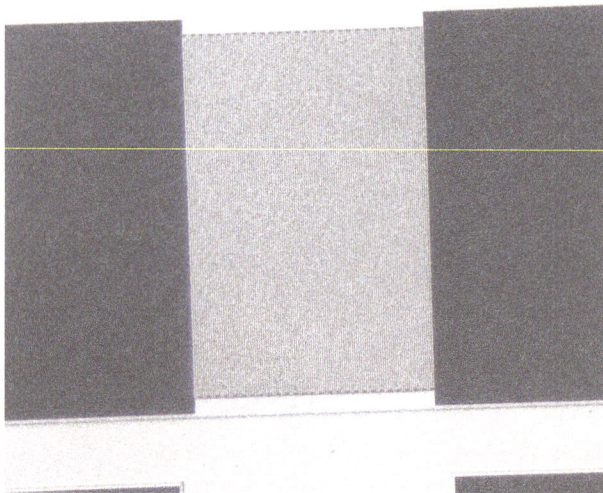

Figure 21. Structure before treatment, centre structure, 200 times enhanced. The finger area is clean of any droplet.

vices with similar design and equal Langasite and metallization material showed good stability at temperatures above 700 °C (Fig. 7) (Bardong, 2013). Optical analysis of the structures showed considerable recrystallization, dewetting and agglomeration of the metal layers as can be seen in Fig. 20. For comparison, an unaged device is shown in Fig. 21. The IDT area shows no dotted pattern.

Mass spectroscopy examination in the temperature range from room temperature up to 600 °C showed several mass counts matching with carbon oxides and carbon alone at the beginning of the treatment (Fig. 22). As this chemical compound can lead to a catalytic degradation of aluminum and its oxide, especially if platinum is present, which by itself is

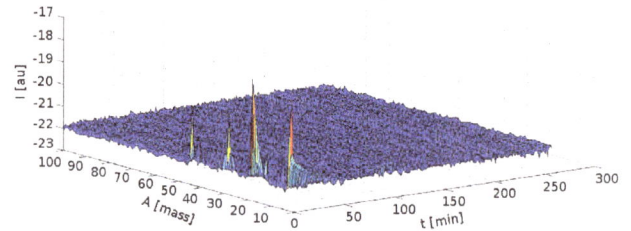

Figure 22. Effusion test graph of one device from room temperature up to 600 °C. At the very beginning of the examination, several mass counts occur, but during annealing, no additional counts are registered, indicating a volatile behaviour of the contaminants.

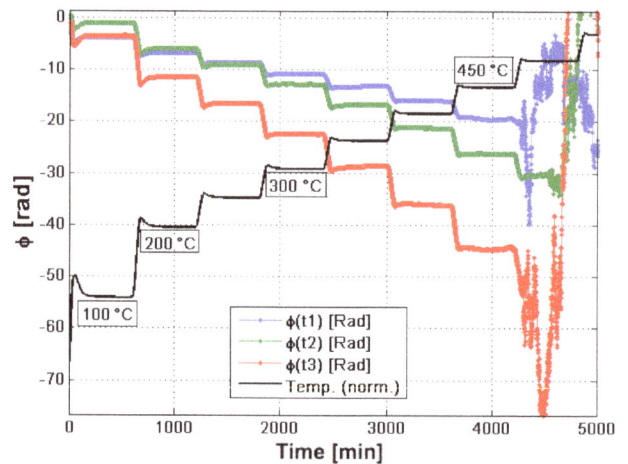

Figure 23. Phase values of a device from the second batch. Device failure above 450 °C has been reproduced.

a very good catalyst, a first assumption leads to reduce the contamination of the system with carbon-based compounds.

5 Further measurements and analysis

Based on these results and assumptions, a second batch of LTCC-housed Langasite devices was produced and measured.

To eliminate possible residual organic compounds inside the housing, the mounted devices were pre-annealed with open lids at 650 °C under oxygen for at least 2 h. The sealing of the lid was done afterwards at 550 °C after application of the glass seal paste under clean room conditions.

However, the previous results were reproduced: all devices failed at temperatures above 450 °C. In Fig. 23, the phase values of an exemplary device from this batch are shown. Again, agglomeration phenomena on the metal layer were observed.

Elementary analysis showed a considerable amount of Bi in the metal system. This explains the reduction of the metal system's melting point as bismuth (Bi) is added to reduce ac-

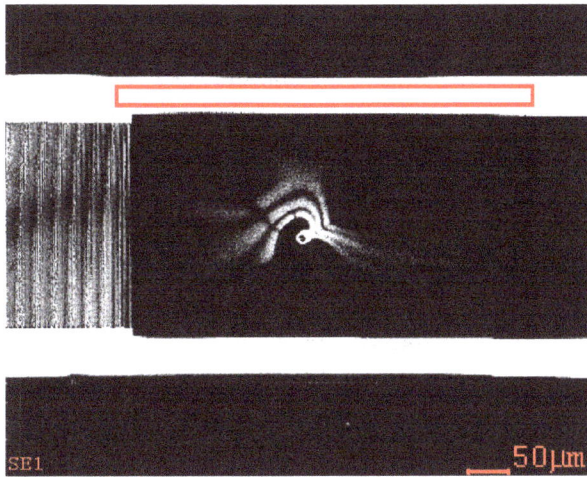

Figure 24. SEM photography of a new device. An area is marked for the elementary (FIB) scan. The IDT structure can be seen on the left between the two bright bus bars. The moiré-like pattern in the centre of this figure results from surface charges due to the SEM scan on this non-conductive, piezoelectric area.

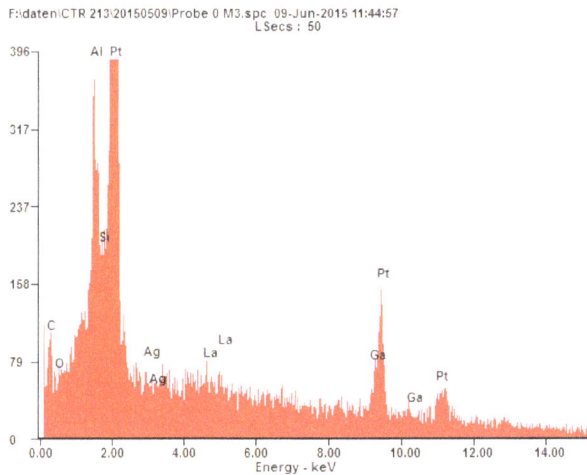

Figure 25. Distribution of elements along the area marked in red in Fig. 24. Mostly Pt and Al, the layer components, are detected and spurious signals of the crystal components like gallium (Ga), lanthanum (La), or silicon (Si).

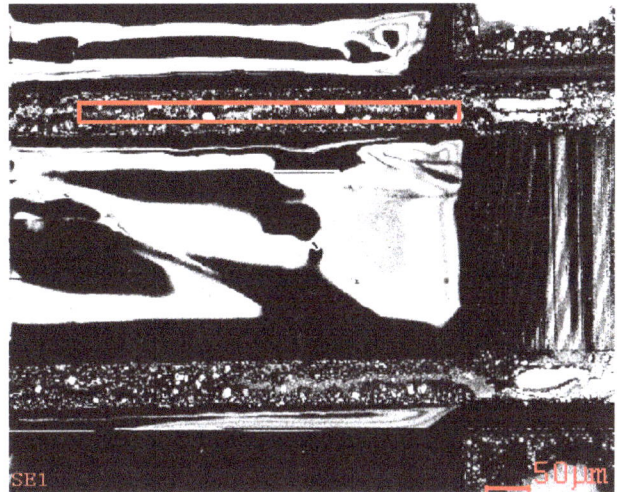

Figure 26. SEM photography of a device after heat treatment. The metallic bus bars show considerable increase in surface roughness. The area in the red box was subject to elementary analysis.

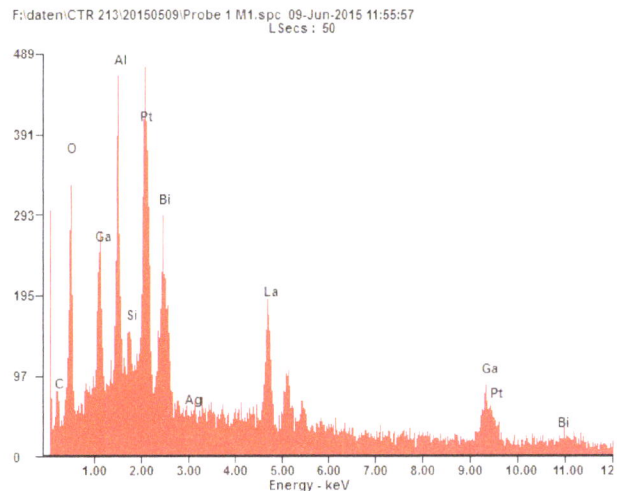

Figure 27. Distribution of elements of the marked area from Fig. 26. The contaminant Bi is detected.

tivation energies of soldering and sintering processes, mostly as a replacement of lead (Pb).

Figure 24 shows a SEM picture of a new chip. The structured metal layer shines bright white, whereas the Langasite shows some minor scattering patterns between the metallic bus bars in the centre of the figure. A FIB scan was performed inside the red marked area, the results are shown in Fig. 25. No bismuth was found.

Figures 26 and 27 show a device after heat treatment and the distribution of elements from the marked area, respectively.

Here, Bi is detected. This indicates that the Bi source is either the housing itself or an assembly component inside it after the seal is applied.

Further FIB scans of housing parts revealed the glass sealant as Bi source. Figure 28 shows the elementary distribution and the corresponding composition in weight and atomic percent.

6 Conclusions and outlook

The limiting factor of the devices considering the maximum temperature was not, as expected, an increase in conductivity of the LTCC leading to an electrical shorting of the signals. It was a contamination of the housing's atmosphere with

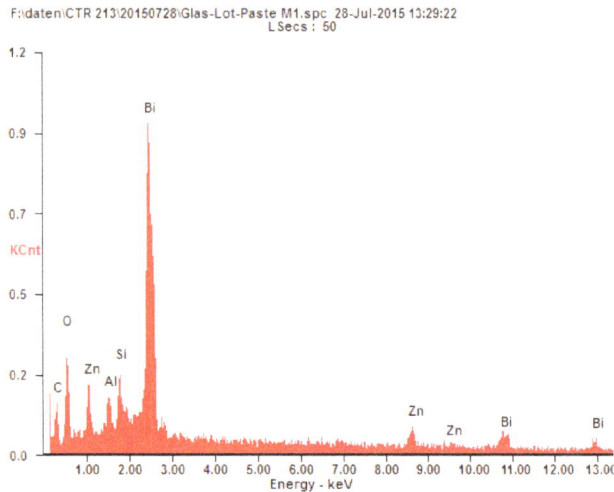

Figure 28. Elementary distribution of glass sealing paste. Bi shows the most exalted peak of all components.

evaporated Bi leading to agglomeration effects of the chip's metal layer and electrical shorting of the IDT fingers. As the sealing containing Bi as a sintering additive to lower the sintering temperature is heated to its glass temperature of about 550 °C, the material becomes more and more liquid, allowing the Bi to enter the gas phase inside the housing. Due to the high volatility of Bi, this element might be able to leave the compound already at lower temperatures. As the temperature profile is always completely executed, the device is treated with temperatures up to 650 °C, allowing the contaminant to leave its containing glass seal area.

This leads to the assumption that without this contaminant, the devices are operable above 450 °C.

Therefore, an alternative sealing strategy has to be tested. To avoid Bi and similar components that are likely to evaporate and contaminate the Pt–Al layer of the chips, metal sealants should be taken into account, e.g. silver–copper (Ag–Cu) based active solders.

The next steps will be tests of several active solders based on Ag and Cu and alternate sealing strategies like ceramic gluing without sintering additives like Bi and Pb. Taking this into account, the devices have potential to be used at temperatures of 500 °C and above.

Acknowledgements. The authors would like to thank M. Gillinger of TU Vienna for providing the effusion measurements.

This work is part of the ENIAC JU project EPPL – Enhanced Power Pilot Line, which is financially supported by Austria, Germany, The Netherlands, France, Italy, Portugal, and the ENIAC Joint Undertaking. The Austrian part of the project is co-funded within the programme "Forschung, Innovation und Technologie für Informationstechnologie" by the Austrian Ministry for Transport, Innovation and Technology (BMVIT).

References

Bardong, J.: Untersuchungen platinbasierter Metallschichtsysteme auf Langasit und Lithiumniobat, Dissertation, FAM – ALU – Freiburg, IMTEK, Dept. of Electrical Instrumentation, 105–116, ISBN: 978-3-86247-377-9, 2013.

Bardong, J., Bruckner, G., Franz, G., Erlacher, A., and Fachberger, R.: Characterisation setup of SAW Devices at high temperatures and ultra high frequencies, IEEE Frequency Control Symposium, Besancon, France, 20–24 April 2009, 28–32, 2009.

Bardong, J., Bruckner, G., and Fachberger, R.: Platinum based metal layer systems for high temperature SAW sensor devices on langasite, Sensoren und Messsysteme 2010 – 15. ITG/GMA Fachtagung, Nürnberg, Germany, 18–19 May, 2010.

Boashash, B.: Time-Frequency Signal Analysis and Processing, Academic Press, Elsevier, London, UK, 2nd Edn., 1056 pp., ISBN 978-0-12-398499-9, 2015.

Butz, T.: Fouriertransformation für Fußgänger, 3rd Edn., Teubner, Stuttgart, Germany, 180 pp., ISBN 3-519-20202-6, 2003.

da Cunha, M. P., Moonlight, T., Lad, R. J., Bernard, G., and Frankel, D. J.: Enabling Very High Temperature Acoustic Wave Devices for Sensor & Frequency Control Applications, Proceedings on IEEE 2007 International Ultrasonics Symposium, 2107–2110, doi:10.1109/ULTSYM.2007.530, 2007.

Hiebel, M.: Grundlagen der vektoriellen Netzwerkanalyse, Rhode&Schwarz GmbH & Co. KG, München, Germany, 1st Edn., PW 0002.3159.00, 2006.

Hornsteiner, J.: Oberflächenwellen – Bauelemente für Hochtemperaturanwendungen: Vom Material zum fertigen Bauteil, Dissertation, TU München, München, 1999.

Malocha, D. C., François-Saint-Cyr, H., Richardson, K., and Robert, C. R.: Measurements of LGS, LGN and LGT Thermal Coefficients of Expansion and Density, IEEE Trans. On Ultrasonics, Ferroelectrics and Frequency Control 49, 50–355, 2002.

Naumenko, N. and Solie, L.: Optimal Cuts of Langasite, La$_3$Ga$_5$SiO$_{14}$ for SAW Devices, IEEE Trans. On Ultrasonics, Ferroelectrics and Frequency Control, 48, 530–537, 2001.

Smith, S.: Digital Signal Processing: A Practical Guide for Engineers and Scientists, Newnes, Oxford, UK, ISBN:0-750674-44-X, 2013.

Toskov, S., Maric, A., Blaz, N., Miskovic, G., and Radosavljevic, G.: Properties of LTCC Dielectric Tape in High Temperature and Water Environment, Int. J. Mater. Mech. Man., 1, 332–336, doi:10.7763/IJMMM.2013.V1.72, 2013

Wall, B., Gruenwald, R., Klein, M., and Bruckner, G.: A 600 °C Wireless and Passive Temperature Sensor Based on Langasite SAW-Resonators, SENSOR 2015, 390–395, doi:10.5162/sensor2015/C3.3, 2015.

Zinke, O. and Brunswig, H.: Hochfrequenztechnik 2 – Elektronik und Signalverarbeitung, 5th Ed. Springer, ISBN 3-540-64728-7, 1999.

26

Platform to develop exhaust gas sensors manufactured by glass-solder-supported joining of sintered yttria-stabilized zirconia

F. Schubert, S. Wollenhaupt, J. Kita, G. Hagen, and R. Moos

University of Bayreuth, Bayreuth Engine Research Center (BERC), Department of
Functional Materials, 95440 Bayreuth, Germany

Correspondence to: F. Schubert (functional.materials@uni-bayreuth.de)

Abstract. A manufacturing process for a planar binary lambda sensor is shown. By joining the heating and the sensing components via glass soldering with a joining temperature of 850 °C, a laboratory platform has been established that allows the manufacturing of two independent parts in high-temperature co-fired ceramics technology (HTCC) with electrodes that are post-processed at lower temperatures, as is required for mixed-potential sensors. The final device is compared to a commercial sensor with respect to its sensing performance. Important processes and possible origins of problems as well as their solutions during sensor development are shown, including heater design and joining process.

1 Introduction

Gas sensors are indispensable parts in modern automotive exhaust gas aftertreatment systems. Driven by stringent emission limits and the need for fuel efficiency and environmental protection, such systems are getting more and more complex. Sensors are required for continuous exhaust gas monitoring to control the combustion process as well as to monitor the conversion of the catalysts (Twigg, 2007).

The most used material in exhaust gas sensing is yttria-stabilized zirconia (YSZ). It provides several advantages like relevant functional properties combined with mechanical strength and high temperature stability to address the requirements arising from application in the harsh exhaust environment (Riegel, 2002; Ramamoorthy et al., 2003; Badwal, 1992). Here, sensors have to withstand temperatures up to 1000 °C; poisoning impacts from oils, ashes, or gases; and mechanical vibrations. As an example, the switching-type lambda probe (or binary lambda probe) shall be briefly introduced in the following to illustrate the major challenges in exhaust gas sensing. Both working principle and manufacturing technology are of relevance for this study and are discussed in the following section.

In general, a lambda probe sensor has to monitor the oxygen content which results from the thermodynamic equilibrium of all gases in the exhaust. If there is oxygen excess in the tailpipe emissions, the combustion is called "lean" and the nitric oxides (NO_x) cannot be converted by classical three-way catalysts (TWCs). In contrast to that, reduced exhaust components (unburned hydrocarbons, HC, or carbon monoxide, CO) prevail in the exhaust in the "rich" engine operation mode. The quotient of the actual air / fuel ratio in the combustion to the stoichiometric ratio is called lambda (λ). An efficient catalytic conversion of NO_x as well as HC and CO in a TWC is only possible when the combustion is controlled to a stoichiometric ratio of $\lambda \approx 1$. The measuring principle of a switching-type lambda probe fits the need to detect deviations from $\lambda = 1$. By comparing the oxygen content in the exhaust with that of a reference atmosphere (e.g., ambient air, $pO_2 \approx 0.2$) via a solid-state oxygen ion conductor, a voltage between exhaust side and reference side can be measured following the Nernst equation. This sensor signal depends directly on the logarithmic quotient of the partial pressures of oxygen in the exhaust and in the reference air. The resulting characteristic curve of the sensor has a binary characteristic. It shows a relatively high voltage of about 1 V in rich atmospheres (large difference between both partial pressures), a voltage below 200 mV in lean conditions, and an almost abrupt transition in between, indicating the stoichiometric point. Such sensors are realized in a planar high-temperature co-fired ceramics technology (HTCC), using YSZ green tapes as base material (Baunach et al., 2006; Riegel, 2002). Functional elements such as electrodes and an internal heating element are manufactured via screen printing using high-temperature-stable noble metal platinum as the conductor material. Pt in that case provides the needed electrical as well as catalytic properties and is stable during the high-temperature sintering of the ceramic substrate at a temperature on the order of 1400 °C. The lambda probe is usually operated at about 700 °C for several reasons: the ceramic carrier must be oxygen ion conducting (the higher the temperature, the better), the sensing electrode – and its covering – must be catalytically active to achieve a thermodynamic equilibrium of all gas components in that local area (the higher the temperature, the better), and soot deposition has to be avoided (soot burns off at about 600 °C). The upper operating temperature is limited by the increasing electronic conductivity of the YSZ to about 1000 °C; see for instance Burke et al. (1971).

Altogether, a YSZ ceramic-based sensor combines all advantages for potentiometric exhaust gas sensors. Therefore, the described lambda probe has been well established for more than 40 years in serial application. On that basis, further developments for example of wide-range lambda sensors, NO_x sensors (Brosha, 2002; Kato et al., 1996; Alkemade and Schumann, 2006) or NH_3 sensors (Wang et al., 2009) were possible and YSZ became the standard material in exhaust gas sensing.

As mentioned above, future exhaust gas aftertreatment systems are getting more complex, and therefore the need for additional gas sensors in that area is increasing. Furthermore, the use of sensors is unavoidable for on-board-diagnostic (OBD) purposes required by law. That means, all emission-relevant parts have to be monitored continuously during operation to indicate a malfunction and trigger a warning to the driver. The sensing challenges will be NO_x (distinguishing between NO and NO_2 at its best), hydrocarbons (HCs), or ammonia (NH_3). Therefore, other sensing principles, e.g., mixed-potential-based systems (Fergus, 2007; Miura et al., 1998, 2014; Schönauer et al., 2011; Wang et al., 2014; Xiong and Kale, 2006; Zosel et al., 2002), are under study for application in the harsh exhaust environment (Guth and Zosel, 2004; Liu et al., 2014; Ménil et al., 2000). Recent studies deal also with novel approaches like pulse polarization methods (Fischer et al., 2010) or direct conversion rate sensors (Hagen et al., 2015a).

The major problem of all these principles is the need to use other electrode materials. In mixed-potential applications, complex alloys, metal-oxide cermets, etc., are of interest (Moos et al., 2009; Miura et al., 2014; Zosel et al., 2004; Zhuikov and Miura, 2007). Such materials may be long-term stable in exhaust atmospheres. But mostly, it is not possible to integrate them into an HTCC setup due to the required high sintering temperatures. Gold for instance, which serves as an electrode in a classical YSZ-based mixed-potential sensor, has a melting point of 1064 °C and cannot be co-fired with zirconia at (typical sintering temperatures of) 1400 °C.

The aim of this contribution is to establish a YSZ-based sensing platform which overcomes this drawback. We present a possibility to manufacture a planar switching-type lambda probe made of two separate produced parts, which are joined together by the new process at moderate temperatures of about 850 °C. This offers the integration of "low" sintering complex electrode materials inside the sensor element (which can be, e.g., in contact with the reference atmosphere).

As the first step we present the manufacturing process for a monolithic sensor device, reproducing the state of the art of a planar switching-type lambda probe. All parts of the sensor are processed in the green state, laminated together and sintered in one step. Here, we could identify the "crux of the matters", meaning key aspects in processing vias, conduction paths, electrical insulation of the heater, lamination parameters, and the firing profile. As a constant temperature is important for accuracy of certain measurement principles (Mahendraprabhu et al., 2013), we integrated a four-wire heating structure that can be controlled to a certain temperature by its four-wire resistance.

For the second part of the work, we divide up the manufacturing process: two parts of the sensor were sintered separately and joined together with a glass ceramic material using lower sintering temperatures. Finally, we compare the "joined" lambda probe with a commercial device in a measurement.

With this setup, a laboratory platform has been established that allows two independent parts to be manufactured in HTCC technology. Later the top side can have electrodes that are post-processed at lower temperatures. As the final step, both parts are joined together at 850 °C.

2 Design and modeling of the new sensor

For better understanding and optimization of the HTCC fabrication process, a switching-type lambda probe was designed in the first step (Fig. 1). The sensor can be divided into the following parts: heater base with the contact pads, where the heater, which was covered by alumina insulation layers, is deposited; heater cover; part with the reference channel; and electrolyte part with screen-printed platinum electrodes on both sides. Each part contains a different number of zirconia layers. Thus the thicknesses of different functional elements can be varied: the thickness of the heater base is 0.4 mm, the heater cover is 0.3 mm, the reference channel is 0.1 mm, and the electrolyte is 0.1 mm thick.

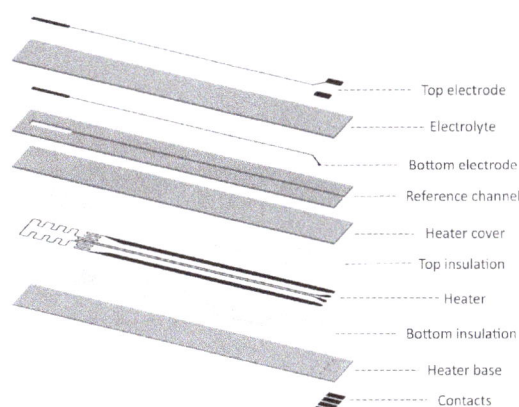

Figure 1. Exploded view of the sensor.

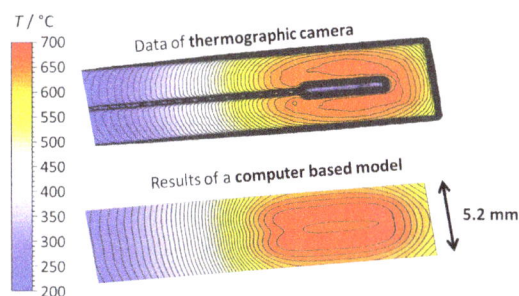

Figure 2. The temperature distribution as calculated by the FEM model and compared with data obtained by infrared camera.

The construction of the heater requires special attention in the here-presented application. The heater should exhibit a homogeneous temperature distribution to maximize the sensor response. Because of the special properties of zirconia, the heater has to be insulated for blackening reasons (Janek, 1999). Therefore, it was insulated from both sides using an alumina thick-film layer. Moreover, we decided to use a four-wire solution. Thus, the precise control of heater resistance and its temperature is possible during heating.

The first design of the heater structure drew upon heater design experience from previous work by Kita et al. (2005). Before realizing the structure, the geometry was optimized using COMSOL Multiphysics®. Properties of materials in the model were fitted to the commercial tape system of Electroscience Lab (ESL) that was chosen for fabricating the sensor (ESL, 2015). It contains all required materials for producing a planar sensor structure, including the 5 mol% yttria-stabilized zirconia tape, an aluminum insulator paste, a platinum conductor paste for the heater and filling vias, and a cermet paste for the electrodes. A sacrificial tape, which is proven to help maintain dimensional stability of the reference channel, is also part of the system (Khoong et al., 2010).

Special attention has been given to the temperature dependency of material data, as stressed by Kita et al. (2015). Therefore, at first the thermal conductivity of zirconia was measured using the LFA 1000 Laser Flash Apparatus (Linseis Messgeräte GmbH, Selb, Germany). Other data for thermal conductivity were taken from the VDI Heat Atlas (2010). Boundary conditions were applied as follows:

– The contact area of the sensor-surrounding temperature was set to 100 °C as the sensor was to be screwed into a heated pipe for testing.

– Boundary conditions for natural convection were applied according to the VDI Heat Atlas (2010).

– Additionally, a thermal radiation was defined with emissivity according to Tanaka et al. (2001).

The optimization results are shown in Fig. 2. The model shows the uniform temperature distribution within the heater. The temperature of the electrode is about 660 °C, with a temperature uniformity better than 10 °C.

After preparation of initial samples, the results of the simulation were compared with an infrared camera measurement of a heated sensor. The outer electrode can be clearly seen due to the lower emissivity of the platinum/cermet paste. The temperature distribution is also in the range of ±5 °C. However, the area of uniform temperature distribution is about 2 mm shorter. This may be a result of the measurement conditions – the sensor had to be measured without protective cover. This increases the convection around the sensor and lowers the temperature of the surrounding gas compared to the model. Nevertheless, the obtained simulation results agree with the data measured in the real samples.

3 Fabrication of the sensor

Platinum structures were screen-printed on unfired YSZ tapes using 325 mesh screens (platinum) or 200 mesh screens (alumina). The widths of the conductor lines in the heater structure were 150 μm; the widths of the feed lines were about 0.9 mm. The insulation for alumina was printed twice to ensure sufficient density of the layer. The alumina layer was set to be 150 μm broader than the conduction paths to allow for some tolerance in the printing process. The reference channel was laser-structured. Heater contacts were deposited on the reverse side of the heater base.

The embedded heater is connected to the contacts on the reverse side of the sensor using vias. Similarly, one of the electrodes on the top needs to be connected by vias. To ensure functionality of the sensor, the vias should be reliably produced. Therefore, a test of the via-filling process was conducted. Pre-laminated green sheets with laser-cut vias were placed on a porous ceramic plate. Vacuum was applied, which caused the deposited paste to be sucked into the via. Two different via-filling methods were tested. In the first attempt, the vacuum was kept low to prevent paste smearing out from the vias at the other side and contaminating the reverse side of the laminate. The vias were filled from both

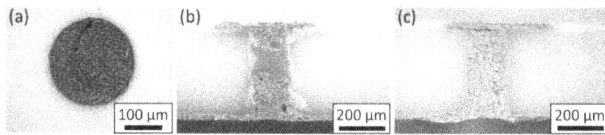

Figure 3. Top view of an unsintered filled via (**a**), cross section of a sintered via filled from both sides (**b**), cross section of a sintered via repeatedly filled from one side (**c**).

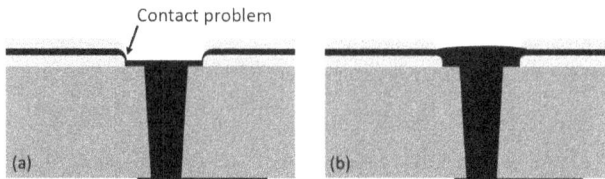

Figure 4. Contact between via and conduction path without (**a**) and with (**b**) additional paste applied.

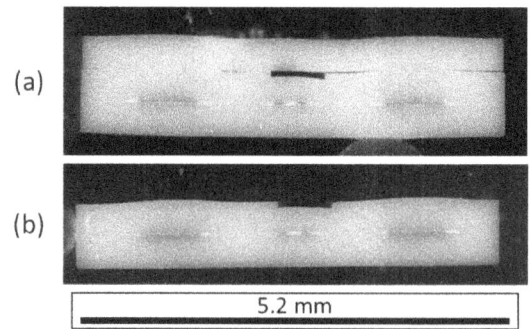

Figure 5. Cross section of a sintered sample with an electrolyte sheet pressed for 12 min at 20 mMPa (**a**) and a sample without electrolyte sheet pressed for 20 min at 25 MPa (**b**).

sides of the laminate. In the second test, the vacuum was increased and the vias were filled repeatedly from one side. Afterwards the residual paste was removed.

Via-filling results are shown in Fig. 3. Generally, filling of the vias in unfired tapes or laminates is more difficult than the via-filling process known from conventional hybrid thick-film technology (Ortolino et al., 2011). The main problem is that, when filling the laminated green sheets, solvents of pastes diffuse rapidly into the green sheets and the paste dries out before filling is completed. Thus, paste from the first attempt very often clogged the via openings and hindered vacuum suction and thus prevented filling the second part of the via. Whereas the via seems to be completely filled (Fig. 3a), the cross section shows unfilled regions (Fig. 3b). Repeatedly filling from one side seems to be a better solution (Fig. 3c). In the first filling procedure, the via walls are covered; in the further steps the vias can be completely filled to achieve full density.

As important as vias are also their connections to the conductive lines. After screen printing of conductive lines, some problems were observed at the edges of the alumina insulation. For a better understanding, the problem was sketched in Fig. 4a. It was found that the relatively thin platinum line placed between thick alumina layers is exposed to high stress during lamination, and the contact between via and conductive line can break up. However, this can easily be remedied when applying an additional drop of paste on top of the vias, as shown in Fig. 4b.

To further increase reliability, the conduction paths have to be improved. The screen-printed alumina layer is generally very rough. Therefore, the minimal line width of the platinum lines should be at least $200 \mu m$ when printed on the dried alumina layer.

One of the key steps to form structural integrity of the sensor is the lamination process. As initial cracks in the ceramic

are detrimental to its strength, special care has to be taken to minimize them, especially in those areas of the heater with its insulation layers.

Generally, two lamination techniques are possible (Imanaka, 2005). For forming cavities, uniaxial pressing is preferred, as the compression of material occurs only on the z axis, where no cavity is to be found between the pressing tools. Thus, cavities can be realized without use of a sacrificial layer if the heater is pre-laminated (Rabe et al., 2007). The other type of lamination is the isostatic one. As the sample is laminated in a liquid, pressure and temperature will be distributed uniformly and thus compensate for the unevenness of the sample. For this reason, isostatic pressing was selected for our experiments. However, when applying this process, cavities have to be filled with a sacrificial tape, which is an additional step of work.

To improve tape handling, each single sensor part was pre-laminated prior to screen printing. Parts with a reference channel and parts with a heater were separately pre-laminated at 10 MPa and 70 °C for 10 min. Applying a relatively low pressure causes single tapes to stick together without achieving final density and stability, meaning that both parts can be further laminated together. Lamination for 12 min at 20 MPa or 20 min at 25 MPa and firing at 1550 °C in a chamber furnace concluded the ceramic process.

Figure 5a shows a cross section of the sintered sensor. Whereas the lamination at the plane of the heater, covered by alumina layers, is very good, lamination of the electrolyte part on an uneven heater required further improvement.

The delamination is visible not only at the edges of the structure but also in the area of the reference channel. The electrolyte sheet made contact only at the protruding areas of the heater. We found that the best results can be achieved if the lamination is divided into three parts: pre-lamination prior to printing, lamination of the heater, and lamination heater and reference channel. When time and temperature of the lamination process were increased, best results were achieved for the heater sheets (Fig. 5b). However, the sacrificial tape in the reference channel is not able to withstand the

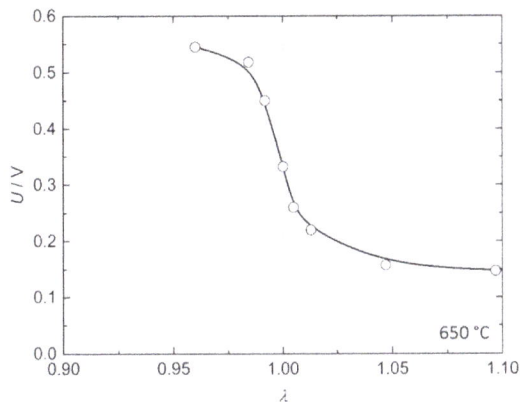

Figure 6. Response of a one-step sintered sensor to lambda variations around the stoichiometric point.

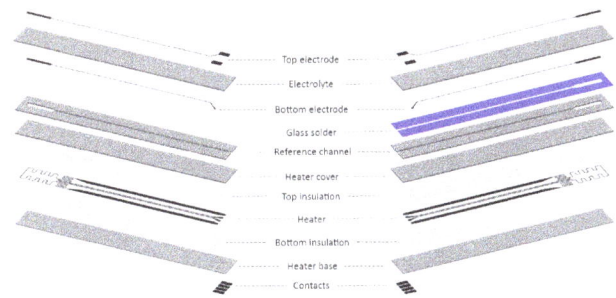

Figure 7. Exploded view of the sensor without (left) and with (right) glass-soldering layer.

pressure and collapses. One solution can be to reduce the insulation layer size and thereby improve the lamination characteristics.

4 First measurement

The prepared one-step sintered sensor was measured at 650 °C. Prior to the measurement, the characteristic curve of the heater temperature vs. the heater resistance was calibrated with a pyrometer. During the measurement, the temperature of the sensor was controlled by the heater resistance. A digital multimeter (Keithley 2700) was used to log the potentials between the electrodes. The gas atmosphere contained CO, H_2, H_2O, and N_2 with varying O_2 contents to adjust λ. A commercial ETAS Lambda Meter was used to measure the λ values. The sensor was mounted onto a gas test bench. To achieve a gas-tight sealing, it was cast integrally into a stainless-steel tube by use of an epoxy resin. A catalyst was applied by brush coating on the electrode surface with a thickness of about $100\,\mu$m to establish the equilibrium even further. The catalyst coating was prepared by impregnating porous Al_2O_3 powder (Baikowski CR10, BET specific surface area: $7\,m^2\,g^{-1}$; agglomerate size d_{90}: $1.25\,\mu$m) with platinum in the form of the platinum complex $[(NH_3)_4\,Pt]\,Cl_2 \cdot H_2O$ (Sigma-Aldrich). The platinum complex was dissolved in deionized water, and the alumina was mixed into the solution, with a mass ratio of 100 : 1 ($m_{alumina} : m_{platinum\,complex}$). Finally, the powder was reduced in a vertical reactor at 500 °C for 1 h under forming gas (5 % H_2 in N_2). For coating, the catalyst was processed to a paste by mixing it with terpineol and ethyl cellulose as an organic binder. The exact properties of the catalyst were determined by Hagen et al. (2015b). Measurements of the sensor are shown in Fig. 6.

The sensor exhibits the desired lambda switch, characterized by the drop of the signal when the partial pressure of oxygen rises over several decades as the atmosphere changes

from rich to lean. The sensor shows a relatively low signal and an offset in the lean area. Further upcoming work will address these issues. However, the general function of the sensor was proven by this test.

5 Joined sensor

Since the results of the one-step sintered sensors were very promising, a modular platform was fabricated in the second step. As mentioned above, such a platform can be utilized to build ample different sensors. To join the two parts of the sensor, we chose a layer of joining glass, a material formulation that has been described by Dev et al. (2014) and Lessing (2007).

For enhanced clarification, the new design is depicted in Fig. 7. The difference between both depicted structures can be seen in the glass-soldering paste that joins the Nernst element to the reference channel instead of sintering them together.

For fabrication of the joining glass paste two different types of glass powder were taken into account (type A and type B). The glass transition point T_G for type A was around 660 °C and for type B around 600 °C. Both powders were sieved with a mesh width of $90\,\mu$m to remove the coarsest particles. Additionally, the glass powder was dried at 80 °C for 8 h.

The solder in the form of a thick-film paste was obtained by mixing commercial recrystallizing glass powder with the thermal expansion coefficient (TCE) adjusted to YSZ ceramics with a binder made of ethyl cellulose and terpineol (ratio 1 : 11) using a three-roll mill (EXAKT).

The prepared paste was screen-printed and leveled for 15 min. In first experiments, the paste was not dried before both sides were joined so that the capillary force could reduce the space between the surfaces and compensate for flatness imperfections that may arise during joining. When the parts adhered, the sensor was dried at 100 °C for 2 h.

Type A was sintered at 1000 °C for 15 min, and type B was sintered for 15 min at 850 °C. Both samples were weighted down to improve adhesion. The resulting samples were cut with a diamond saw and polished to investigate the joining

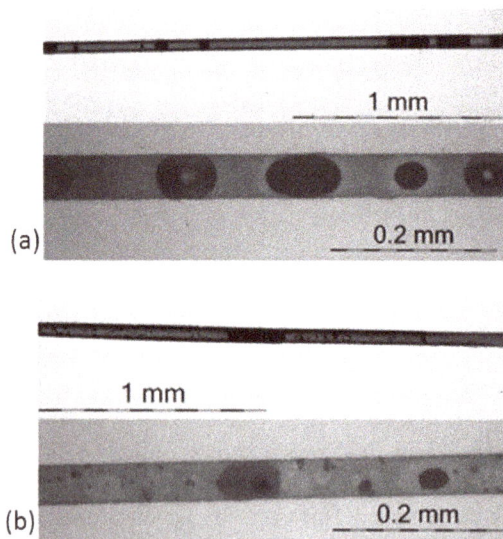

Figure 8. Microscope images of cross sections of the joining layers of type A (**a**) and type B (**b**). The parts were joined when the solders were in the wet state.

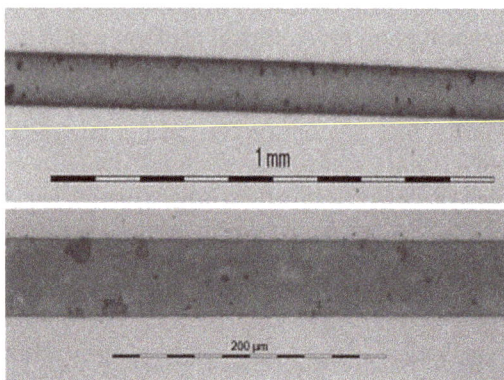

Figure 9. Microscope images of a cross section of the joining layers of type B. The parts were joined when the solder was in the dried state.

layer by a light optical microscope. Figure 8 shows microscope images of cross sections of joined layers for both types. The layers are mechanically stable even when they have been cut for inspection.

Type A exhibits many pores, with sizes ranging from 50 to 500 μm. Type B also shows many pores; however the number of big pores is reduced. Whether this can be attributed to the glass itself is not clear yet. Both glasses seem to be suitable to form a stable joint. Mainly due to its lower sintering temperature, type B was chosen for further investigation as this gives more freedom for different electrode materials.

In the second experiment, the type B binder was dried before joining the parts. With respect to flatness imperfections of the joining surfaces, the thickness of the glass solder was doubled by printing on both joining surfaces. The obtained

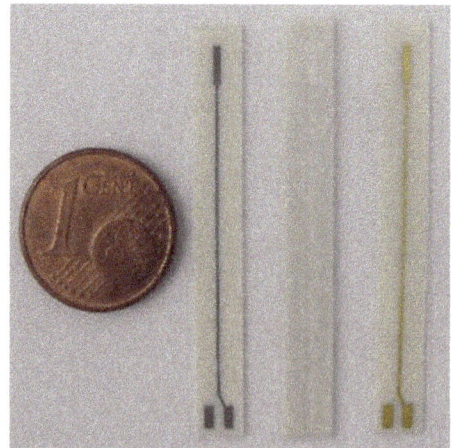

Figure 10. Heating element of a sensor with electrolyte with platinum and gold electrodes.

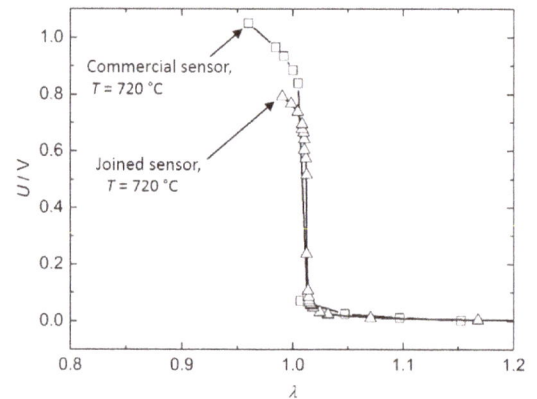

Figure 11. Sensor response of a commercial lambda probe and of the joined sensor. λ was varied around the stoichiometric point and to some extent to lean exhaust.

sensor was checked for tightness by dye penetrant inspection.

As shown in Fig. 9, pores can be greatly reduced in number and size by drying the paste before joining the layers. The pore size is reduced for almost all pores to below 30 μm.

Figure 10 shows sintered parts ready for joining. The platinum electrode is standard for the HTCC process. The gold electrode, however, is only feasible for this design due to the newly developed joining process. Both electrodes can be joined with the heating element with the reference channel.

After the tightness of the joined sensor devices was confirmed by a dye impregnation test, measurements of glass-joined structures were conducted using the same setup as for the one-step sintered sensor. On the electrodes exposed to the test gas, the same platinum catalyst on alumina carrier as mentioned above was applied by brush coating (thickness about 100 μm) to enhance chemical equilibrium kinetics. The sensor response was compared with a commercial type lambda probe with respect to its sensitivity to lambda changes (at around the stoichiometric point at $\lambda = 1$).

Both sensors were operated at approximately 720 °C. The sensor characteristics of the commercial lambda probe and the joined sensor are compared in Fig. 11. As in the case of the one-step sintered sensor, the lambda switch is clearly visible; the commercial type sensor and the joined sensor agree quite well. The lower values for the sensor voltage in rich gases may originate from the more advanced catalyst coating on the electrode of the commercial sensor.

6 Conclusion

In this study we show a method to manufacture a low-temperature-joined lambda sensor. The initial step was to build a monolithic sensor. Special attention was given to the heater layout, where the design process was supported by means of FEM modeling. The resulting sensor showed a uniform temperature distribution in the range of ± 10 °C at 660 °C. During manufacturing of the sensor, several key processes were found to be critical for reliability and were therefore improved. Vias have to be filled completely from one side, ensuring electrical connection. Roughness of the insulation layer is detrimental to the quality of the conduction paths; the reliability can be increased by printing wider conduction paths. At the area of the sensor where conduction paths have to cross the start of the insulation layer, additional paste has to be applied to prevent failure. Tests in gas atmospheres show that the monolithic sensor shows the desired binary lambda behavior. Later on, a low-temperature joining process was applied. It enabled us to seal tightly two parts of the YSZ sensor. Best results for tightness can be obtained if the solder is applied to the parts to be joined and dried before bonding them together. Tests at a gas test bench proved the functionality of the sensor.

It is now possible to vary the material of both electrodes on a switching-type lambda sensor. This may lead to many interesting prospects in future research and application as the variety of possible electrode materials increases drastically with the possibility of using materials with a sintering temperature as low as 850 °C.

References

Alkemade, U. and Schumann, B.: Engines and exhaust after treatment systems for future automotive applications, Solid State Ionics, 177, 2291–2296, doi:10.1016/j.ssi.2006.05.051, 2006.

Badwal, S.: Zirconia-based solid electrolytes: microstructure, stability and ionic conductivity, Solid State Ionics, 52, 23–32, doi:10.1016/0167-2738(92)90088-7, 1992.

Baunach, T., Schänzlin, K., and Diehl, L.: Sauberes Abgas durch Keramiksensoren, Physik Journal, 5, 33–38, 2006.

Brosha, E.: Development of ceramic mixed potential sensors for automotive applications, Solid State Ionics, 148, 61–69, doi:10.1016/S0167-2738(02)00103-0, 2002.

Burke, L. E., Rickert, H., and Steiner R.: Elektrochemische Untersuchungen zur Teilleitfähigkeit, Beweglichkeit und Konzentration der Elektronen und Defektelektronen in dotiertem Zirkondioxid und Thoriumdioxid, Z. Phys. Chem. Neue. Fol., 74, 146–167, 1971.

Dev, B., Walter, M. E., Arkenberg, G. B., and Swartz, S. L.: Mechanical and thermal characterization of a ceramic/glass composite seal for solid oxide fuel cells, J. Power Sources, 245, 958–966, doi:10.1016/j.jpowsour.2013.07.054, 2014.

ESL ElectroScience: Materials for planar oxygen sensors and other HTCC multilayer structures, available at: http://www.electroscience.com/sensormaterials.html, last access: 12 December 2015.

Fergus, J.: Materials for high temperature electrochemical NO_x gas sensors, Sensor Actuat. B-Chem., 121, 652–663, doi:10.1016/j.snb.2006.04.077, 2007.

Fischer, S., Pohle, R., Farber, B., Proch, R., Kaniuk, J., Fleischer, M., and Moos, R.: Method for detection of NO_x in exhaust gases by pulsed discharge measurements using standard zirconia-based lambda sensors, Sensor Actuat. B-Chem., 147, 780–785, doi:10.1016/j.snb.2010.03.092, 2010.

Guth, U. and Zosel, J.: Electrochemical solid electrolyte gas sensors – hydrocarbon and NO_x analysis in exhaust gases, Ionics, 10, 366–377, doi:10.1007/BF02377996, 2004.

Hagen, G., Burger, K., Wiegärtner, S., Schönauer-Kamin, D., and Moos, R.: A mixed potential based sensor that measures directly catalyst conversion – A novel approach for catalyst on-board diagnostics, Sensor Actuat. B-Chem., 217, 158–164, doi:10.1016/j.snb.2014.10.004, 2015a.

Hagen, G., Leupold, N., Wiegärtner, S., and Moos, R.: Sensor Tool for Fast Catalyst Material Light-off Characterization, CAPOC10 – 10th International Congress on Catalysis and Automotive Pollution Control, 28–30 October 2015, Brussels, Belgium, Vol. 2, 283–293, 2015b.

Imanaka, Y.: Multilayered Low Temperature Cofired Ceramics (LTCC) Technology, Springer, New York, USA, 2005.

Janek, J.: Electrochemical blackening of yttria-stabilized zirconia – morphological instability of the moving reaction front, Solid State Ionics, 116, 181–195, doi:10.1016/S0167-2738(98)00415-9, 1999.

Kato, N., Nakagaki, K., and Ina, N.: Thick film ZrO_2 NO_x sensor, SAE Technical Paper, 960334, doi:10.4271/960334, 1996.

Khoong, L. E., Tan, Y. M., and Lam, Y. C.: Overview on fabrication of three-dimensional structures in multi-layer ceramic substrate, J. Eur. Ceram. Soc., 30, 1973–1987, doi:10.1016/j.jeurceramsoc.2010.03.011, 2010.

Kita, J., Rettig, F., Moos, R., Drüe, K.-H., and Thust, H.: Hot Plate Gas Sensors-Are Ceramics Better?, Int. J. Appl. Ceram. Tec., 2, 383–389, doi:10.1111/j.1744-7402.2005.02037.x, 2005.

Kita, J., Engelbrecht, A., Schubert, F., Groß, A., Rettig, F., and Moos, R.: Some practical points to consider with respect to thermal conductivity and electrical resistivity of ceramic substrates for high-temperature gas sensors, Sensor Actuat. B-Chem., 213, 541–546, doi:10.1016/j.snb.2015.01.041, 2015.

Lessing, P. A.: A review of sealing technologies applicable to solid oxide electrolysis cells, Journal of Material Science, 42, 3465–3476, doi:10.1007/s10853-006-0409-9, 2007.

Liu, Y., Parisi, J., Sun, X., and Lei, Y.: Solid-state gas sensors for high temperature applications – a review, J. Mater. Chem., 2, 9919, doi:10.1039/C3TA15008A, 2014.

Mahendraprabhu, K., Miura, N., and Elumalai, P.: Temperature dependence of NO2 sensitivity of YSZ-based mixed potential type sensor attached with NiO sensing electrode, Ionics, 19, 1681–1686, doi:10.1007/s11581-013-0894-1, 2013.

Ménil, F., Coillard, V., and Lucat, C.: Critical review of nitrogen monoxide sensors for exhaust gases of lean burn engines, Sensor Actuat. B-Chem., 67, 1–23, doi:10.1016/S0925-4005(00)00401-9, 2000.

Miura, N., Lu, G., and Yamazoe, N.: High-temperature potentiometric/amperometric NO_x sensors combining stabilized zirconia with mixed-metal oxide electrode, Sensor Actuat. B-Chem., 52, 169–178, doi:10.1016/S0925-4005(98)00270-6, 1998.

Miura, N., Sato, T., Anggraini, S. A., Ikeda, H., and Zhuiykov, S.: A review of mixed-potential type zirconia-based gas sensors, Ionics, 20, 901–925, doi:10.1007/s11581-014-1140-1, 2014.

Moos, R., Sahner, K., Fleischer, M., Guth, U., Barsan, N., and Weimar, U.: Solid state gas sensor research in Germany – a status report, Sensors, 9, 4323–4365, doi:10.3390/s90604323, 2009.

Ortolino, D., Kita, J., Wurm, R., Blum, E., Beart, K., and Moos, R.: Investigation of the short-time high-current behavior of vias manufactured in hybrid thick-film technology, Microelectron. Reliab., 51, 1257–1263, doi:10.1016/j.microrel.2011.02.025, 2011.

Rabe, T., Kuchenbecker, P., Schulz, B., and Schmidt, M.: Hot Embossing: An Alternative Method to Produce Cavities in Ceramic Multilayer, Int. J. Appl. Ceram. Tec., 4, 38–46, doi:10.1111/j.1744-7402.2007.02117.x, 2007.

Ramamoorthy, R., Dutta, P. K., and Akbar, S. A.: Oxygen sensors: Materials, methods, designs and applications, J. Mater. Sci., 38, 4271–4282, doi:10.1023/A:1026370729205, 2003.

Riegel, J.: Exhaust gas sensors for automotive emission control, Solid State Ionics, 152–153, 783–800, doi:10.1016/S0167-2738(02)00329-6, 2002.

Schönauer, D., Nieder, T., Wiesner, K., Fleischer, M., and Moos, R.: Investigation of the electrode effects in mixed potential type ammonia exhaust gas sensors, Solid State Ionics, 192, 38–41, doi:10.1016/j.ssi.2010.03.028, 2011.

Tanaka, H., Sawai, S., Morimoto, K., and Hisano, K.: Measurement of Spectral Emissivity and Thermal Conductivity of Zirconia by Thermal Radiation Calorimetry, J. Therm. Anal. Calorim., 64, 867–872, doi:10.1023/A:1011538022439, 2001.

Twigg, M. V.: Progress and future challenges in controlling automotive exhaust gas emissions, Appl. Catal. B-Environ., 70, 2–15, doi:10.1016/j.apcatb.2006.02.029, 2007.

VDI Heat Atlas, 2nd ed, VDI-Buch, Springer, Berlin, Germany and London, UK, 2010.

Wang, D. Y., Yao, S., Shost, M., Yoo, J.-H., Cabush, D., Racine, D., Cloudt, R., and Willems, F.: Ammonia Sensor for Closed-Loop SCR Control, SAE Int. J. Passeng. Cars – Electron. Electr. Syst., 1, 323–333, doi:10.4271/2008-01-0919, 2009.

Wang, D., Racine, D., Husted, H., and Yao, S.: Sensing Exhaust NO_2 Emissions Using the Mixed Potential Principle, SAE Paper 2014-01-1487, 2014, doi:10.4271/2014-01-1487, 2014.

Xiong, W. and Kale, G. M.: Novel high-selectivity NO2 sensor incorporating mixed-oxide electrode, Sensor Actuat. B-Chem., 114, 101–108, doi:10.1016/j.snb.2005.04.010, 2006.

Zhuiykov, S. and Miura, N.: Development of zirconia-based potentiometric NO_x sensors for automotive and energy industries in the early 21st century: What are the prospects for sensors?, Sensor Actuat. B-Chem., 121, 639–651, doi:10.1016/j.snb.2006.03.044, 2007.

Zosel, J., Westphal, D., Jakobs, S., Müller, R., and Guth, U.: Au–oxide composites as HC-sensitive electrode material for mixed potential gas sensors, Solid State Ionics, 152–153, 525–529, doi:10.1016/S0167-2738(02)00355-7, 2002.

Zosel, J., Ahlborn, K., Müller, R., Westphal, D., Vashook, V., and Guth, U.: Selectivity of HC-sensitive electrode materials for mixed potential gas sensors, Solid State Ionics, 169, 115–119, doi:10.1016/S0167-2738(03)00082-1, 2004.

27

Qualification concept for optical multi-scale multi-sensor systems

A. Loderer and T. Hausotte

Institute of Manufacturing Metrology, Friedrich-Alexander-Universität Erlangen-Nürnberg (FAU), Erlangen, Germany

Correspondence to: A. Loderer (andreas.loderer@fau.de)

Abstract. This article describes a new qualification concept for dimensional measurements on optical measuring systems. Using the example of a prototypical multi-scale multi-sensor fringe projection system for production-related inspections of sheet-bulk metal-formed parts, current measuring procedures of the optical system are introduced. Out of the shown procedures' deficiencies, a new concept is developed for determining the orientations and positions of the sensors' measuring ranges in a common coordinate system. The principle element of the concept is a newly developed flexible reference artefact, adapted to the measuring task of the fringe projection system. Due to its dull surface, the artefact is optimized for optical measuring systems, like the used fringe projection sensors. By measuring the reference artefact with each fringe projection sensor and aligning the resulting data sets on a digital reference model of the artefact, sensor-specific transformation matrices can be calculated which allow transformation of the sensors' data sets into a common coordinate system, without the need for any overlapping areas. This approach is concluded in an automated measuring procedure, using alignment algorithms from commercial available software where necessary. With the automated measuring procedure, geometrical relations between individual measured features can be determined and dimensional measuring beyond the measuring range of a sensor became possible. Due to a series of experiments, the advantages of the new qualification concept in comparison with the current measuring procedures are finally revealed.

1 Introduction

New production technologies, like sheet-bulk metal forming (Schaper et al., 2011), involve new challenges for dimensional measurements of the manufactured parts. In the case of sheet-bulk metal forming, metrological requirements of a production-related inspection arise from the short cycle time, the complex and filigree geometry, and varying surface roughness due to the high, irregularly distributed forming forces (Merklein et al., 2012). The challenges of inspecting complex workpieces can be explained by considering the "golden rule of measuring metrology" (Berndt et al., 1968). In 1968 Georg Berndt developed a rule for selecting appropriate measurement systems. Therefore, the measurement uncertainties of the measurement systems have to be known. Following the recommendation of the golden rule, the measurement uncertainty should be at least less than a fifth, and better less than a tenth, of the tolerance width. If

this minimum requirement can be met, it ensures that the measurement results are accurate enough (Loderer et al., 2013). To achieve these requirements, a prototype of a multi-scale multi-sensor fringe projection system was developed, designed for a production-related environment (see Fig. 1).

The main parts of the prototype systems are three different types of fringe projection sensors with varying measuring ranges and resolutions (see Table 1).

To get an overview of the workpiece and also to measure large features simultaneously, an exchangeable fringe projection sensor with a measuring range of the size of the workpiece is installed (Ohrt et al., 2012). For the measurement of filigree elements, two other types of fringe projection sensors are used as detail sensors which can be arranged around the workpiece. Each of these two sensors captures only one feature, but at a resolution adapted to the feature's size. Whereas there is only one overview sensor available, there are up to

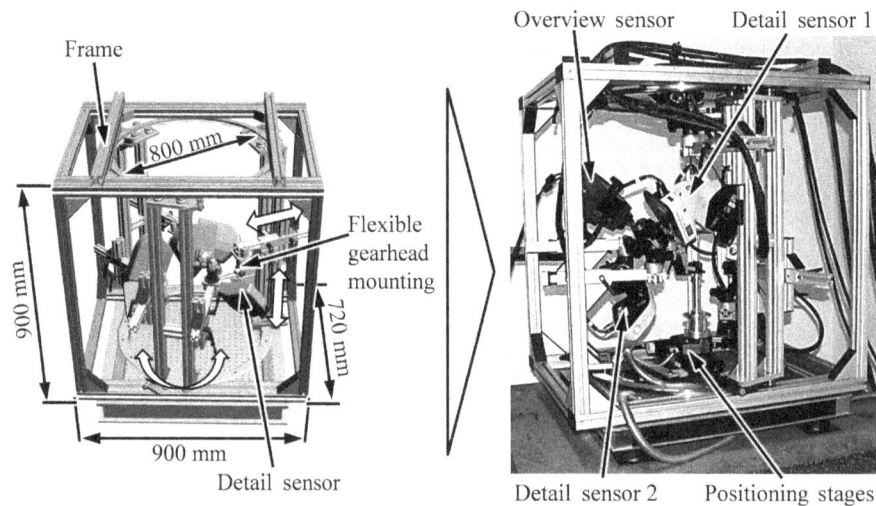

Figure 1. Technical sketch of the idea without an overview sensor (left side) and realized set-up of the prototypical multi-scale multi-sensor fringe projection system (right side).

Table 1. Technical specifications of the fringe projection sensors.

Sensor type	Available	Measuring range in mm	Resolution in μm
Overview sensor (GOM ATOS Compact Scan 2M)	1 $\times$	$115 \times 88 \times 92$	80 (mean point spacing)
Detail sensor 1 (GFM MicroCAD 1.0 μm)	8 $\times$	$13 \times 10 \times 3$	17 (lateral); 1 (vertical)
Detail sensor 2 (GFM MicroCAD 0.3 μm)	4 $\times$	$4 \times 3 \times 1$	2.5 (lateral); 0.3 (vertical)

eight detail sensors 1 and up to four detail sensors 2 (Loderer et al., 2015).

The process of gathering measuring results out of multi-scale data sets is divided into four main steps (see Fig. 2): firstly, measurements were done by all selected fringe projection sensors automatically. Then the sensors' data sets have to be transformed into a common coordinate system and combined into one holistic data set by a merging process. For this purpose the data sets are roughly aligned in a coarse registration by selecting corresponding points in each data set manually. Only if there is a large overlapping area with at least one significant feature can an automatic algorithm, e.g. presented in Shaw et al. (2013), be considered, but often this requirement cannot be met. In contrast to this for the following fine registration, various automatic algorithms are available. For the presented procedures, the best-fit algorithm of the commercial available Polyworks IMInspect 2014 software is used, which provides numerous settings for multi-data-set alignments. Next to point cloud data sets, polygonal models can also be aligned. The last step of the standard measuring procedure is the evaluation which can also be performed automatically. That procedure requires at least a small overlapping area. However, even if enough cor-

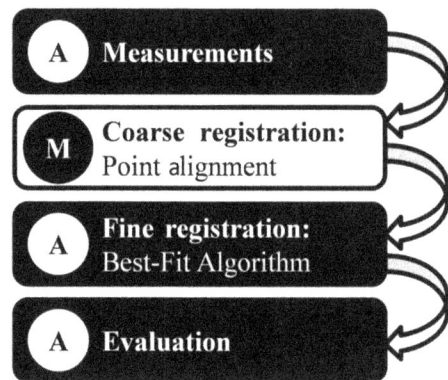

Figure 2. Automatic (A) and manual (M) steps of the standard measuring procedure.

responding points are available, the procedure is neither fast nor accurate enough to benefit from the high accuracy of the fringe projection sensors, as can be seen in the results presented in Fig. 8.

In order to get a reliable geometrical relation between individual measured features and the ability to transform data sets without overlapping areas into a common coordinate

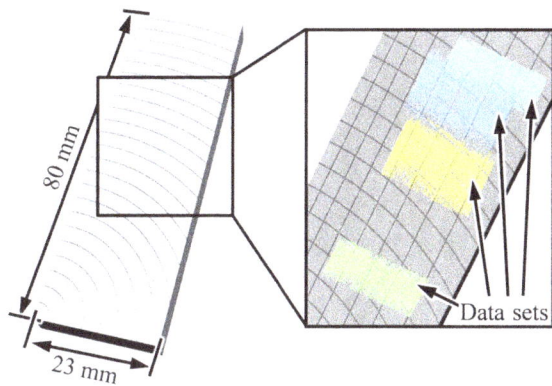

Maximum qualification field size: ≈ 1.800 mm²

Figure 3. Flat reference artefact for testing the qualification principle.

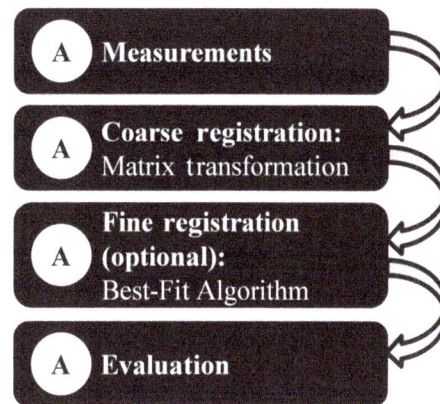

Figure 4. Automated (A) measuring procedure by using transformation matrices.

system, a qualification concept, adapted to the properties of the prototypical multi-sensor multi-scale measuring system, has to be developed.

For the underlying research the shown measuring procedures have been selected as the most suitable approaches for further adaptations of the multi-scale multi-sensor fringe projection system. Besides the explained four steps, there are other approaches for combining measurement data, e.g. presented in Puente León and Kammel (2003), Komander et al. (2014) and Keck et al. (2014).

2 Qualification principle

To prove the basic idea of the qualification of optical multi-scale multi-sensor systems, a flat reference artefact was used (see Fig. 3). On the reference artefact, surface lines of differing distances as well as radii of differing sizes were milled in, and thereby a unique surface structure was created (Kästner et al., 2013).

The measuring ranges of the considered optical sensors are significantly smaller than the artefact's size. Setting up a multi-sensor measurement, the sensors' measuring ranges are positioned onto a measuring object. Subsequently, the measuring object is replaced by the reference artefact. Each sensor now measures a part of the reference artefact's surface and, due to the unique surface structure, the position of each data set can be allocated. By a manually coarse registration using point alignments and a following automatically fine registration, each data set can be aligned to a CAD (computer-aided design) model of the reference artefact. All necessary transformations to get the data sets in the correct positions can be expressed in transformation matrices. These matrices represent the sensor orientations and have to be saved. Replacing the reference artefact by a measuring object, the data sets of each sensor can be transformed in the correct position again by using the transformation matrices of the sensor. If the measuring range of a fringe projection

sensor is changed, the qualification procedure has to be done once more.

With the flat reference artefact, a qualification field size of about 1800 mm² can be used which is equal to the surface size. Due to the flat design, only a lateral qualification is possible, whereas all sensors have a similar vertical position.

The important advantage of the qualification principle is the loss of need for corresponding areas. Even data sets that do not overlap can be located correctly, and thereby dimensional measurements with optical multi-scale multi-sensor systems are enabled. Moreover, the time-consuming manual coarse registration has to be done only in the qualification procedure. Once all transformation matrices are available, the steps of the measuring procedure run automatically (see Fig. 4).

3 Flexible qualification concept

A crucial disadvantage is the flat shape of the reference artefact. Sheet-bulk metal-formed objects and the prototypical multi-scale multi-sensor fringe projection system designed for measuring often are of round shapes with varying diameters (see Fig. 5) (Merklein et al., 2015). With a flat reference artefact, the fringe projection sensors can only be qualified if their measuring ranges are positioned at the same height and oriented similarly. However, complex features like cylinders require differently positioned sensors with differing heights of their measuring ranges. The flat reference artefact is not capable of fulfilling these demands.

Thus, a flexible qualification concept was worked out to also allow dimensional measurement of complex features by using optical multi-sensor systems. This concept is mainly based on a new flexible and, adapted to the demands of sheet-bulk metal forming, reference artefact (see Fig. 6). The basic principle of the reference artefact, which is a unique surface structure as well, can be found on cylindrical "reference heads". These heads are mounted on "adjustment arms",

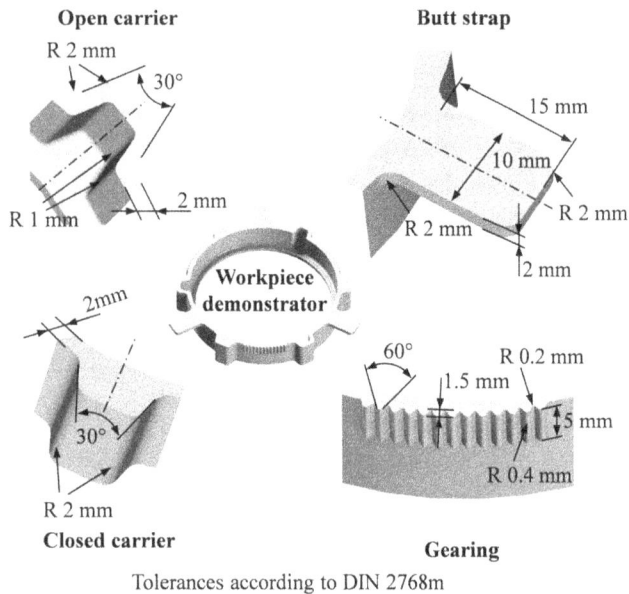

Figure 5. Workpiece demonstrator of sheet-bulk metal-forming processes with its differing sizes and shapes.

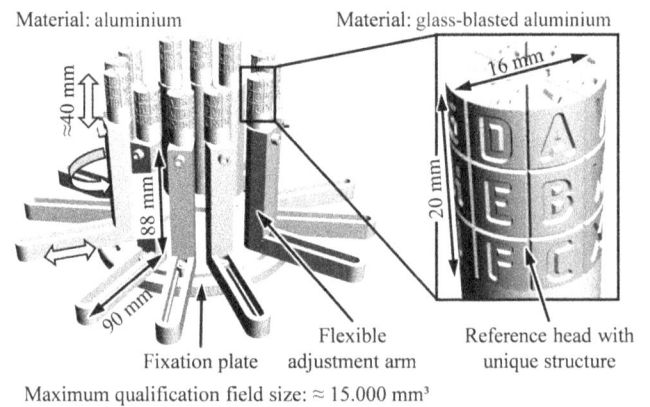

Figure 6. Flexible reference artefact adapted on sheet-bulk metal-formed parts.

which are adjustable in the lateral and vertical directions. Thereby, measuring ranges do not have to be set up at the same height, but rather can be oriented freely. In order to optimize the reference heads for optical measuring systems, the surfaces are glass-blasted to generate dull and very measurable surface structures.

With the flexible reference artefact, a qualification field size of about $15\,000\,\mathrm{mm}^3$ can be used. Due to the vertical adjustment of the reference heads, a lateral as well as vertical qualification is possible.

According to the qualification concept, ten main steps have to be considered when setting up a complete multi-sensor measurement (see Fig. 7). Firstly, the measuring ranges of the fringe projection sensors have to be positioned on the measuring object. Then the measuring object is replaced with the reference artefact and the reference heads are positioned into the measuring ranges. At least one reference head has to be inside the measuring range of each sensor. In the next step, measurements of the fringe projection sensors are triggered. To generate a reference polygonal model of the reference artefact, it is digitized by using an optical sensor with a bigger measuring range. The quality of the measurement with the overview sensor is crucial for the qualification concept. The more accurate the overview sensor and the higher the quality of the digitization, the more accurate the qualification procedure's result. In the shown experiments, the overview sensor is used to digitize the complete reference artefact.

Next the data sets of the fringe projection sensors are aligned to the digital reference polygonal model of the reference artefact. From these alignments, transformation matri-

ces for each data set are calculated, which express the orientation of each fringe projection sensor in a common coordinate system. The qualification procedure finishes by replacing the reference artefact with the measuring object. With measurements of the measuring object, the following measuring procedure starts. Using the transformation matrices, the data sets of the fringe projection sensors can be transformed into the common coordinate system and combined into one common data set by a merging process. By repeating the measuring procedure, this qualification concept enables holistic dimensional measurements of complex features.

4 Comparison

To detect the advantages provided by the developed qualification concept, a comparison between the standard measuring procedure and the automated measuring procedure with the new qualification concept is worked out.

Therefore, the height (1 mm) and width (3 mm) of a step height standard have to be measured (see Fig. 8). This measuring task represents the need for multi-sensor measurements: due to optical effects like shadowing and technical limitations, e.g. the maximum thread angle, the measurement of both parameters by using only one fringe projection sensor is not possible. Only by changing the position of the sensor or step height standard and performing more measurements can the features be tediously detected by a single sensor. For a fast and reliable measurement, more sensors with differing measuring positions are needed. In order to compare both measuring procedures, the deviations of height and width of the qualified values are considered as parameters. The calculation of heights and width is done with Polyworks IMInspect 2014. A consideration of DIN EN ISO 5436-1 for calculating step heights is not possible due to software restrictions. Contrary to the standard, Polyworks IMInspect 2014 calculates two Gaussian plains and evaluates the vectorial distance between both as the step height. This approach is not

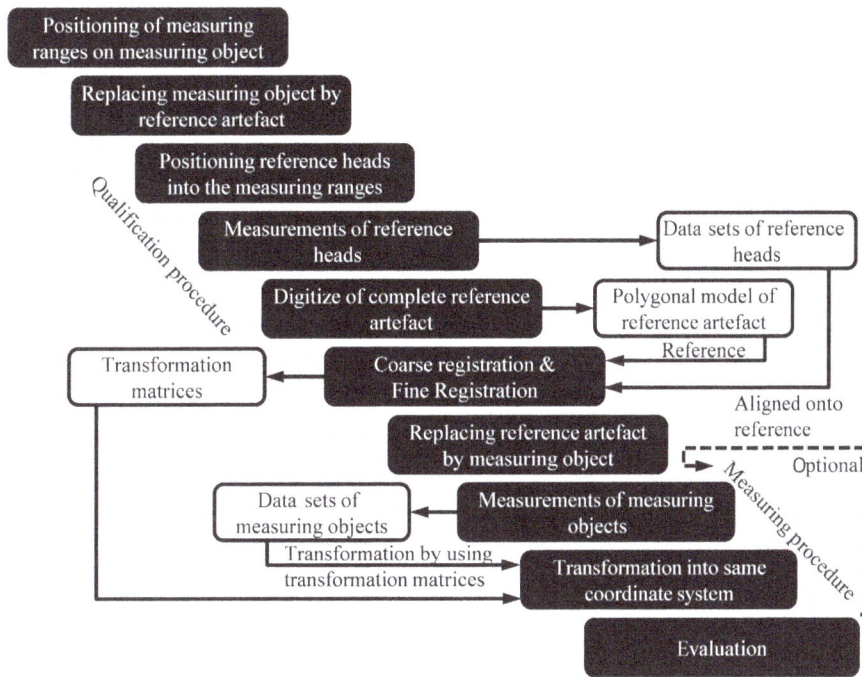

Figure 7. Complete steps for the qualification and measuring procedure.

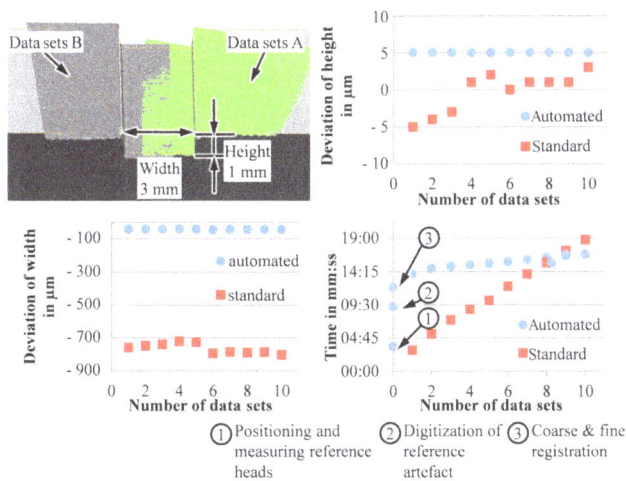

Figure 8. Considered parameters for comparing the standard measuring procedure (standard) and the automated measuring procedure (automated) with the new qualification concept (upper left corner). Results for deviations of height (upper right corner), width (lower left corner) and comparison of needed time (lower right corner).

standardized and only generates valid results when the two plains are parallel. Due to the use of a step height standard, the parallelism is ensured, and thus the Polyworks IMInspect 2014 approach is permissible.

In addition to the deviations of height and width of the qualified values, the time needed for performing all necessary steps is evaluated, too.

The results for the deviation of width show a significant difference between the automated measuring procedure with the corresponding new qualification concept and the former standard measuring procedure. Whereas the measuring results, gathered by using the standard measuring procedure, are between 0.7 and 0.8 mm smaller than the qualified value, the deviation averages 0.42 μm using the automated measuring procedure. Considering the deviation of height, there is also a difference between both procedures. The deviation averages 5.0 μm for the automated procedure and -0.3 μm for the standard procedure. Although this difference seems to be small, its statistical significance is proven by using a Student's t test. However, focusing on the results' distributions, the reliability of the automated measuring procedure becomes obvious. The automated procedure provides continuously the same value for results, which is caused by using the same sensors' transformation matrices for all ten data sets, whereas the values of the standard procedure are spread between 5 and 3 μm.

Comparing the duration needed for performing all required steps, the automated measuring procedure takes less time. Even though the detected difference is only about 2 min, the automated process is of benefit the more data sets are used.

When performing the standard measuring process, the needed time increases nearly linearly, and one time step rep-

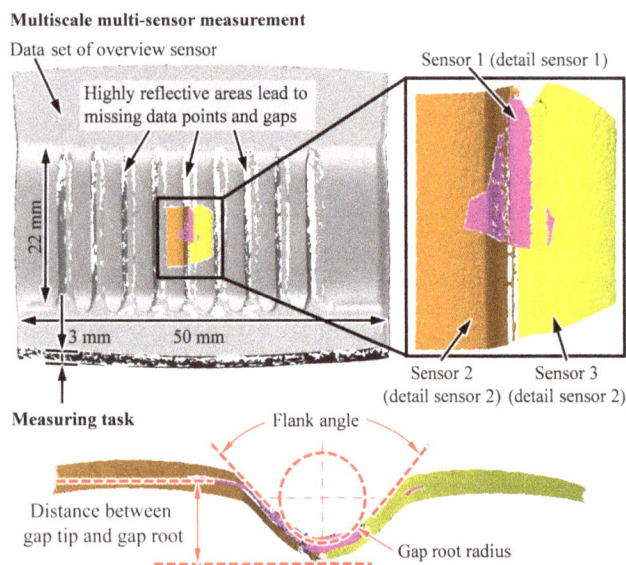

Figure 9. Multi-scale multi-sensor measurement of a sheet-bulk metal-formed multiple gap structure (upper side) and the corresponding measuring task (lower side).

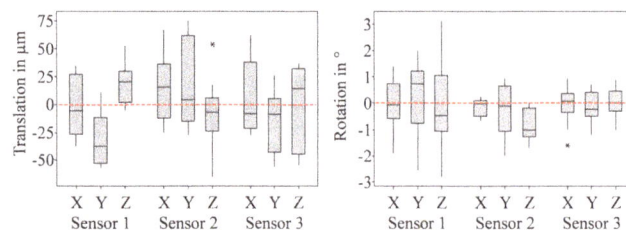

Figure 10. Ranges of data sets' translations and rotations done by a manually coarse registration (boxplots) in comparison to best-fit fine registration (pointed line).

resents one completed data set. In contrast, the most time-consuming steps of the automated procedure are steps 1 to 3. These steps belong to the qualification procedure and include the positioning and measuring of the reference heads, the digitalization of the reference artefact and the registration of the reference heads' data sets in order to calculate the transformation matrices. Once the matrices are available and a script for an automated measuring process is created, which is done in the fourth step, the needed time for the following data sets is significantly shorter.

5   Application

With the comparison of the standard and automated measuring procedures by using the qualified step height standard, the advantages of the automated measuring procedure could be shown. In order to prove the advantages in a measuring task similar to a task for which the multi-scale multi-sensor fringe projection system was developed, the automated measuring procedure is applied in an inspection of a sheet-bulk metal-formed part.

With a process of the sheet-bulk metal forming, a multiple-gap structure is formed in DC04 sheet metal (see Fig. 9). In order to have precise data for evaluating and further improving the process, a holistic detection of the relevant middle section of the multiple-gap structure is necessary. Such detection is not possible by using only one conventional fringe projection sensor with a large enough measuring range, e.g. the overview sensor. The smooth and thereby highly reflective surface of the formed section in combination with the structure's flank angle leads to missing data points and gaps

in the data set (Loderer et al., 2015). Moreover, due to small dimensions of the multiple-gap structure, the fringe projection sensor's resolution would be not accurate enough. Using only one fringe projection sensor with a smaller measuring range but instead an appropriate resolution, the whole relevant section cannot be detected at once.

This conflict leads to a multi-scale multi-sensor measuring set-up, similar to the measurement of the step height standard, where one side of the structure is detected by one separate sensor of the type of detail sensor 2 and the gap root's detection is done by a third, more accurate sensor of the type of detail sensor 1. In order to allow dimensional measurements of flank angle, gap root radius and distance between gap tip and gap root, a qualification procedure has to ensure a precise and also robust alignment of the three resulting data sets.

To evaluate the stability and robustness of the automated measuring procedure, one data set of each sensor is run through the procedure 10 times, again using Polyworks IMInspect 2014. This means each data set is aligned manually onto a reference data set of the reference artefact for a coarse registration, and subsequently a fine registration by Polyworks IMInspect's best-fit algorithm is done. The resulting transformation matrices were used to transform the data sets correctly in a common coordinate system and the feature evaluation was done finally. Due to using the same data set for each sensor 10 times, the same results for the evaluated features should always be calculated. If there are any variations, these can be clearly matched as deviations in the automated measuring procedure.

The large spread of the manually coarse registration becomes obvious when comparing the data sets' translations and rotations to the translations and rotations of the best-fit algorithm, which were considered as a reference (see Fig. 10). Picking corresponding points in the sensors' data sets and in the reference artefact's reference data sets in the manually coarse registration leads to differing translations and rotations. Obviously, by manual registration only, a stable alignment is not ensured. Nevertheless, when performing a best-fit fine registration after the manually coarse registration, the spreads can be eliminated. Thereby the best-fit fine registration always provides the same translation and rota-

Table 2. Results of feature evaluation.

Feature	Result	Variation	Repetitions
Flank angle	90.83°	0.00°	10
Gap root radius	0.819 mm	0.000 mm	10
Distance: gap tip–gap root	0.825 mm	0.000 mm	10

tion in the ten repetitions without any spread, whatever the manual input registration deviation was.

After the fine registration, the data sets were aligned and the features could be evaluated (see Table 2). Due to always having the same values for translations and rotations for each sensor data set of the best-fit algorithm, the results for the considered features are also identical every time without any spread or deviation. In this manner the robustness of the automated measuring process could be proven in a measuring task for which the multi-scale multi-sensor fringe projection system was designed. In a real inspection set-up, the manual registration step is necessary only once in the qualification process for allowing a subsequently fine registration. With the transformation matrices generated then, the data sets of the following measurements are aligned correctly and no manual interaction is required.

6 Conclusion

In this article a new qualification concept, optimized for optical multi-scale multi-sensor measuring systems, was introduced, which allows dimensional measurements of features larger than the measuring range of the optical sensor. The basic principle was proven by a flat-shaped reference artefact using the example of a prototypical multi-scale multi-sensor fringe projection system. A unique surface structure of the reference artefact ensures that the measured data sets can be aligned on a polygonal reference model, and thus transformation matrices for each fringe projection sensor can be calculated. These matrices contain the positions and orientations of each sensor, expressed in a common coordinate system. Thereby the correct transformation of all measurement data sets in a common coordinate system is enabled in order to generate a holistic data set. The basic principle was transferred to a flexible reference artefact, adapted on the shape of sheet-bulk metal-formed parts for whose inspection the prototypical multi-scale multi-sensor measuring system was designed. Together with the new qualification and measuring procedures, an automated and reliable measurement of complex workpieces is possible now. Comparing the new qualification concept with the former standard measuring procedure by setting up a series of experiments, the gathered advantages become obvious.

In order to test the automated measuring procedure, which contains the new qualification concept, a measuring task of a sheet-bulk metal-formed multiple gap structure was set up.

Here, large variations of the manually coarse registration became obvious, which were corrected by the subsequent best-fit fine registration. By repeating the qualification procedure 10 times with the same data sets, variations in the features' results can be matched to deviations in the procedure itself. However, the final feature evaluation provides the same result for each feature in every trail, and thus the stability and robustness of the qualification procedure could be proven for dimensional measurements with optical multi-scale multi-sensor measuring systems.

Acknowledgements. The authors are grateful to the German Research Foundation (DFG) for supporting the investigations in research project Manufacturing of complex functional components with variants by using a new metal forming process – Sheet-Bulk metal forming (SFB/TR 73; online: https://www.tr-73.de). In addition, special thanks should be expressed to Michael Harant of the Friedrich-Alexander-Universität Erlangen-Nürnberg for his contribution, as well as to Thomas Schneider, Daniel Gröbel, Philipp Hildenbrand, and Johannes Koch from the Institute of Manufacturing Technology of the Friedrich-Alexander-Universität Erlangen-Nürnberg, and to Michael Gröne from the Institute of Forming Technology and Machines of the Leibniz Universität Hannover.

References

Berndt, G., Hultzsch, E., and Weinhild, H.: Funktionstoleranz und Meßunsicherheit, Wissenschaftliche Zeitschrift der Technischen Universität Dresden, 17, 465–471, 1968.

Kästner, M., Hausotte, T., Reithmeier, E., Loderer, A., Ohrt, C., and Sieczkarek, P.: Fertigungsnahe Qualitätskontrolle von Werkzeug und Werkstück, Tagungsband zum 2. Erlanger Workshop Blechmassivumformung 101–118, 2013.

Keck, A., Böhm, M., Knierim, K. L., Sawodny, O., Gronle, M., Lyda, W., and Osten, W.: Multisensorisches Messsystem zur dreidimensionalen Inspektion technischer Oberflächen, Technisches Messen, 81, 280–288, 2014.

Komander, B., Lorenz, D., Fischer, M., Petz, M., and Tutsch, R.: Data fusion of surface normals and point coordinates for deflectometric measurements, J. Sens. Sens. Syst., 3, 281–290, doi:10.5194/jsss-3-281-2014, 2014.

Loderer, A., Galovskyi, B., Hartmann, W., and Hausotte, T.: Measurement strategy for a production-related multi-scale inspection of formed work pieces, Proceedings of the 11th Global Conference on Sustainable Manufacturing – GCSM 2013, 23–25 September 2013, Berlin, 148–153, 2013.

Loderer, A., Timmermann, M., Matthias, S., Kästner, M., Schneider, T., Hausotte, T., and Reithmeier, E.: Measuring systems for sheet-bulk metal forming, Key Engineering Materials, 639, 291–298, 2015.

Merklein, M., Allwood, J. M., Behrens, B.-A., Brosius, A., Hagenah, H., Kuzmann, K., Mori, K., Tekkaya, A. E., and Weckenmann, A.: Bulk forming of sheet metal, Annals of the CIRP, 61, 725–745, 2012.

Merklein, M., Gröbel, D., Löffler, M., Schneider, T., and Hildenbrand, P.: Sheet-bulk metal forming forming of functional components from sheet metals, Proceedings of the 4th International Conference on New Forming Technology, MATEC Web of Conferences, 01001, 1–12, 2015.

Ohrt, C., Hartmann, W., Kästner, M., Weckenmann, A., Hausotte, T., and Reithmeier, E.: Holistic measurement in the sheet-bulk metal forming process with fringe projection, Key Engineering Materials, 504, 1005–1010, 2012.

Puente León, F. and Kammel, S.: Image fusion techniques for robust inspection of specular surfaces, in: Multisensor, Multisource Information Fusion: Architectures, Algorithms and Applications, edited by: Dasarathy, B. V., Proceedings of SPIE, 5099, 77–86, 2003.

Schaper, M., Lizunkova, Y., Vucetic, M., Cahyono, T., Hetzner, H., Opel, S., Schneider, T., Koch, J. and Plugge, B.: Sheet-bulk Metal Forming - A New Process for the Production of Sheet Metal Parts with Functional Components, Metallurgical and Mining Industry, 7, 53–58, 2011.

Shaw, L., Ettl, S., Mehari, F., Weckenmann, A., and Häusler, G.: Automatic registration method for multisensor datasets adopted for dimensional measurements on cutting tools, Measurement Science and Technology, 24, 8 pp., 2013.

Permissions

All chapters in this book were first published in JSSS, by Copernicus Publications; hereby published with permission under the Creative Commons Attribution License or equivalent. Every chapter published in this book has been scrutinized by our experts. Their significance has been extensively debated. The topics covered herein carry significant findings which will fuel the growth of the discipline. They may even be implemented as practical applications or may be referred to as a beginning point for another development.

The contributors of this book come from diverse backgrounds, making this book a truly international effort. This book will bring forth new frontiers with its revolutionizing research information and detailed analysis of the nascent developments around the world.

We would like to thank all the contributing authors for lending their expertise to make the book truly unique. They have played a crucial role in the development of this book. Without their invaluable contributions this book wouldn't have been possible. They have made vital efforts to compile up to date information on the varied aspects of this subject to make this book a valuable addition to the collection of many professionals and students.

This book was conceptualized with the vision of imparting up-to-date information and advanced data in this field. To ensure the same, a matchless editorial board was set up. Every individual on the board went through rigorous rounds of assessment to prove their worth. After which they invested a large part of their time researching and compiling the most relevant data for our readers.

The editorial board has been involved in producing this book since its inception. They have spent rigorous hours researching and exploring the diverse topics which have resulted in the successful publishing of this book. They have passed on their knowledge of decades through this book. To expedite this challenging task, the publisher supported the team at every step. A small team of assistant editors was also appointed to further simplify the editing procedure and attain best results for the readers.

Apart from the editorial board, the designing team has also invested a significant amount of their time in understanding the subject and creating the most relevant covers. They scrutinized every image to scout for the most suitable representation of the subject and create an appropriate cover for the book.

The publishing team has been an ardent support to the editorial, designing and production team. Their endless efforts to recruit the best for this project, has resulted in the accomplishment of this book. They are a veteran in the field of academics and their pool of knowledge is as vast as their experience in printing. Their expertise and guidance has proved useful at every step. Their uncompromising quality standards have made this book an exceptional effort. Their encouragement from time to time has been an inspiration for everyone.

The publisher and the editorial board hope that this book will prove to be a valuable piece of knowledge for researchers, students, practitioners and scholars across the globe.

List of Contributors

S. Baldo and L. Tripodi
Istituto per la Microelettronica e Microsistemi, CNR, VIII Strada 5, 95121, Catania, Italy
Dipartimento di Fisica e Astronomia, Università degli Studi di Catania, Via S. Sofia, 95125, Catania, Italy

V. Scuderi, A. La Magna and S. Scalese
Istituto per la Microelettronica e Microsistemi, CNR, VIII Strada 5, 95121, Catania, Italy

S.G. Leonardi, N. Donato and G. Neri
Dipartimento di Ingegneria Elettronica, Chimica e Ingegneria Industriale, Università degli Studi di Messina, Contrada di Dio, Salita Sperone 31, Messina, Italy

S. Filice
Dipartimento di Chimica e Tecnologie Chimiche, Università della Calabria, Via P. Bucci, cubo 14/D, 87036, Arcavacata di Rende (CS), Italy

M. Klenner, T. Abels, C. Zech, A. Hülsmann and M. Schlechtweg
Fraunhofer Institute for Applied Solid State Physics (IAF), Freiburg im Breisgau, Germany

O. Ambacher
Fraunhofer Institute for Applied Solid State Physics (IAF), Freiburg im Breisgau, Germany
Department of Microsystems Engineering (IMTEK), University of Freiburg, Freiburg im Breisgau, Germany

A. König and K. Thongpull
Institute of Integrated Sensor Systems, TU Kaiserslautern, 67663 Kaiserslautern, Germany

F. Schmaljohann, D. Hagedorn and F. Löffler
Physikalisch-Technische Bundesanstalt, Bundesallee 100, 38116 Braunschweig, Germany

Z. Zelinger, P. Janda, P. Kubát and S. Civiš
J. Heyrovský Institute of Physical Chemistry AS CR, Prague, Czech Republic

J. Suchánek and M. Dostál
J. Heyrovský Institute of Physical Chemistry AS CR, Prague, Czech Republic
Faculty of Safety Engineering, VŠB – Technical University of Ostrava, Ostrava, Czech Republic

V. Nevrlý and P. Bitala
Faculty of Safety Engineering, VŠB – Technical University of Ostrava, Ostrava, Czech Republic

A. Weiss, M. Bauer and C.-D. Kohl
Institute of Applied Physics, JLU Giessen, Giessen, Germany

S. Eichenauer and E. A. Stadlbauer
Competence Centre for Energy and Environmental Engineering, University of Applied Science THM, Campus Giessen, Giessen, Germany

F. Roth, E. Ionescu, N. Nicoloso and R. Riedel
Technical University Darmstadt, Institute of Material Science, Jovanka-Bontschits-Strasse 2, 64287 Darmstadt, Germany

C. Schmerbauch and O. Guillon
Forschungszentrum Jülich, Institute of Energy and Climate Research, Wilhelm-Johnen-Strasse, 52425 Jülich, Germany

A. Dickow and G. Feiertag
Munich University of Applied Sciences, Munich, Germany

P. Fremerey
Department of Functional Materials, University of Bayreuth, Bayreuth, Germany
Department of Chemical Engineering, University of Bayreuth, Bayreuth, Germany

A. Jess
Department of Chemical Engineering, University of Bayreuth, Bayreuth, Germany

R. Moos
Department of Functional Materials, University of Bayreuth, Bayreuth, Germany

A. Henseleit, C. Pohl, Th. Bley and E. Boschke
Institute of Food Technology and Bioprocess Engineering, Technische Universität Dresden, 01062 Dresden, Germany

S. G. Nedilko, S. L. Revo, V. P. Chornii and V. P. Scherbatskyi
Taras Shevchenko National University of Kyiv, Kyiv, Ukraine

M. S. Nedielko
E. O. Paton Electric Welding Institute of NASU, Kyiv, Ukraine

C. M. Zimmer, K. T. Kallis and F. J. Giebel
Intelligent Microsystems Institute, Faculty of Electrical Engineering and Information, Technische Universität Dortmund, Dortmund, Germany

L. Ebersberger and G. Fischerauer
Chair of Metrology and Control Engineering, Universität Bayreuth, Bayreuth, Germany

G. Dumstorff, E. Brauns and W. Lang
Institute of Microsensors, -actuators, and -systems (IMSAS), Microsystems Center Bremen, University of Bremen, Bremen, Germany

H. S. Wasisto, A. Waag and E. Peiner
Institut für Halbleitertechnik (IHT), Technische Universität Braunschweig, Braunschweig, Germany
Laboratory for Emerging Nanometrology (LENA), Braunschweig, Germany

S. Merzsch
Institut für Halbleitertechnik (IHT), Technische Universität Braunschweig, Braunschweig, Germany
Infineon Technologies AG, Munich, Germany

E. Uhde
Material Analysis and Indoor Chemistry Department (MAIC), Fraunhofer-WKI, Braunschweig, Germany

J. C. B. Fernandes and E. V. Heinke
The Municipal University of São Caetano do Sul, USCS, Centre Campus, 50, Santo Antônio Street, 09521-160, São Caetano do Sul – SP, Brazil

M. Großklos
Institut Wohnen und Umwelt, Darmstadt, Germany

Derssie D. Mebratu and Charles Kim
Electrical and Computer Engineering, Howard University, Washington DC, 20059, USA

W. Minkina and D. Klecha
The Faculty of Electrical Engineering, Częstochowa University of Technology, Częstochowa, 42–200, Poland

M. Brüne, J. Spiegel and A. Pflitsch
Department of Geography, Ruhr-Universität Bochum, Universitätsstraße 150, 44801 Bochum, Germany

K. Potje-Kamloth
Fraunhofer ICT-IMM, Carl-Zeiss-Straße 18–20, 55129 Mainz, Germany

C. Stein
smartGAS Mikrosensorik GmbH, Kreuzenstraße 98, 74076 Heilbronn, Germany

Marc-Peter Schmidt, Aleksandr Oseev, Andreas Brose and Bertram Schmidt
Institute of Micro and Sensor Systems, Otto von Guericke University Magdeburg, Universitätsplatz 2, 39106 Magdeburg, Germany

Christian Engel
TEPROSA GmbH, Paul-Ecke-Str. 6, 39114 Magdeburg, Germany

Sören Hirsch
Department of Engineering, University of Applied Sciences Brandenburg, Magdeburger Str. 50, 14770 Brandenburg an der Havel, Germany

A. Tempelhahn, H. Budzier, V. Krause and G. Gerlach
Technische Universität Dresden, Electrical and Computer Engineering Department, Solid-State Electronics Laboratory, Dresden, Germany

J. Schilm, A. Goldberg, U. Partsch and A. Michaelis
Fraunhofer IKTS, Fraunhofer Institute for Ceramic Technologies and Systems, Winterbergstraße 28, 01277 Dresden, Germany

W. Dürfeld and D. Arndt
ALL IMPEX GmbH, Bergener Ring 43, 01458 Ottendorf-Okrilla, Germany

A. Pönicke
Modine Europe GmbH, Arthur-B.-Modine-Straße 1, 70794 Filderstadt, Germany

M. Windisch, J. Lienig and L. Schulze
Institute of Electromechanical and Electronic Design, Dresden University of Technology, 01062 Dresden, Germany

K.-J. Eichhorn
Leibniz Institute of Polymer Research Dresden, Hohe Str. 6, 01069 Dresden, Germany

G. Gerlach
Solid-State Electronics Laboratory, Dresden University of Technology, 01062 Dresden, Germany

Jochen Bardong and Alfred Binder
Carinthian Tech Research, Europastraße 4/1, Villach, Austria

Sasa Toskov, Goran Miskovic and Goran Radosavljevic
Institute of Sensor and Actuator Systems, Gusshausstrasse 27–29, Vienna, Austria

F. Schubert, S. Wollenhaupt, J. Kita, G. Hagen and R. Moos
University of Bayreuth, Bayreuth Engine Research Center (BERC), Department of Functional Materials, 95440 Bayreuth, Germany

A. Loderer and T. Hausotte
Institute of Manufacturing Metrology, Friedrich-Alexander-Universität Erlangen-Nürnberg (FAU), Erlangen, Germany

Index

www.ingramcontent.com/pod-product-compliance
Lightning Source LLC
Chambersburg PA
CBHW080622200326
41458CB00013B/4474

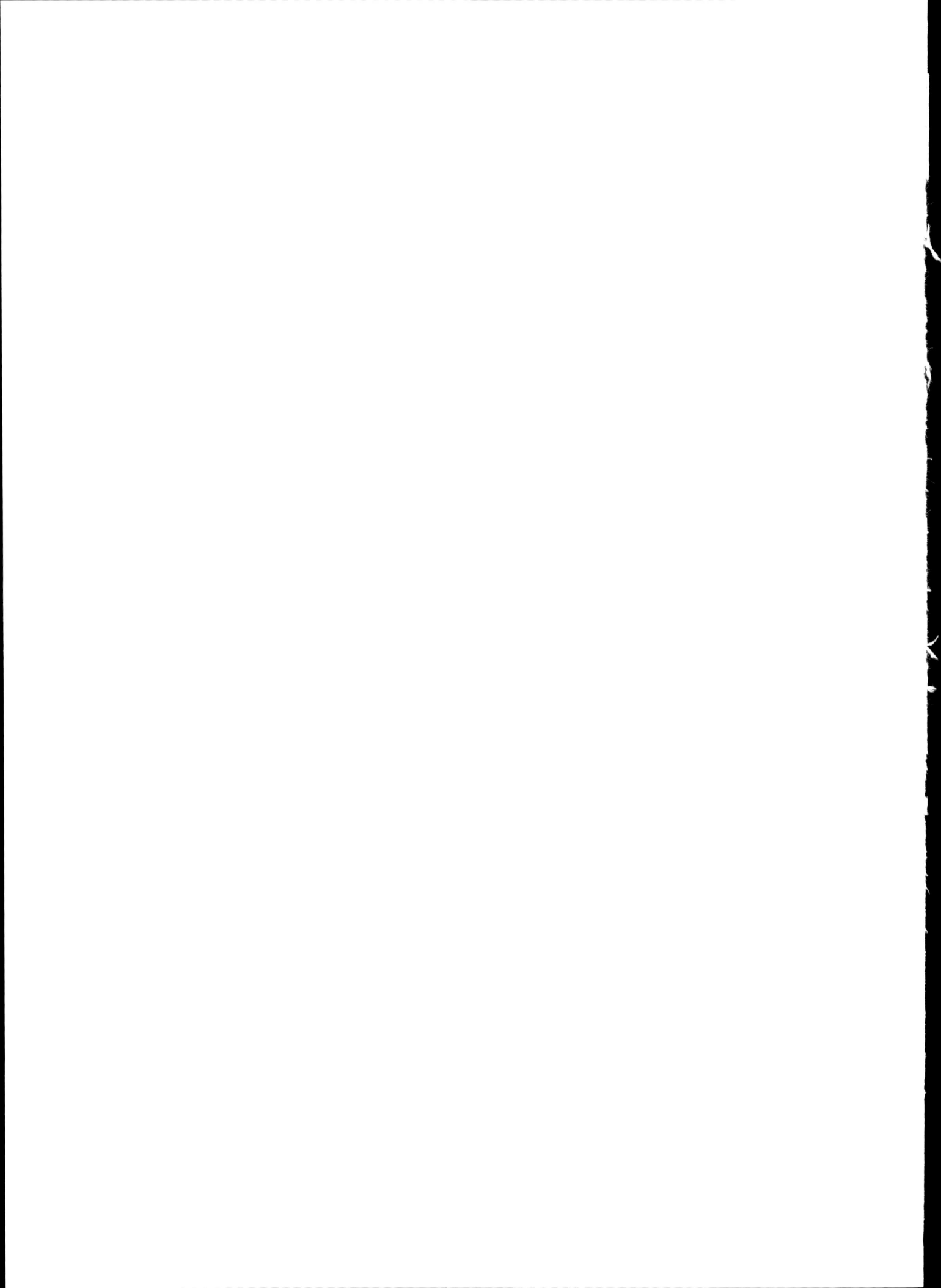